Ultrathin Magnetic Structures III

J.A.C. Bland · B. Heinrich (Eds.)

Ultrathin Magnetic Structures III

Fundamentals of Nanomagnetism

With 128 Figures, Including 28 in Color

Springer

J. Anthony C. Bland
The Cavendish Laboratory
Department of Physics
University of Cambridge
Madingley Road
CB3 0HE Cambridge
United Kingdom
e-mail: jacb1@phy.cam.ac.uk

Bretislav Heinrich
Physics Department
Simon Fraser University
Burnaby, BC, V5A 1S6
Canada
e-mail: bheinric@sfu.ca

ISBN 978-3-642-06021-2 e-ISBN 978-3-540-27163-5

Springer is a part of Springer Science+Business Media

springeronline.com

Softcover reprint of the hardcover 1st edition 2005

Cover production: Erich Kirchner, Heidelberg

Printed on acid-free paper 57/3141/YL - 5 4 3 2 1 0

Preface

The field of magnetic nanostructures is now an exciting and central area of modern condensed matter science, which has recently led to the development of a major new direction in electronics – so called 'spintronics'. This is a new approach in which the electron spin momentum plays an equal role to the electrical charge, and these radical ideas have galvanised the efforts of previously disparate research communities by offering the promise of surpassing the limits of conventional semiconductors. Clearly the world of magnetism has now entered electronics in a very fundamental manner. This is a very fast growing and exciting field which attracts a steadily increasing number of researchers, bringing a constant stream of new ideas. Both spintronics and magnetic nanostructures are already household names in the broad scientific community and we are now, as a result, at the important stage of beginning to develop entirely new approaches to electronics and information technology. 50 Gigabyte/sq inch storage densities in hard drive disks are now a reality. Magnetic Random Access Memories are being introduced commercially and they will soon change the operation of PC's and laptops. Computer logic architectures based on spintronics are already being widely discussed.

Spintronics spreads beyond the traditional boundaries of physics research, device applications and electronics. Researchers in biology and the medical sciences find this approach equally exciting. In this background it is obvious that a deplorable absence of magnetism teaching within University curricula, which started with the advent of an enormous growth of semiconductor physics, and electronics in the early sixties, is now a complete anachronism. There is a pressing need to have books suitable for lecturers in advanced undergraduate and postgraduate courses. Teaching staff at Universities need such literature to quickly incorporate the field of magnetic nanostructures and spintronics into the University teaching program. Scientists working in spintronics applications come from a very broad science and technology background. They also need access to literature which addresses fundamentals and which helps to achieve a broader understanding of this field.

We addressed the basic topics of magnetic multilayers in Volumes I and II which still underpin many of these developments today. In the early nineties, Giant Magnetoresistance and new materials based on the unique properties of interfaces of ultrathin films structures were already in place, but applications were only a promise and the 'engineering' of new magnetic materials using nanostructures was still not well known to the wider community. Since that time the field has moved way ahead and undergone a complete transformation. This is indeed a true success story of modern materials science based on nanostructures, which has led to very powerful and far reaching developments in information storage and device technologies. In view of these developments we have been encouraged by our fellow scientists to update the information base started by the earlier volumes and to provide in Vols. III and IV a new perspective on both nanomagnetism and spintronics, aiming at the reader who needs a concise coverage of the underlying phenomena. These volumes have been written keeping in mind that the prime purpose of these books is to educate and help to eliminate gaps in the understanding of the complex phenomena which magnetic nanostructures manifest. This is highly multidisciplinary science where the enormous and rapid growth currently occurring is hard to follow without having access to a treatment which aims to encompass both the present knowledge and direction of the field, so providing insight into its likely future development.

In preparing these volumes we were fortunate to be able to enlist many of the leading experts in this field. Not only have authors come from leading scientific Institutions and made pioneering contributions but they have often played a role as scientific ambassadors of this fast developing science and technology, often encouraging young scientist to bring their talents to this exciting and demanding research endeavour. We hope that this treatment, based at it is on such wide experience, will therefore be particularly attractive to readers already working in, or planning a career in nanoscience.

We would like to express our thanks to all participating authors for their willingness to put aside an appreciable amount of time to write and keep updating their chapters and to cross-correlate their writing with other contributions. We appreciate all the authors' sharing the experience and expertise which has allowed them to contribute so successfully and fundamentally to magnetic nanostructures and spintronics. Finally we hope that the reader will find these two new volumes a pleasure to read and that the material presented will enrich the reader's understanding of this truly fascinating and revolutionary field of science.

Burnaby and Cambridge
September 2004

B. Heinrich
J. A. C. Bland

Contents

Contributors

J. A. C. Bland
University of Cambridge
Department of Physics
The Cavendish Laboratory
Madingley Road
CB3 0HE Cambridge
UK

W. H. Butler
Center for Materials for Information
Technology University of Alabama
Tuscaloosa, AL 35487-0209
USA

B.-C. Choi
Department of Physics & Astronomy
University of Victoria
Victoria, BC V8W 3P6
Canada

M. R. Freeman
Department of Physics
University of Alberta
Edmonton, AB T6G 2J1
Canada

E. Fullerton
San Jose Research Center, E3
Hitachi Global Storage Technologies
650 Harry Road
San Jose, CA 95120
USA

B. Heinrich
Physics Department
Simon Fraser University
8888 University Drive
Burnaby, BC, V5A 1S6
Canada

P. LeClair
NW14-2126
170 Albany Str.
Cambridge, MA 02139
USA

J.-S. Moodera
Francis Bitter Magnet Laboratory
MIT
Cambridge, MA 02139
USA

S. K. Sinha
Department of Physics
University of California San Diego
9500 Gilman Drive
La Jolla, CA 92093-0354
USA

M. Stiles
National Institute of Standards
and Technology
100 Bureau Dr. Stop 8412
Gaithersburg, MD 20899-8412
USA

H. J. M. Swagten
Eindhoven University of Technology
Applied Physics
Physics of Nanostructures
PO Box 513, NLe 1.08
5600 MB Eindhoven
The Netherlands

C. A. F. Vaz
University of Cambridge
Department of Physics
Cavendish Laboratory
Madingley Rd
Cambridge CB3 0HE
UK

X.-G. Zhang
Computational Science and Mathematics
Division
Oak Ridge National Laboratory
PO Box 2008
Oak Ridge, TN 37831-6114
USA

Acronyms

2D	two-dimensional
AES	Auger electron spectroscopy
AFM	atomic force microscopy
AP	antiparallel state
BCS	Bardeen–Cooper–Schrieffer
BEEM	ballistic electron emission microscopy
BLS	Brillouin light scattering
CA	crystal analyser
CIP	current in-plane geometry
CLO	$Ce_{0.69}La_{0.31}O_{1.845}$
CPP	current perpendicular to plane
DOS	density of states
DWBA	distorted wave Born approximation
EELS	electron energy loss spectroscopy
FERPS	electrons with random point-like scatterers
FMAR	ferromagnetic antiresonance
FMER	ferromagnetic elastic resonance
FMR	ferromagnetic resonance
FTIR	Fourier-transform infrared spectroscopy
GMR	giant magnetoresistance effect
IBZ	Brillouin zone
L.L.G.	L.L. Gilbert equation of motion
L.L.	Landau Lifshitz equation of motion
LCMO	$La_{0.7}Ca_{0.3}MnO_3$
LSDA	local-spin-density approximation
LSMO	$La_{0.67}Sr_{0.33}MnO_3$
M.B.B.	modified Bloch–Bloembergen relaxation term
MBE	molecular-beam epitaxy
MFM	magnetic force microscopy
ML	monolayer
MOKE	magneto-optic Kerr effect

MRAM	magnetic RAM
MRFM	resonance force microscopy
MR	magnetoresistance
MTJ	magnetic tunnel junctions
NEXI	non-equilibrium exchange interaction
NMR	nuclear magnetic resonance
P_2	low resistance parallel state
PIMM	pulsed inductive microwave magnetometer
PNR	polarised neutron reflection
PSD	power spectral density
RBS	Rutherford backscattering
RHEED	reflection high energy electron diffraction
RKKY	Ruderman–Kittel–Kasuya–Yosida
RPA	random phase approximation
SEMPA	scanning electron microscopy with polarization analysis
SEM	scanning electron microscopy
SHMOKE	second-harmonic magneto-optic Kerr effect
SPEEL	spin polarized electron energy loss
SPT	spin-polarized tunneling technique
SQUID	superconducting quantum interference device
STM	scanning tunneling microscopy
STO	$SrTiO_3$
SWASER	spin wave amplification by stimulated emission of radiation
TEM	transmission electron microscopy
TMR	tunneling magnetoresistance
TOF	time-of-flight
TR-SKM	time-resolved scanning Kerr microscope
UHV	ultra-high vacuum
UPS	ultraviolet photoelectron spectroscopy
VSM	vibrating sample magnetometry
WKB	Wentzel–Kramer–Brillouin
XAS	x-ray absorption edge spectroscopy
XMCD	x-ray magnetic circular dichroism
XPS	x-ray photoelectron spectroscopy

1

Introduction

J.A.C. Bland and B. Heinrich

Since the publication of Vols. I and II in this series 10 years ago, there has been an explosion of interest and activity in the subject of thin film magnetism. Much of this activity has been stimulated by the use of giant magnetoresistance read heads in hard disc drives and by the continuing advances in storage densities achievable in thin film media. Such applications are now almost as familiar as those of the semiconductor transistor, while 10 years ago, the phenomenon of giant magnetoresistance was largely unknown outside the research laboratory.

As early as the 1950s, researchers had already recognised the enormous technological potential of thin magnetic films for use as sensors and information storage devices. Louis Néel identified the importance of the surface in leading to modified switching fields, the role of finite thickness in modifying the domain structure of a thin ferromagnetic film and the role of interface roughness in mediating interlayer dipole coupling. Many researchers recognised the possibilities of using such modified magnetic properties to create technologically useful devices. However it was soon recognised that difficulties in controlling sample quality, often due to the inevitable chemical contamination resulting from the inadequate vacuum available for thin film growth, frustrated attempts to control thin film properties and to perform reliable experiments in the search for modified properties. Despite advances in surface science techniques and the widespread use of molecular beam epitaxy in the 1980s it was only in the late 1990s that the early dreams of a new technology have begun to be truly fulfilled.

The very success of the giant magnetoresistance spin valve structure has led to increased efforts to develop magnetic tunnel junction devices based on metal/insulator/metal structures. Spintronic devices based on spin polarised electron injection and detection in all semiconductor or hybrid metal/semiconductor structures are now being very actively developed. Such devices rely for their operation on the manipulation of the electron spin rather the electron charge and momentum as in conventional semiconductor devices. Ultimately it is believed that by controlling the spin polarised transport channels it may be possible to engineer complete suppression of one of the spin conduction channels in the presence of an applied magnetic or electric field, leading to infinite magnetoresistance ratios in future spintronic devices. Advances in our understanding of spin polarised electron transport in magnetic multilayers have emphasised the role of the microscopic spin polarised electron scattering processes in magnetotransport and have led to the beginnings of a theoretical understanding of the reciprocal effect, current induced magnetic reversal, in which the electron current induces a reversal of the magnetisation in magnetic nanostructures. This phenomenon would allow magnetic switching in nanoscale devices by all electrical means without the need to apply external magnetic fields.

In the earlier two volumes, UMS I and II, we described many of the fundamental properties of thin magnetic films and techniques used to investigate them. These properties largely underpin the remarkable technological developments of the last decade. However the last decade has seen considerable progress and refinement in our understanding of magnetotransport and interlayer coupling but also the blurring of the boundaries between metals and semiconductors research in the quest for new spin polarised phenomena: it is largely these developments which form the focus of the present volumes. Here in Vol. III, the first of the two new volumes, we present further advances in the fundamental understanding of thin film magnetic properties and of methods for characterising thin film structure which underpin the present spintronics revolution.

The success of spintronics depends on our fundamental understanding of spin polarised electron transport. In Chap. 2, Butler and Zhang describe computational studies of electron transport in magnetic multilayers using a semiclassical approach based on solution of the Boltzmann equation with realistic Fermi surface properties. These results emphasise the link between the spin split band structure and the resulting electron transport properties and, in particular emphasise the distinction between diffusive processes and ballistic processes. The latter mechanism is important not only for tunnel magnetoresistance but is likely to be the key to understanding polarised electron transport on the nanoscale. The spin dependent quantum tunneling between two ferromagnetic layers first proposed in 1975 but only demonstrated experimentally in the last few years, provides a larger magnetoresistive effect than giant magnetoresistance. Consequently it offers great promise in the field of spintronics. Magnetic tunnel junctions formed from two ferromagnetic films separated by an insulating barrier layer are used in random access memory arrays (see the chapters by Katti and also by Shi). However the fundamental physics behind these devices is only beginning to be understood. In Chap. 3, LeClair, Moodera, and Swagten describe developments in the understanding of the fundamental physics of magnetic tunnelling based on a wide range of experimental studies in planar structures and finally consider the wider outlook for spintronic applications based on spin dependent tunneling. The phenomenon of indirect exchange coupling between ferromagnetic films was first identified experimentally in 1986. The effect proved to be crucial in the development of spin valve devices leading to the development of 'spin engineering'. By the early 90's several models had been proposed but the fundamental understanding of indirect coupling effect was still in a state of development and many experimental results could not be fully explained. In Chap. 4 Stiles describes the early development of these models and the subsequent theoretical advances which led to a fuller understanding of interlayer exchange coupling based on precision experimental measurements on near defect-free structures in the late 90's. The subject of the magnetisation dynamics is intimately linked to the need to switch magnetic nanostructures at ultrahigh rates for information storage applications. In Chap. 5 Heinrich describes ferromagnetic resonance studies of spin relaxation processes in magnetic metallic layers and multilayers and in Chap. 6 Choi and Freeman describe experimental studies of nonequilibrium spin dynamics in laterally defined nanostructures. They present a detailed description of an experimental method for imaging nonequilibrium magnetic phenomena in the picosecond temporal regime and with sub-micrometer spatial resolution based on stroboscopic scanning Kerr microscope. They present exemplar data illustrating dynamic micromagnetic processes during magnetization switching and spontaneous magnetic domain pattern formation in small magnetic elements. It is now recognised that ultimately the atomic scale structure of interfaces need to be described to properly account for the magnetic properties. Fundamental properties of ultrathin structures such as the magnetic moment and magnetic anisotropy ultimately have an atomic scale origin and these properties can differ markedly from the corresponding bulk properties. Probes of buried interface structures are therefore pivotal in characterising magnetic multilayer structures. While there is an abundance of surface sensitive structural probes there are few techniques which allow completed multilayer structures to be probed. In Chap. 7 Bland and Vaz describe the use of polarised neutron reflection for layer selective magnetometry in thin (nm scale) film structures and show that the layer dependent magnetisation vector and total layer magnetisation vector can be very accurately determined. In Chap. 8 Fullerton and Sinha discuss the basic concepts of X-ray scattering studies from ultrathin metallic structures and show that the average structure and the atomic scale roughness can be determined with very high precision.

In Vol. IV we deal with the fundamentals of spintronics: magnetoelectronic materials, spin injection and detection, micromagnetics and the development of magnetic random access memory based on giant magnetoresistance and tunnel junction devices.

The reader is encouraged to use these volumes not only as an introduction to recent developments in thin film magnetism and to the new field of spintronics but to see this work as part of a continuing evolution in a subject which continues to grow in importance, both technologically and scientifically. By focusing on fundamental issues we hope that the material we have covered will continue to be of value as a tutorial guide for some time. Inevitably we have not been able to cover all important topics in the present volumes, many of which are still in a state of rapid development. Nevertheless we hope that the present volumes will serve to help interest grow still further in a fascinating field.

2

Electron Transport in Magnetic Multilayers

W.H. Butler and X.-G. Zhang

2.1 Introduction

Almost all electronic devices depend for their operation on the response of the electron's *charge* to applied electric fields. Until very recently, however, no use had been made of another degree of freedom which electrons possess, their *spin*. This situation has changed dramatically, however, during the past 15 years. The giant magnetoresistance effect was discovered in 1988 [2.1]. This was followed quickly by the rediscovery of tunneling magnetoresistance [2.2–4]. Today, the combination of charge and spin transport in heterostructures offers almost unlimited opportunity for new discoveries and applications.

Giant magnetoresistance (GMR) is a change (usually a significant decrease) in the electrical resistance of a magnetically inhomogeneous metallic system that occurs when an applied magnetic field aligns the magnetic moments in different regions of the system. Because of this moment alignment, one of the two spin currents is able to traverse the system relatively unimpeded compared to the other. Although GMR may be observed in many geometries, applications typically employ ultrathin magnetic multilayers. GMR is now important commercially because it can be used to make very small and sensitive magnetic field sensors that have numerous applications most notably as read sensors in magnetic disk drives.

Tunneling magnetoresistance (TMR) may be observed when ferromagnetic electrodes are separated by a thin insulating layer that serves as a tunneling barrier. A significant change in the tunneling conductance (usually an increase) is often observed when an applied magnetic field aligns the moments in the two ferromagnetic electrodes. TMR is likely also to soon become commercially important. Two potential applications are read sensors for disk drives and non-volatile magnetic random access memory devices.

Future directions of a field changing so rapidly are difficult to predict. There is much interest presently in spin-polarized current induced switching [2.5, 6]. In this phenomenon, the roles of current and magnetic moments are reversed compared to

GMR. In GMR, the relative orientation of the magnetic moments in the ferromagnetic layers affect the current through the film. Here, the current can change the relative orientation of the magnetization of the layers. Another area of increasing interest is magnetic semiconductors and the injection of polarized currents into semiconductors [2.7]. This would allow us to combine the new ability to manipulate electrons with their spin as well as their charge with sophisticated semiconductor technology, and allow rapid and practical development of new devices.

In the following exposition we have shamelessly concentrated on our own work on transport in magnetic multilayers. We apologize to colleagues whose excellent work has been slighted. Under no circumstances should this be viewed as a comprehensive review of the work in this area. A recent review article on GMR [2.8] may be useful for a broader perspective on some of the topics covered here. We hope that our approach has, at least, the advantage of a single coherent point of view.

2.2 Transport Theory for Inhomogeneous Materials

Theoretical approaches to the study of electron transport in magnetic multilayers range from fully quantum mechanical linear response theory based on the Kubo formula, to simpler models based on free-electron bands and semiclassical assumptions. Simple free-electron models, however, often fail to capture some of the essential physics in spin-dependent transport. We will first briefly review the free-electron based models in next section, then in Sect. 2.5, discuss the role of electronic structure in transport. That will be followed by discussions of transport theory based on first-principles band structures, with the emphasis on the semiclassical Boltzmann approach for diffusive transport and the Landauer approach for ballistic transport.

2.2.1 Quantum Theory of Linear Response

Fully quantum mechanical expressions for transport coefficients can be derived from linear response theory . Consider a system of noninteracting electrons moving in the presence of a random potential. Kubo [2.9] and Greenwood [2.10] have shown that the zero temperature dc conductivity may be written as

$$\sigma_{\mu\nu} = \frac{\pi\hbar}{V}\left\langle \sum_{\alpha,\alpha'} \langle\alpha|j_\mu|\alpha'\rangle\langle\alpha'|j_\nu|\alpha\rangle \right\rangle \delta(E_\mathrm{F}-E_\alpha)\delta(E_\mathrm{F}-E_\alpha)\ , \tag{2.1}$$

where j_μ is the current operator,

$$\langle\alpha|j_\mu|\alpha'\rangle = \frac{-\mathrm{i}e\hbar}{2m}\int \mathrm{d}\boldsymbol{r}\left[\psi^*_\alpha(\boldsymbol{r})\nabla_\mu\psi_{\alpha'}(\boldsymbol{r}) - \nabla_\mu\psi^*_\alpha(\boldsymbol{r})\psi_{\alpha'}(\boldsymbol{r})\right]\ , \tag{2.2}$$

and V is the volume. The quantum states $|\alpha\rangle$ are the exact eigenfunctions of a particular configuration of the random potential and the large angle brackets indicate an average over configurations. We will find it useful to write (2.1) in terms of the Green function which is defined as $G = [E - H]^{-1}$. It is related to the sum over states in (2.1) through

$$\sum_{\alpha} |\alpha\rangle\langle\alpha|\delta(E - E_{\alpha}) = \frac{-1}{2i\pi}(G^{+} - G^{-}) \equiv -\mathcal{G}/\pi \ . \tag{2.3}$$

Here $G^{+} = [E + i\eta - H]^{-1}$ and $G^{-} = [E - i\eta - H]^{-1}$ with η a positive infinitesimal, are the retarded and advanced Green functions, respectively. Using these definitions, (2.1) can be written in the form

$$\sigma_{\mu\nu} = \frac{-e^{2}\hbar^{3}}{4\pi V m^{2}} \int \mathrm{d}\boldsymbol{r} \int \mathrm{d}\boldsymbol{r}' \big\langle \nabla_{\mu}\mathcal{G}(\boldsymbol{r},\boldsymbol{r}')\nabla'_{\nu}\mathcal{G}(\boldsymbol{r}',\boldsymbol{r}) + \nabla'_{\nu}\mathcal{G}(\boldsymbol{r},\boldsymbol{r}')\nabla_{\mu}\mathcal{G}(\boldsymbol{r}',\boldsymbol{r}) \\ -\nabla_{\mu}\nabla'_{\nu}\mathcal{G}(\boldsymbol{r},\boldsymbol{r}')\mathcal{G}(\boldsymbol{r}',\boldsymbol{r}) - \mathcal{G}(\boldsymbol{r},\boldsymbol{r}')\nabla'_{\nu}\nabla_{\mu}\mathcal{G}(\boldsymbol{r}',\boldsymbol{r})\big\rangle \ . \tag{2.4}$$

This allows the definition of a non-local kernel for the conductivity of the form [2.11],

$$\sigma_{\mu\nu}(\boldsymbol{r},\boldsymbol{r}') = \frac{-e^{2}\hbar^{3}}{4\pi V m^{2}} \big\langle \nabla_{\mu}\mathcal{G}(\boldsymbol{r},\boldsymbol{r}')\nabla'_{\nu}\mathcal{G}(\boldsymbol{r}',\boldsymbol{r}) + \nabla'_{\nu}\mathcal{G}(\boldsymbol{r},\boldsymbol{r}')\nabla_{\mu}\mathcal{G}(\boldsymbol{r}',\boldsymbol{r}) \\ -\nabla_{\mu}\nabla'_{\nu}\mathcal{G}(\boldsymbol{r},\boldsymbol{r}')\mathcal{G}(\boldsymbol{r}',\boldsymbol{r}) - \mathcal{G}(\boldsymbol{r},\boldsymbol{r}')\nabla'_{\nu}\nabla_{\mu}\mathcal{G}(\boldsymbol{r}',\boldsymbol{r})\big\rangle \ . \tag{2.5}$$

This result emphasizes the fact that the current at a point $\boldsymbol{r}$ depends not just on the electric field at that point but on the field at points $\boldsymbol{r}'$ within the vicinity (approximately the electronic mean free path) of $\boldsymbol{r}$,

$$J_{\mu}(\boldsymbol{r}) = \int \mathrm{d}\boldsymbol{r}' \sum_{\nu} \sigma_{\mu\nu}(\boldsymbol{r},\boldsymbol{r}')\mathcal{E}(\mathbf{r}') \ . \tag{2.6}$$

Thus $\sigma_{\mu\nu}(\boldsymbol{r},\boldsymbol{r}')$ is the current in direction μ at point $\boldsymbol{r}$ induced by an electric field of unit strength in direction ν that exists at point $\boldsymbol{r}'$. The realization that the conductivity is non-local, i.e. that the current at one point depends on fields applied at other points is key to understanding giant-magnetoresistance for the technologically important case in which the current flows parallel to the planes of the multilayer.

2.3 Free Electrons with Random Point Scatterers

In order to get a better understanding of this non-local conductivity let us evaluate it for the simple case of free electrons with random point-like scatterers (FERPS). The Green function, $G^{+}(\boldsymbol{r},\boldsymbol{r}')$ is defined by,

$$\left[\frac{\hbar^{2}}{2m}\nabla^{2} + E - \sum_{i} v\left(\boldsymbol{r} - \boldsymbol{R}_{i}\right)\right] G^{+}(\boldsymbol{r},\boldsymbol{r}') = \delta(\boldsymbol{r} - \boldsymbol{r}') \ . \tag{2.7}$$

The "+" superscript on the Green function indicates that it is a "retarded" or causal Green function. For our purposes, this means that the energy, E, has an infinitesimal imaginary part. In the FERPS model, the scatterers are assumed to be located at random positions, R_i.

The FERPS Green function can be expanded as follows,

$$G^{+}(\boldsymbol{r},\boldsymbol{r}') = G_0^{+}(\boldsymbol{r},\boldsymbol{r}') + \int \mathrm{d}\boldsymbol{r}_1 G_0^{+}(\boldsymbol{r},\boldsymbol{r}_1) \sum_i v(\boldsymbol{r}_1 - \boldsymbol{R}_i) G^{+}(\boldsymbol{r}_1,\boldsymbol{r}') \,, \tag{2.8}$$

where $G_0^{+}(\boldsymbol{r},\boldsymbol{r}')$ is the Green function in the absence of the the random scatterers,

$$G_0^{+}(\boldsymbol{r},\boldsymbol{r}') = \frac{2m}{\hbar^2} \frac{\mathrm{e}^{\mathrm{i}\kappa|\boldsymbol{r}-\boldsymbol{r}'|}}{4\pi|\mathbf{r}-\boldsymbol{r}'|} \,, \tag{2.9}$$

where $\kappa = \sqrt{2mE}/\hbar$. The integral equation for the Green function, (2.8) is known as the Lippmann-Schwinger equation and can be verified by substituting it into (2.7).

Let us write the Lippmann-Schwinger expression for the Green function including the random point scatterers using the simplified notation, $G = G_0 + \sum_i G_0 v_i G$ in which integration over "internal" variables is suppressed. Then we can expand by substituting the entire expression for the G on the right hand side,

$$G = G_0 + \sum_i G_0 v_i G_0 + \sum_{i,j} G_0 v_i G_0 v_j G_0 + \sum_{i,j,k} G_0 v_i G_0 v_j G_0 v_k G_0 + \cdots \,. \tag{2.10}$$

If we use angle brackets to denote an average over configurations, i.e. over the possible positions of the scatterers then we can assume that $\langle v_i \rangle = 0$, since a shift in the average potential can be accommodated as a shift in the energy zero; then $\Delta v_i = v_i - \langle v_i \rangle$ and we can write,

$$\begin{aligned}\langle G \rangle = G_0 &+ \sum_i G_0 \langle \Delta v_i G_0 \Delta v_i \rangle G_0 + \sum_i G_0 \langle \Delta v_i G_0 \Delta v_i G_0 \Delta v_i \rangle G_0 \\ &+ \sum_{ij} G_0 \langle \Delta v_i G_0 \Delta v_i \rangle G_0 \left\langle \Delta v_j G_0 \Delta v_j \right\rangle G_0 + \cdots \,.\end{aligned}$$

Then we can write,

$$\langle G \rangle = G_0 + G_0 \Sigma G_0 + G_0 \Sigma G_0 \Sigma G_0 + \cdots \,, \tag{2.11}$$

where

$$\Sigma = \langle \Delta v_i G_0 \Delta v_i \rangle + \langle \Delta v_i G_0 \Delta v_i G_0 \Delta v_i \rangle + \cdots \,. \tag{2.12}$$

If the potential differences, Δv_i, are very short ranged, relatively weak, and randomly distributed in space, the electron self-energy, Σ, will be approximately independent of the electron momentum and position. It will, however, generally be a function of energy[1].

[1] If the scatterers are literally delta functions, *e.g.* if $< \Delta v(\boldsymbol{r}) \Delta v(\boldsymbol{r}') >= \gamma\delta(\boldsymbol{r}-\boldsymbol{r}')$, then $\Sigma \approx n\gamma G(0)$ (where n is the density of scatterers) is formally divergent because the equal argument free electron Green function is divergent. Fortunately, it is usually the imaginary part of the self-energy that enters expressions for the conductivity and this is well defined. The imaginary part of Σ is negative for G^{+} and positive for G^{-}.

Finally, we have made enough approximations to be able to write down the average Green Function in the FERPS approximation,

$$[\frac{\hbar^2}{2m}\nabla^2 + E - \Sigma]\langle G^+(\boldsymbol{r},\boldsymbol{r}')\rangle = \delta(\boldsymbol{r}-\boldsymbol{r}') , \tag{2.13}$$

$$\langle G^+(\boldsymbol{r},\boldsymbol{r}')\rangle = \frac{1}{(2\pi)^3}\int \mathrm{d}^3\boldsymbol{k}\frac{\mathrm{e}^{\mathrm{i}\boldsymbol{k}\cdot(\boldsymbol{r}-\boldsymbol{r}')}}{E-\Sigma-\dfrac{\hbar^2k^2}{2m}} = \frac{2m}{\hbar^2}\frac{\mathrm{e}^{\mathrm{i}\kappa|\boldsymbol{r}-\boldsymbol{r}'|}}{4\pi|\boldsymbol{r}-\boldsymbol{r}'|} , \tag{2.14}$$

where $\kappa = \sqrt{2m(E-\Sigma)}/\hbar$. In the following we omit the angle brackets to simplify the notation, but it should be remembered that we are concerned with the average of the green function over the atomic configurations.

The expression for the conductivity, (2.4), involves the average of two Green functions, $\langle \mathcal{G}(\boldsymbol{r},\boldsymbol{r}')\mathcal{G}(\boldsymbol{r}',\boldsymbol{r})\rangle$. It is very common, however, to average the Green functions independently. This is called the neglect of "vertex corrections". It is an approximation that can be made in both the quantum and in the semi-classical approaches to transport. In the latter case, this approximation is called "neglect of the scattering-in terms". We shall show that whether or not these terms can be neglected for a layered systems depends on the geometry. For the case in which the current is perpendicular to the layers (CPP), we shall show in Sect. 2.6.2 that at least an approximate treatment of the vertex corrections is necessary for a consistent theory. However for the case in which the current is in the plane of the layers (CIP) or for a homogeneous system these terms do not contribute to the current if the scattering is isotropic. Since we will be primarily concerned with the latter case, and since they greatly complicate the calculations, we will neglect them for the time being.

We can now evaluate the conductivity for a homogeneous system by integrating over $\boldsymbol{r}'$ and averaging over $\boldsymbol{r}$ and directions (μ). Thus $\boldsymbol{J} = \sigma_0\mathcal{E}$ where,

$$\begin{aligned}\sigma_0 &= \frac{1}{3V}\sum_{\mu}\int \mathrm{d}^3\boldsymbol{r}\int \mathrm{d}^3\boldsymbol{r}'\sigma_{\mu\mu}(\boldsymbol{r},\boldsymbol{r}') \\ &= -\frac{1}{3}\frac{e^2\hbar^3}{\pi m^2 V}\int \mathrm{d}^3\boldsymbol{r}\int \mathrm{d}^3\boldsymbol{r}'\nabla\mathrm{Im}G(\boldsymbol{r},\boldsymbol{r}')\cdot\nabla'\mathrm{Im}G(\boldsymbol{r}',\boldsymbol{r}) .\end{aligned} \tag{2.15}$$

Here we have taken advantage of the fact that the Green function for a homogeneous system is symmetric in $\boldsymbol{r}$ and $\boldsymbol{r}'$. This implies that $\mathcal{G}(\boldsymbol{r},\boldsymbol{r}') = \mathrm{Im}G^+(\boldsymbol{r},\boldsymbol{r}')$. It also allows us to equate the first and second as well as the third and fourth terms in (2.5) which defines the non-local conductivity. Finally, (2.15) is obtained by equating the first and third terms of (2.5). This is justified in this case because the system is homogeneous.

Letting $\boldsymbol{R} = \boldsymbol{r}-\boldsymbol{r}'$ and $|\boldsymbol{R}| = R$ in (2.17), we have.

$$\sigma_0 = \frac{e^2\hbar^3}{3\pi m^2}\int \mathrm{d}^3\boldsymbol{R}[\nabla\mathrm{Im}G(R)]^2 . \tag{2.16}$$

This integration can be performed exactly using elementary techniques and yields,

$$\sigma_0 = \frac{e^2}{12\pi^2\hbar}\frac{\kappa_R^2}{\kappa_I}\,, \tag{2.17}$$

where $\kappa_R = \mathrm{Re}[\sqrt{2m(E-\Sigma)}/\hbar]$ and $\kappa_I = \mathrm{Im}[\sqrt{2m(E-\Sigma)}/\hbar]$. This is the same as the usual expression for free electrons if we identify $\kappa_R = k_\mathrm{F}$ and $2\kappa_I = 1/\lambda$ where λ is the mean free path. The factor of two arises because the electron probability decays twice as fast as the electron amplitude. Thus,

$$\sigma_0 = \frac{e^2 k_\mathrm{F}^2\lambda}{6\pi^2\hbar} = \frac{k_\mathrm{F}^3}{6\pi^2}\frac{e^2\tau}{m} = \frac{Ne^2\tau}{m}\,, \tag{2.18}$$

where we used $\lambda = v_\mathrm{F}\tau = \hbar k_\mathrm{F}\tau/m$. Here λ is the electron mean free path, $v_\mathrm{F} = \hbar k_\mathrm{F}/m$ is the Fermi velocity, $\tau = \lambda/v_\mathrm{F}$ is the electron lifetime and $N = k_\mathrm{F}^3/(6\pi^2)$ is the number of free electrons per unit volume for a single spin channel.

The above exercise shows that the Kubo-Greenwood quantum mechanical linear response formalism applied to the FERPS model gives familiar results for a homogeneous system. In preparation for dealing with layered systems, let us treat a homogeneous system as an artificial layered system by calculating the non-local conductivity that would arise if we could apply an electric field in a plane of vanishing thickness. Thus the current density $J_\mu(z)$ induced in direction μ in plane z due to an electric field, $\mathcal{E}_\nu(z')$ applied in direction ν to plane z' (see Fig. 2.1) are related through

$$J_\mu(z) = \int \mathrm{d}z' \sum_\nu \sigma_{\mu\nu}(z,z')\mathcal{E}_\nu(z')\,. \tag{2.19}$$

Non-local Conductivity

The functions $\sigma_{zz}(z-z')$ (2.29) and $\sigma_{xx}(z-z')$ (2.21) defined in this section give the current induced in the z and x directions respectively for electric fields applied a distance Z away in the z direction from the point where the current is induced. The currents are in the same direction as the applied fields. The geometry for σ_{zz} and σ_{xx} is indicated in Fig. 2.1. σ_{xx} is instructive concerning electron transport for the current in the plane (CIP) geometry, while σ_{zz} is instructive concerning transport in the current perpendicular to the planes (CPP) geometry.

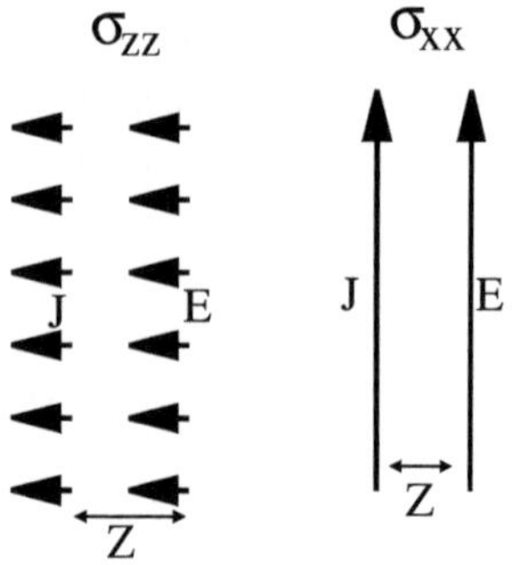

Fig. 2.1. Geometry for longitudinal non-local conductivity, σ_{zz}, and transverse non-local conductivity, σ_{xx}

Transverse Non-local Conductivity

This non-local conductivity transverse to the layers is given by,

$$\sigma_{xx}(z,z') = \sigma_{yy}(z,z') = -\frac{e^2\hbar^3}{\pi m^2 A}\int \mathrm{d}x\,\mathrm{d}y \int \mathrm{d}x'\,\mathrm{d}y' \nabla_x \mathrm{Im}G(\boldsymbol{r},\boldsymbol{r}')\nabla'_x \mathrm{Im}G(\boldsymbol{r}',\boldsymbol{r}) \,, \quad (2.20)$$

which also can be evaluated exactly. The details of the evaluation are given in reference [2.12]. The results for the current and field parallel to the planes z and z' can be written (using $Z = |z - z'|$) as,

$$\begin{aligned}\sigma_{xx}(Z) = \frac{e^2}{16\pi^2\hbar} &\times [\kappa^2 E_{1,3}(-2\mathrm{i}\kappa Z) + \kappa^{*2} E_{1,3}(2\mathrm{i}\kappa^* Z) + 2|\kappa|^2 E_{1,3}(2\kappa_I Z) \\ &+ \frac{2\mathrm{i}\kappa}{Z} E_{2,4}(-2\mathrm{i}\kappa Z) - \frac{2\mathrm{i}\kappa^*}{Z} E_{2,4}(2\mathrm{i}\kappa^* Z) + \frac{4\kappa_I}{Z} E_{2,4}(2\kappa_I Z) \\ &- \frac{1}{Z^2} E_{3,5}(-2\mathrm{i}\kappa Z) - \frac{1}{Z^2} E_{3,5}(2\mathrm{i}\kappa^* Z) + \frac{2}{Z^2} E_{3,5}(2\kappa_I Z)] \,. \quad (2.21)\end{aligned}$$

Here the functions, $E_{n,m}(x)$ are combinations of exponential integrals defined by

$$E_{n,m}(x) = \int_1^\infty \mathrm{e}^{-xt}[t^{-n} - t^{-m}]\,\mathrm{d}t = E_n(x) - E_m(x) \,. \quad (2.22)$$

Note that the transverse non-local conductivity contains integrals over terms that oscillate on the scale of one half of the electron wavelength. These terms arise from the quantum nature of the transport and are absent in the semiclassical approximation. However, as shown in Fig. 2.2 they give only a relatively small modification to the monotonic semiclassical result. They do however, remove the logarithmic singularity at $Z = 0$ in the semiclassical result for σ_{xx}.

Longitudinal Non-local Conductivity

For the case in which the current and fields are perpendicular to the z and z' planes, the first and third terms of (2.5) are not equivalent and must be treated separately. In this case it is convenient to take advantage of the fact that the Green function can be can be represented either in real space or in reciprocal space. For layered systems that are homogeneous in two dimensions it is often convenient to represent the Green function using a hybrid representation; reciprocal space for the variation parallel to the layers and real space for the variation perpendicular. Thus,

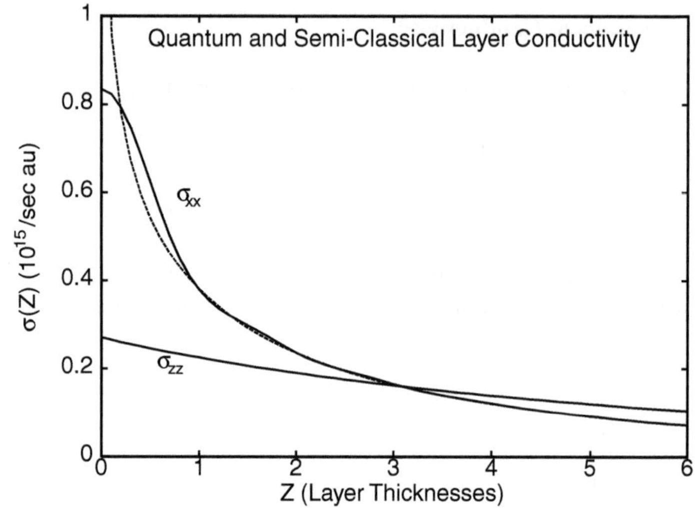

Fig. 2.2. Quantum and semi-classical non-local layer conductivity as a function of layer separation, Z, for a homogeneous free electron system. Solid lines are the quantum conductivity. The dashed line for σ_{xx} is the semi-classical approximation. For σ_{zz}, the quantum and semiclassical expressions are the same. In this example the lattice constant is that of copper (0.3615 nm); the Fermi momentum corresponds to 0.5 electrons per spin channel; and Z is measured in terms of the thickness of (111) layers of copper (0.209 nm). $\sigma(Z)$ is measured in units of 10^{15}/sec au where 1 au = 0.0529 nm. Z is the distance between the plane at which the field is applied and the plane at which the current is induced

$$\begin{aligned} G^+(\boldsymbol{r},\boldsymbol{r}') &= \frac{1}{(2\pi)^2}\int \mathrm{d}k_x\,\mathrm{d}k_y\,\mathrm{e}^{\mathrm{i}(k_x(x-x')+k_y(y-y'))} \\ &\quad\times\frac{1}{2\pi}\int \mathrm{d}k_z\frac{\mathrm{e}^{\mathrm{i}k_z(z-z')}}{E-\Sigma-\dfrac{\hbar^2}{2m}(k_x^2+k_y^2+k_z^2)} \\ &= \frac{1}{(2\pi)^2}\int \mathrm{d}^2\boldsymbol{k}_\parallel\,\mathrm{e}^{\mathrm{i}\boldsymbol{k}_\parallel\cdot\boldsymbol{R}_\parallel}G(k_\parallel,z,z') \end{aligned} \qquad (2.23)$$

where $G(k_\parallel,z,z') = \frac{2m}{\hbar^2}\,\mathrm{e}^{\mathrm{i}kZ}/2\mathrm{i}k$. Here $k(k_\parallel) = \sqrt{2m(E-\Sigma)/\hbar^2 - k_\parallel^2}$ is the component of the electron momentum perpendicular to the layers.

The Kubo formula for $\sigma_{zz}(Z)$ in terms of $G(k_\parallel,z,z')$ is

$$\begin{aligned} \sigma_{zz}(Z) &= \frac{e^2\hbar^3}{2\pi m^2}\frac{1}{(2\pi)^2}\int \mathrm{d}^2\boldsymbol{k}_\parallel \\ &\quad\times\left[\left(\frac{\partial}{\partial Z}\mathrm{Im}G(k_\parallel,Z)\right)^2-\frac{\partial^2}{\partial Z^2}\mathrm{Im}G(k_\parallel,Z)\mathrm{Im}G(k_\parallel,Z)\right]. \end{aligned} \qquad (2.24)$$

Defining $f = \frac{\hbar^2}{2m}\mathrm{Im}G^+(k_\parallel, Z) = \frac{-1}{4}(\frac{\mathrm{e}^{ikZ}}{k} + \frac{\mathrm{e}^{ik^*Z}}{k^*})$, we have $(f')^2 - f''f = \frac{k_R^2}{|k|^2}$, where $k = k_R + ik_I$ and $k^2 = \frac{2m}{\hbar^2}(E - \Sigma) - k_\parallel^2 = k_\mathrm{F}^2 - k_\parallel^2 + i\Delta$.

Thus $\sigma_{zz}(Z)$, where $Z = z - z'$ is given by

$$\sigma_{zz}(Z) = \frac{e^2}{2\pi^2\hbar}\int_0^\infty k_\parallel\,\mathrm{d}k_\parallel \frac{k_R^2}{|k|^2}e^{-2k_I Z}\;. \tag{2.25}$$

Using

$$2k_R^2(k_\parallel) = \sqrt{(k_\mathrm{F}^2 - k_\parallel^2)^2 + \Delta^2} + (k_\mathrm{F}^2 - k_\parallel^2) = \frac{\Delta^2}{2k_I^2(k_\parallel)} \tag{2.26}$$

and

$$2k_I^2(k_\parallel) = \sqrt{(k_\mathrm{F}^2 - k_\parallel^2)^2 + \Delta^2} - (k_\mathrm{F}^2 - k_\parallel^2)\;, \tag{2.27}$$

we obtain $dk_I = \frac{k_I k_\parallel dk_\parallel}{|k|^2}$ and

$$\sigma_{zz}(Z) = \frac{e^2\Delta^2}{8\pi^2\hbar}\int_{\kappa_I}^\infty \frac{\mathrm{d}k_I}{k_I^3}e^{-2k_I Z}\;, \tag{2.28}$$

where for consistency with (2.17) and (2.21) we define $\kappa_I = k_I(0)$ and $\kappa_R = k_R(0)$. Substituting $t\kappa_I = k_I(k_\parallel)$ and using $\frac{\Delta^2}{\kappa_I^2} = 4\kappa_R^2$ we obtain our final expression for the longitudinal non-local conductivity,

$$\sigma_{zz}(Z) = \frac{e^2\kappa_R^2(0)}{2\pi^2\hbar}\int_1^\infty \frac{\mathrm{d}t}{t^3}e^{-2t k_I Z}\;. \tag{2.29}$$

In [2.12] we obtained an incorrect result for the quantum expression for $\sigma_{zz}(Z)$ which is corrected here.

The non-local conductivities, $\sigma_{xx}(Z)$ and $\sigma_{zz}(Z)$ are shown in Fig. 2.2 for an electron density approximately equal to that of copper. The CIP and CPP non-local conductivity both decrease monotonically as a function of the distance between applied field and induced current.

2.3.1 Semiclassical Limit

The semiclassical limit of the non-local conductivities can be obtained by replacing the damped oscillatory functions with exponentially decaying functions that have the same volume integral, thus

$$
\begin{aligned}
-2\mathrm{i}\kappa t\ \mathrm{d}Z\,\mathrm{e}^{2\mathrm{i}\kappa Zt} &\rightarrow 2\kappa_I t\ \mathrm{d}Z\,\mathrm{e}^{-2k_I Zt} \\
\frac{\mathrm{d}Z}{Z}\mathrm{e}^{2\mathrm{i}\kappa Zt} &\rightarrow \frac{\mathrm{d}Z}{Z}\mathrm{e}^{-2\kappa_I Zt} \\
\frac{\mathrm{d}Z}{-\mathrm{i}\kappa Z^2}\mathrm{e}^{2\mathrm{i}\kappa Zt} &\rightarrow \frac{\mathrm{d}Z}{\kappa_I Z^2}\mathrm{e}^{-2\kappa_I Zt}\ .
\end{aligned}
\tag{2.30}
$$

Using this substitution in (2.21) yields

$$
\sigma_{xx}(Z) = \frac{e^2\kappa_R^2}{8\pi^2\hbar}E_{1,3}(2Z\kappa_I) = \frac{e^2\kappa_R^2}{8\pi^2\hbar}\int_1^\infty [t^{-1} - t^{-3}]e^{-|z-z'|/\lambda}\ . \tag{2.31}
$$

The semiclassical result for the transverse non-local conductivity is compared to the quantum result in Fig. 2.2. For $\sigma_{zz}(Z)$ the semiclassical and quantum results for the non-local conductivity of a homogeneous FERPS system are equivalent,

$$
\sigma_{zz}(Z) = \frac{e^2\kappa_R^2}{4\pi^2\hbar}E_3(2Z\kappa_I) = \frac{e^2\kappa_R^2}{4\pi^2\hbar}\int_1^\infty t^{-3}e^{-|z-z'|/\lambda}\ . \tag{2.32}
$$

These expressions for the non-local conductivity, (2.31 and 2.32) allow a generalization for the case in which the electron mean free path, λ, varies slowly with z,

$$
e^{-|z-z'|/\lambda} \rightarrow e^{-\int_{z<}^{z>}\mathrm{d}z''/\lambda(z'')}\ . \tag{2.33}
$$

In this section we have obtained the semiclassical expressions for the non-local conductivity as an average of the quantum expressions. They are usually derived from a classical approach that is outlined in the next section.

2.4 The Semiclassical Approach to Transport

In the semi-classical approach to transport, the electrons are assumed to behave like classical particles. The only concessions made to their quantum nature is the use of Fermi statistics (which implies that it is the Fermi energy electrons that are important for transport) and the use of quantum mechanics to calculate the relation between electron energy and momentum and to calculate the transmission and reflection probabilities at interfaces.

Semiclassical transport theory begins with the concept of the electron distribution function, $f^s(\boldsymbol{k},\boldsymbol{r},t)$ which is defined as the number of electrons with given values of wavevector, $\boldsymbol{k}$ and spin s, at position $\boldsymbol{r}$ at time t. It is a 7 dimensional function (for each spin) measured in dimensionless units. In the absence of applied fields, the electrons will be at equilibrium and the distribution function will be the equilibrium distribution function $f_0(\mathrm{e}_k^s - \mu_0) = \left[1+\exp(\mathrm{e}_k^s - \mu_0)/k_\mathrm{B}T)\right]^{-1}$. We now imagine that a field has been applied but that we have waited long enough that the system is in a steady state, *i.e.* the distribution function is no longer changing so that $\mathrm{d}f/\mathrm{d}t = 0$. The Boltzmann equation is obtained by balancing the changes in the distribution function caused by the applied field against processes that act to bring it back towards equilibrium.

Thus at steady state, the time rate of change of the distribution function is given by,

$$\frac{\mathrm{d}f}{\mathrm{d}t} = 0 = \left.\frac{\partial f}{\partial t}\right|_{\mathrm{drift}} + \left.\frac{\partial f}{\partial t}\right|_{\mathrm{field}} + \left.\frac{\partial f}{\partial t}\right|_{\mathrm{scatt}} , \tag{2.34}$$

where the three contributions to the time derivative of the distribution function are due to electron drift, i.e. the movement of electrons in space due to their velocity, acceleration of the electrons due to the applied field, and the scattering of the electrons by imperfections (i.e. deviations from periodicity) in the lattice.

The drift term can be evaluated from the fact that the electrons entering a volume near point $\boldsymbol{r}$ at time t were previously at position $\boldsymbol{r} - \boldsymbol{v}\,\mathrm{d}t$ at time $t - \mathrm{d}t$. Thus

$$\left.\frac{\partial f(\boldsymbol{r},\boldsymbol{k},t)}{\partial t}\right|_{\mathrm{drift}} = -\boldsymbol{v}(\boldsymbol{k}) \cdot \nabla_{\boldsymbol{r}} f(\boldsymbol{r},\boldsymbol{k},t) . \tag{2.35}$$

Similarly, the field term can be evaluated as a drift term in momentum space because the electrons entering a volume of momentum space near point $\boldsymbol{k}$ at time t were previously located in momentum space at $\boldsymbol{k} - (\mathrm{d}\boldsymbol{k}/\mathrm{d}t)\,\mathrm{d}t$ at time $t - \mathrm{d}t$. Then, using Newton's second law to relate the force from the applied field to the rate of change of the electron momentum, $-\mathrm{e}\boldsymbol{E} = \hbar \mathrm{d}\boldsymbol{k}/\mathrm{d}t$ we have,

$$\left.\frac{\partial f(\boldsymbol{r},\boldsymbol{k},t)}{\partial t}\right|_{\mathrm{field}} = -\frac{\partial f(\boldsymbol{r},\boldsymbol{k},t)}{\partial \boldsymbol{k}}\frac{\partial \boldsymbol{k}}{\partial t} = \frac{\mathrm{e}}{\hbar}\nabla_{\boldsymbol{k}} f(\boldsymbol{r},\boldsymbol{k},t) \cdot \boldsymbol{E} , \tag{2.36}$$

where the symbol e represents the *magnitude* of the electronic charge.

The scattering term can be written in terms of the probability, $P_{\boldsymbol{k}\boldsymbol{k}'}$, for an electron to scatter between momentum states $\boldsymbol{k}$ and $\boldsymbol{k}'$. It will be the sum of the probabilities for an electron to scatter into state $\boldsymbol{k}$ from some other momentum state less the probability for an electron to scatter out of state $\boldsymbol{k}$,

$$\begin{aligned}\left.\frac{\partial f(\boldsymbol{r},\boldsymbol{k},t)}{\partial t}\right|_{\mathrm{scatt}} &= \sum_{\boldsymbol{k}'} P_{\boldsymbol{k}\boldsymbol{k}'} \left\{ f(\boldsymbol{r},\boldsymbol{k}',t)\,[1 - f(\boldsymbol{r},\boldsymbol{k},t)] \right. \\ &\quad \left. - f(\boldsymbol{r},\boldsymbol{k},t)\left[1 - f(\boldsymbol{r},\boldsymbol{k}',t)\right]\right\} \\ &= \sum_{\boldsymbol{k}'} P_{\boldsymbol{k}\boldsymbol{k}'} \left[f(\boldsymbol{r},\boldsymbol{k}',t) - f(\boldsymbol{r},\boldsymbol{k},t)\right] .\end{aligned} \tag{2.37}$$

In principle, the scattering probability should also include processes that scatter electrons from one spin-channel to another, but we neglect those here because we are concentrating on the two current model.

Assembling the three terms, and assuming steady state, we obtain the Boltzmann equation,

$$-\boldsymbol{v}(\boldsymbol{k}) \cdot \nabla_{\boldsymbol{r}} f(\boldsymbol{r},\boldsymbol{k}) + \frac{\mathrm{e}}{\hbar}\nabla_{\boldsymbol{k}} f(\boldsymbol{r},\boldsymbol{k}) \cdot \boldsymbol{E} + \sum_{\boldsymbol{k}'} P_{\boldsymbol{k}\boldsymbol{k}'} \left[f(\boldsymbol{r},\boldsymbol{k}') - f(\boldsymbol{r},\boldsymbol{k})\right] = 0 . \tag{2.38}$$

We are attempting to calculate a linear response that is proportional to the field so we write the distribution function as the equilibrium distribution function plus a correction term called the "deviation" function that describes the deviation from equilibrium, $f(\boldsymbol{r},\boldsymbol{k},t) = f_0(\varepsilon_k - \mu_0) + g(\boldsymbol{r},\boldsymbol{k})$. Substituting this form into the Boltzmann equation we obtain,

$$\begin{aligned} & -\boldsymbol{v}(\boldsymbol{k}) \cdot \nabla_{\boldsymbol{r}} g(\boldsymbol{r},\boldsymbol{k}) + \frac{\mathrm{e}}{\hbar} \nabla_{k} f_0(\mathrm{e}_k - \mu_0) \cdot \boldsymbol{E} \\ & + \sum_{\boldsymbol{k}'} P_{\boldsymbol{k}\boldsymbol{k}'} \left[g(\boldsymbol{r},\boldsymbol{k}') - g(\boldsymbol{r},\boldsymbol{k}) \right] = 0 , \end{aligned} \tag{2.39}$$

where we have only retained the lowest order contribution to the field term because the field, E is assumed to be small. The field term can be further simplified using

$$\nabla_{k} f_0(\varepsilon_k - \mu_0) = \frac{\partial f_0(\varepsilon_k - \mu_0)}{\partial \varepsilon_k} \nabla_{k} \varepsilon_k = \frac{\partial f_0(\varepsilon_k - \mu_0)}{\partial \varepsilon_k} \hbar \boldsymbol{v}(\boldsymbol{k}) , \tag{2.40}$$

and the "scattering-out" term can be simplified using

$$\sum_{\boldsymbol{k}'} P_{\boldsymbol{k}\boldsymbol{k}'} = 1/\tau_k ; \tag{2.41}$$

which defines the lifetime as the inverse of the total scattering rate for electrons to scatter out of momentum state $\boldsymbol{k}$. Thus, the Boltzmann equation becomes,

$$\begin{aligned} & -\boldsymbol{v}(\boldsymbol{k}) \cdot \nabla_{\boldsymbol{r}} g(\boldsymbol{r},\boldsymbol{k}) - \frac{g(\boldsymbol{r},\boldsymbol{k})}{\tau_k} + \sum_{\boldsymbol{k}'} P_{\boldsymbol{k}\boldsymbol{k}'} g(\boldsymbol{r},\boldsymbol{k}') \\ & = -\mathrm{e} \frac{\partial f_0(\varepsilon_k - \mu_0)}{\partial \varepsilon_k} \boldsymbol{v}(\boldsymbol{k}) \cdot \boldsymbol{E} , \end{aligned} \tag{2.42}$$

and the current density is given by

$$\boldsymbol{J}(\boldsymbol{r}) = -\frac{\mathrm{e}}{V} \sum_{\boldsymbol{k}} \boldsymbol{v}(\boldsymbol{k}) g(\boldsymbol{r},\boldsymbol{k}) . \tag{2.43}$$

Often, we do not know very much about the details of the scattering probability $P_{\boldsymbol{k}\boldsymbol{k}'}$. In these cases it is popular to make the "lifetime approximation" which consists of dropping the scattering-in term, $\sum_{\boldsymbol{k}'} P_{\boldsymbol{k}\boldsymbol{k}'} f(\boldsymbol{k}')$. The $\boldsymbol{k}$ dependence of the lifetime is also often neglected. If the scattering is isotropic in the sense that $P_{\boldsymbol{k}\boldsymbol{k}'}$ does not depend on the angle between $\boldsymbol{k}$ and $\boldsymbol{k}'$, then one can often argue that the scattering-in term vanishes or is small because of symmetry because $g(\boldsymbol{k}')$ usually vanishes when summed over $\boldsymbol{k}'$. In general, however, the scattering is not isotropic and the neglect of the scattering-in term is an important, non-trivial approximation. This is particularly true of the case in which the system is inhomogeneous in the direction of the applied field and current.

2.4.1 Layered Systems

If we consider a system that is homogeneous in the x and y directions but with properties that may vary in the z direction, we may write (2.42) as

$$-v_z(z,\boldsymbol{k})\frac{\partial h(z,\boldsymbol{k})}{\partial z} - \frac{h(z,\boldsymbol{k})}{\tau_k} + \sum_{k'} P_{kk'} h(z,\boldsymbol{k}') = e\boldsymbol{v}(z,\boldsymbol{k}) \cdot \boldsymbol{E} , \tag{2.44}$$

where we substituted $g(\boldsymbol{r},\boldsymbol{k}) = h(\boldsymbol{r},\boldsymbol{k})\delta(\varepsilon_k - \varepsilon_F)$ because, for metals[2] at room temperature and below, it is the states at the Fermi energy that are involved in transport. We also assumed that the scattering probability, $P_{kk'}$ is energy conserving, i.e. $\varepsilon_k = \varepsilon_{k'}$. This is valid if the scattering is dominated by electron-impurity scattering and is not badly wrong for electron-phonon scattering.

We can take advantage of the fact that it is only electrons on the Fermi surface that participate in transport to explicitly reduce the sums over all momentum states that occur, for example, in (2.43) to two dimensional sums over the transverse momentum. Thus for an arbitrary function of momentum, $y(z,\boldsymbol{k})$,

$$\begin{aligned}\sum_k y(z,\boldsymbol{k})g(z,\boldsymbol{k}) &= \sum_k y(z,\boldsymbol{k})h(z,\boldsymbol{k})\delta(\varepsilon_k - \varepsilon_F) \\ &= \frac{V}{2\pi A}\sum_{k_\parallel}\left[y(z,\boldsymbol{k}_\parallel,k_z^+)\frac{h^+(z,\boldsymbol{k}_\parallel)}{\hbar|v_z^+(\boldsymbol{k}_\parallel)|} + y(z,\boldsymbol{k}_\parallel,k_z^-)\frac{h^-(z,\boldsymbol{k}_\parallel)}{\hbar|v_z^-(\boldsymbol{k}_\parallel)|}\right] .\end{aligned} \tag{2.45}$$

Here we have used the fact that for a given value of transverse momentum, $\boldsymbol{k}_\parallel$, the solutions to $\varepsilon(\boldsymbol{k}_\parallel, k_z) = \varepsilon_F$ occur in pairs with opposite signs for the z-component of the Fermi velocity. Thus

$$\delta(\varepsilon_k - \varepsilon_F) \to \frac{\delta\left(k_z(\boldsymbol{k}_\parallel) - k_z^+\right)}{\hbar|v_z^+(\boldsymbol{k}_\parallel)|} + \frac{\delta(k_z(\boldsymbol{k}_\parallel) - k_z^-)}{\hbar|v_z^-(\boldsymbol{k}_\parallel)|} . \tag{2.46}$$

It should be noted that, in general there may be more than one pair of positive and negative v_z states at any value of $\boldsymbol{k}_\parallel$. It should also be noted that $|v_z^+|$ is in general only equal to $|v_z^-|$ if the lattice has mirror symmetry in a plane perpendicular to the z direction. $h^+(z,\boldsymbol{k}_\parallel)$ and $h^-(z,\boldsymbol{k}_\parallel)$ are respectively the distribution functions for $+z$-going and $-z$-going electrons respectively.

2.4.2 Semiclassical Non-Local Conductivity for FERPS

In order to make contact with the expressions for the non-local conductivity derived in the Sect. 2.3.1 let us evaluate these quantities for the FERPS model using the Boltzmann equation,

[2] Note that this approximation would only be valid for a doped semiconductor at low temperature. It might also be questionable if there is a rapid variation in the electronic structure near E_F (on the scale of $k_B T$).

$$v_z^{\pm}\frac{\partial h^{\pm}(z,\boldsymbol{k}_{\|})}{\partial z}+\frac{h^{\pm}(z,\boldsymbol{k}_{\|})}{\tau}=-e\boldsymbol{v}^{\pm}\cdot\boldsymbol{E}\delta(z-z')\,. \tag{2.47}$$

Here the driving term for the Boltzmann equation is only applied at the plane defined by $z=z'$. Note that we have omitted the scattering-in term. This is a sensible approximation for $\sigma_{xx}(Z)$ (CIP), but not for $\sigma_{zz}(Z)$ (CPP). As explained in Sect. 2.6.2, however, the formula for $\sigma_{zz}(Z)$ is meaningful if the scattering is isotropic and the local electric field that is used to calculate the current is the total field including that arising from charge accumulation.

It is readily verified that the solution to (2.47) is given by

$$\begin{aligned} h^{+}(z,\boldsymbol{k}_{\|})&=-\theta(z-z')\frac{\boldsymbol{v}^{+}(\boldsymbol{k}_{\|})\cdot\boldsymbol{E}}{|v_z^{+}(\boldsymbol{k}_{\|})|}\mathrm{e}^{-\frac{|z-z'|}{|v_z^{+}|\tau}}\\ h^{-}(z,\boldsymbol{k}_{\|})&=-\theta(z-z')\frac{\boldsymbol{v}^{-}(\boldsymbol{k}_{\|})\cdot\boldsymbol{E}}{|v_z^{-}(\boldsymbol{k}_{\|})|}\mathrm{e}^{-\frac{|z-z'|}{|v_z^{-}|\tau}}\,. \end{aligned} \tag{2.48}$$

Substituting for the current using (2.43) and (2.45) yields

$$\begin{aligned} \sigma_{zz}(Z)&=\frac{e^2}{2\pi^2\hbar}\int_0^{k_\mathrm{F}}k_{\|}\,\mathrm{d}k_{\|}\,\mathrm{e}^{-\frac{Zk_\mathrm{F}}{\lambda k_z}}\\ \sigma_{xx}(Z)&=\frac{e^2}{4\pi^2\hbar}\int_0^{k_\mathrm{F}}k_{\|}\,\mathrm{d}k_{\|}\frac{k_{\|}^2-k_z^2}{k_z^2}\mathrm{e}^{-\frac{Zk_\mathrm{F}}{\lambda k_z}} \end{aligned} \tag{2.49}$$

where we used $Z=|z-z'|$, and $v_z\tau=k_z\lambda/k_\mathrm{F}$. Substitution of $t=k_\mathrm{F}/k_z(k_{\|})$ yields (2.32 and 2.31) immediately.

2.4.3 Quantum and Semiclassical Conductivities for Multilayers

If we generalize the FERPS model slightly to allow the self-energy to depend on z, the transverse non-local conductivity will be given by

$$\sigma_{xx}(z,z')=-\frac{e^2\hbar^3}{8\pi^3m^2}\int\mathrm{d}^2\boldsymbol{k}_{\|}\,\nabla_x\mathrm{Im}G(\boldsymbol{k}_{\|};z,z')\nabla_x'\mathrm{Im}G(\boldsymbol{k}_{\|};z',z)\,, \tag{2.50}$$

where the partial Green function, $G(k_{\|};z,z')$, is defined by

$$\left[E+\frac{\hbar^2}{2m}\left(\frac{\partial^2}{\partial z^2}-k_{\|}^2\right)-\Sigma(z)\right]G(\boldsymbol{k}_{\|};z,z')=\delta(z-z')\,. \tag{2.51}$$

The solution to the quantum FERPS model can be obtained in a closed form for a homogeneous thin film [2.12, 13]. For multilayers, however, it can only be obtained numerically. The partial Green function, $G(k_{\|};z,z')$, can be calculated using the general solution [2.14],

$$G(k_{\parallel}; z, z') = \frac{\psi_{\rm L}(z_<)\psi_{\rm R}(z_>)}{W} , \tag{2.52}$$

where $\psi_{\rm L}(z)$ and $\psi_{\rm R}(z)$ are solutions to the homogeneous part of (2.51) which satisfy boundary conditions on the left and right sides respectively of the multilayer, and where W is the Wronskian of $\psi_{\rm L}$ and $\psi_{\rm R}$. Thus the quantum calculation can be performed relatively simply by solving a one-dimensional Schroödinger equation.

Let us compare this result with the generalized semiclassical approach introduced in (2.31) and (2.33). In this case the non-local transverse conductivity is given by,

$$\sigma_{xx}(z, z') = \frac{e^2}{8\pi^2\hbar} k_{\rm F}^2 \int_1^{\infty} {\rm d}t \left(\frac{1}{t} - \frac{1}{t^3}\right) {\rm e}^{-\phi(z,z')t} , \tag{2.53}$$

where $k_{\rm F}$ is the Fermi wavevector and $\phi(z, z')$ is given by an integral over the inverse of the local mean free path

$$\phi(z, z') = \int_{z_<}^{z_>} {\rm d}z'' \frac{1}{\lambda(z'')} . \tag{2.54}$$

In Fig. 2.3 we display a comparison of the current density defined as $\sigma_{xx}(z) = \int {\rm d}z' \sigma_{xx}(z, z')$ calculated for a multilayer film within the model of free electrons with random point scatterers. In this calculation it was assumed that the current flowed parallel to the layers – known as the current in the plane or CIP configuration. This figure indicates that the semiclassical approach described in Sect. 2.4 becomes a good approximation to the quantum theory of transport as the thickness of the layers approaches or exceeds the electronic mean free path. As deposition techniques improve and magnetic multilayers become more perfect, it will become more important to

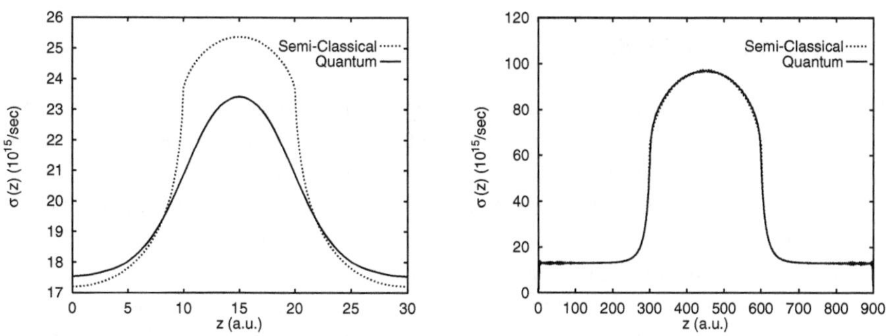

Fig. 2.3. *Left panel:* Local CIP conductivity for a trilayer consisting of a central clean layer between 10 and 20 a.u. surrounded by two dirty layers. The mean free path for the clean layer is $\lambda = 360$ a.u. and for the dirty layer $\lambda = 36$ a.u. *Right Panel:* CIP conductivity for a trilayer consisting of a central clean layer (300 a.u. thick, $\lambda = 36$ a.u.) surrounded by dirty layers (600 a.u. thick, $\lambda = 360$ a.u.)

treat them using the quantum theory. The semiclassical theory, however captures much of the physics, is computationally much faster and is easier to understand.

At this point, let us note two important "classical" limits which provide limiting cases for any theory of transport in multilayers. The first limit is the so-called "thin-layer" limit. This is a limit in which the thickness of each layer is small compared to the electron mean-free-path. In this limit, the effective scattering rate of the electron is the average of the scattering rates in all layers. Thus,

$$\frac{1}{\lambda_{\text{thin}}} = \sum_I \frac{d_I}{d} \frac{1}{\lambda_I} , \tag{2.55}$$

where λ_{thin} is the effective mean-free-path of the system, d_I is the thickness of layer I, λ_I is the mean-free-path in layer I, and d is the total thickness of all the layers. At the other extreme, when the layer thickness is large compared to the mean-free-path, each layer can be considered as a separate resistor in a resistor network, and the limits are different for the current-in-plane (CIP) geometry and for the current-perpendicular-to-plane (CPP) geometry. For CPP, we have a resistors in series network and,

$$\frac{1}{\lambda_{\text{CPP}}} = \sum_I \frac{d_I}{d} \frac{1}{\lambda_I} , \tag{2.56}$$

which is the same as the thin limit. This is often referred to as the "self-averaging" character of the CPP resistance. In the case of CIP, we have a resistors in parallel network and the limit is,

$$\lambda_{\text{thick}} = \sum_I \frac{d_I}{d} \lambda_I . \tag{2.57}$$

2.5 Electronic Structure

2.5.1 Two Current Model

The theoretical approaches described in the preceding two sections have not included the possible differences in the two spin states of the conduction electrons. If the coupling between these two spin states can be ignored then they can be treated as channels that conduct independently. For non-magnetic metals such as copper, the two spin channels are equivalent in the sense that they have the same Fermi energy, density of states, and electron velocities, and therefore carry the same current in response to an applied electric field. Thus the only effect of spin in a non-magnetic material, is the doubling of the number of channels available for conduction and consequently the doubling of the conductance. For the ferromagnetic transition metals and alloys, however, the two channels are quite different[3]. It is often a good approximation to

[3] The most extreme difference is found in a class of conductors called "half-metals" which conduct in one spin channel but have no states at the Fermi energy in the other.

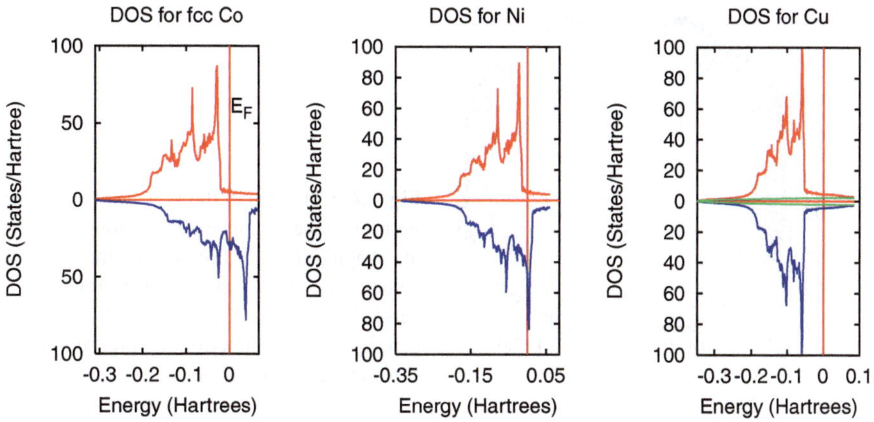

Fig. 2.4. Electronic density of states for fcc-cobalt (*left panel*), nickel (*center*) and copper (*right panel*). The copper DOS also shows a parabolic "free electron" DOS. The departure of the DOS from this parabola is largely due to the *d*-states. The vertical line through $E = 0$ indicates the Fermi energy. Majority (minority) DOS are shown above (below) the Energy axis

assume that two spin channels conduct independently. This approximation is called the "two current" model. In this model each spin channel can be considered separately within transport theory, and the total current is the sum of the currents from each of the two spin channels. The limitations of the two current model and some of the phenomena that it omits will be discussed in Sect. 2.5.5.

Within the "two current" model, the electrons states, $\psi_{j,s,\boldsymbol{k}}(\boldsymbol{r})$, for a periodic system are labelled by wave-vector $\boldsymbol{k}$, band index j and spin s. Associated with each state is an energy, $\varepsilon_{j,s}(\boldsymbol{k})$ called the band energy. Knowledge of the band energy as a function of wave-vector is sufficient to calculate two very important quantities, the density of states as shown, for example, in Figs. 2.4 and 2.5 and the electron velocities (Figs. 2.6 and 2.7).

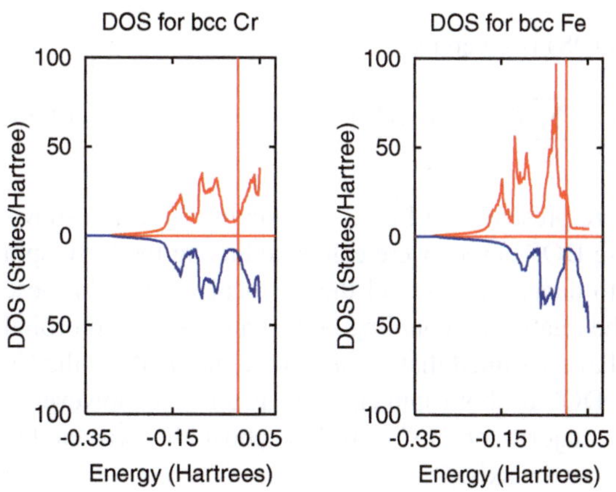

Fig. 2.5. Electronic density of states for nonmagnetic chromium (*left panel*) and for iron (*right panel*)

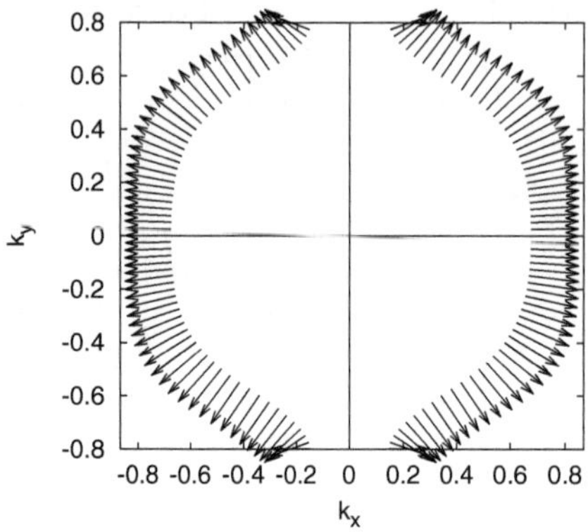

Fig. 2.6. Cut through the $k_z = 0$ plane of the Fermi surface of copper. The arrows indicate the magnitude and direction of the electron velocity. k_x and k_y are given in units of inverse Bohr radii (1.89 Å^{-1}). The k_y direction here is the (111) direction with respect to the conventional cubic axes

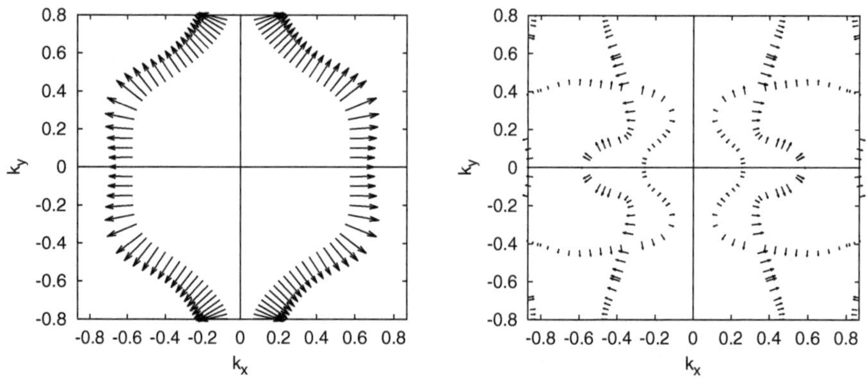

Fig. 2.7. Cuts through the $k_z = 0$ plane of the majority (*left panel*) and minority (*right panel*) Fermi surfaces of cobalt. The arrows representing the velocities are drawn to the same scale as for copper in Fig. 2.6. k_x and k_y are given in units of inverse Bohr radii

2.5.2 Density of States

The electronic density of states (DOS) is given by,

$$n_s(E) = \sum_{j,k} \delta\left[E - \varepsilon_{j,s}(\boldsymbol{k})\right] . \tag{2.58}$$

The DOS of fcc Co, Ni and Cu are presented in Fig. 2.4. Likewise Fig. 2.5 shows the DOS for bcc Cr and Fe. These DOS curves were calculated using the local spin density approximation (LSDA) to density functional theory local [2.15]. In these calculations we have neglected the relativistic coupling between the spin and orbital motions of the electrons and we have assumed that the magnetic moments of the Co and Ni atoms are all aligned. The DOS for both spin channels is positive, however, it has become traditional to plot the majority or "up" spin DOS above the axis and the minority or "down" spin DOS below the axis.

The density of states of the 3-d transition metals is often thought of as a free-electron like parabola ($n \propto \sqrt{E}$). For metals such as Cr, Fe, Co, Ni or Cu this parabola begins approximately 0.3 Hartrees[4] ($\approx$ 8 eV) below the Fermi energy. Because the electrons forming the "free electron" component of the DOS have an effective mass of the same order as the electron mass, a Fermi energy of 0.3 Hartree corresponds to approximately 0.5 electrons per spin channel. The free electron part of the DOS can be barely discerned in the right hand panel of Fig. 2.4 which shows the DOS for Cu. Superimposed upon this free electron DOS is the DOS associated with the "*d*-bands" that derive from the *d*-states of the isolated atoms.

In the isolated atom there would be 10 degenerate[5] states, five for each spin, but in the solid they interact with each other and with the "free electron" states to form a complex of bands containing 5 electrons per spin channel spread over an energy range of 5–10 eV. Actually, the hybridization between the free electron bands and the "*d*-state" derived bands is quite strong so that effects of the "*d*-states" extend well above the nominal top of the "*d*-bands".

The difference between the number of majority and minority valence electrons is the spin magnetic moment per atom measured in Bohr magnetons. This number is zero for Cu as it has identical DOS for both spin states. For Co, the difference is 1.6 (which may be compared to the experimental value of about 1.75 Bohr magnetons for the total moment, which includes both spin and orbital contributions), and for Ni it is 0.6. Note that both the majority and minority *d*-bands are filled for copper. For Co and Ni the majority *d*-bands are filled while and the Fermi energy falls within the minority *d*-bands which are only partially filled.

Figure 2.5 shows the DOS for non-magnetic chromium and for ferromagnetic bcc iron. For bcc Fe and Cr, neither element has either majority or minority *d*-bands filled. Note the similarity of the DOS curves near the Fermi energy for Co, Ni and Cu in the majority channel and for Fe and Cr in the minority channel. These similarities in electronic structure form the basis for the spin-dependent scattering that leads to the GMR effect.

2.5.3 Velocities of Bloch Electrons at the Fermi Energy

The electron velocities at the Fermi energy will be very important for electron transport. The electron velocity is determined from the gradient of the energy bands with respect to the crystal momentum [2.17],

$$\boldsymbol{v}_{j,s,\boldsymbol{k}}(\boldsymbol{k}) \equiv \frac{-\mathrm{i}\hbar}{m} \left\langle \psi_{j,s,\boldsymbol{k}} | \nabla | \psi_{j,s,\boldsymbol{k}} \right\rangle = \nabla_{\boldsymbol{k}} \varepsilon_{j,s}(\boldsymbol{k}) / \hbar \,. \tag{2.59}$$

These velocities vary considerably from point to point on the Fermi surface, both in magnitude and in direction as can be seen in Figs. 2.6 and 2.7. The Fermi surface, defined by $\varepsilon_{j,s}(\boldsymbol{k}) = E_\mathrm{F}$, is the iso-energy surface in $\boldsymbol{k}$-space that separates

[4] 1 Hartree = 27.2 eV.

[5] This assumes that we neglect spin-orbit coupling

the occupied and unoccupied electronic states. It is the states at the Fermi energy that participate in electron transport for modest electrical fields and temperatures.

The Fermi surface and electron velocities are two important ingredients needed for calculating the transport properties of metals in the semiclassical approximation. In order to treat magnetic multilayered systems, however, we will need to understand the electronic structure near interfaces as well.

2.5.4 Electronic Structure Near Interfaces

Figure 2.8 shows the number of valence electrons per atom for each spin-channel and for each atomic layer near the interfaces between permalloy ($Ni_{0.8}Fe_{0.2}$)and cobalt and also between cobalt and copper. For these metallic systems of similar atomic size and electronegativity there are only small perturbations in the number of electrons per atom and spin channel on the layers near the interface. Even when the interfacial perturbations on the electronic structure are significantly larger, they can be incorporated into the the transmission and reflection probabilities because interfacial charge rearrangements are usually limited to a few layers on either side of the interface in good metals.

By putting together the fact demonstrated by Fig. 2.8 that perturbations in the electronic structure at interfaces in magnetic multilayers are confined to a few atomic layers on either side of the interface and the fact demonstrated by Figs. 2.4–2.7 that materials can be found that match much better in one spin channel than the other we can obtain an understanding of the physical origin of giant magnetoresistance. For Cu, Ni, and fcc Co there is good matching in the majority spin channel. For Fe and Cr, there is good matching in the minority channel. Giant magnetoresistance arises from a "short circuit" effect caused by the low resistance in the channel for which this matching occurs.

If the system is composed of layers of different material stacked in the z direction, it is often a good approximation to assume that we have two dimensional periodicity within each layer. If the layers are not too thin, we may also imagine that within

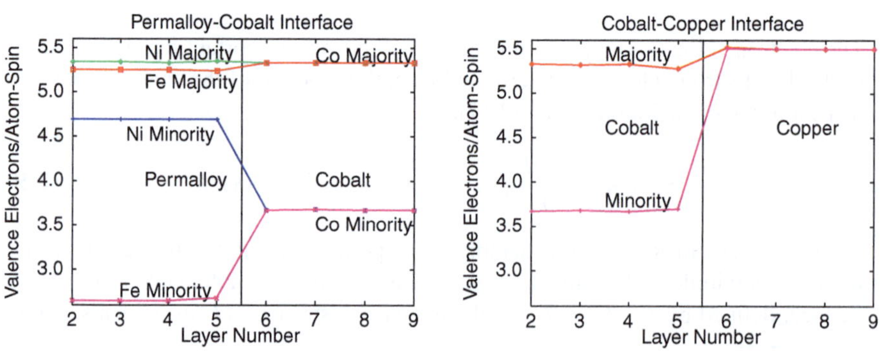

Fig. 2.8. Number of electrons in each spin-channel for a permalloy-cobalt and a cobalt-copper interface

each layer we can use the dispersion relation appropriate to that material in bulk. We would, of course, need to be careful to obtain the correct relative placement of the energy bands because, in general, when two materials are brought together, a dipole layer forms at the interface to balance the electrochemical potentials and allow the materials to have their correct Fermi energies far from the interfaces. These interfacial dipoles can be calculated by modern self-consistent electronic structure codes [2.18]

These approximations lead, then, to a model in which the band energies, $\varepsilon_{n,s,i}(\boldsymbol{k})$, and velocities,$\boldsymbol{v}_{n,s,i}(\boldsymbol{k})$ within each layer are assumed to be those for a perfect (infinite) crystal. Here, an additional index, i, has been added to label the layer. The layers are separated by thin, interfacial regions that can be described by transmission and reflection probabilities as we shall show in a later section.

2.5.5 Corrections to the Two Current Model

It is important to remember that the two current model is an approximation. It is valid in a limit in which spin-orbit coupling is neglected and in which it is assumed that the direction of the magnetic moments of all of the atoms are aligned (parallel or antiparallel) along the same axis. Spin-orbit coupling is a relativistic effect that couples the spin and orbital motions of the electrons. It introduces a small additional term into the non-relativistic Schrödinger equation of the form,

$$H_{\mathrm{so}} = \frac{\hbar^2}{2m^2c^2r}\frac{\mathrm{d}V}{\mathrm{d}r}\boldsymbol{L}\cdot\boldsymbol{S}\,, \tag{2.60}$$

with $\boldsymbol{L}$ and $\boldsymbol{S}$ the orbital and spin angular momentum operators, respectively, and $V(r)$ the effective potential (assumed here to depend only on the distance to the nucleus). Because this term couples the spin and orbital motions of the electrons, their energy bands can no longer be described as purely up or down spin.

If all of the magnetic moments are not aligned parallel or antiparallel, the two current model also breaks down. In a ferromagnet (ignoring spin-orbit coupling) the majority and minority electrons experience different potentials, $V^{\uparrow}$ and $V^{\downarrow}$, respectively. If the moments in two nearby magnetic layers are aligned anti-parallel, the majority electrons in the first layer (where they experience potential $V^{\uparrow}$) will experience potential $V^{\downarrow}$ when they travel to the nearby layer with moments antiparallel to the first. Nevertheless, the two spin channels can be treated separately. If, on the other hand, the moments in the second ferromagnetic layer are aligned at some arbitrary angle relative to the first layer, the two spin channels defined in the first layer will be coupled in the second layer, and they can no longer be treated separately. In this case one can consider the moments in the first layer to be aligned in the z-direction. In the second layer, the atomic potentials may be expressed in the form of a two by two matrix in spin space,

$$\underline{\mathrm{V}}(\boldsymbol{r}) = \frac{1}{2}\left[V^{\uparrow}(\boldsymbol{r}) + V^{\downarrow}(\boldsymbol{r})\right]\underline{1} + \frac{1}{2}\left[V^{\uparrow}(\boldsymbol{r}) - V^{\downarrow}(\boldsymbol{r})\right]\underline{\sigma}\cdot\hat{\boldsymbol{e}} \tag{2.61}$$

Here $\underline{\sigma}$ is a vector constructed from two dimensional Pauli spin $\frac{1}{2}$ matrices [2.19], $\underline{\sigma} = \sigma_x\hat{\boldsymbol{x}} + \sigma_y\hat{\boldsymbol{y}} + \sigma_z\hat{\boldsymbol{z}}$, and $\hat{\boldsymbol{e}}$ is a unit vector in the direction of the moments in the second layer relative to the first. Because σ_x and σ_y are non-diagonal, the two spin channels can only be be treated separately if $\hat{\boldsymbol{e}}$ is in the $\pm z$ direction. Of course, if all moments are collinear, we can choose the z direction to point along the moment direction and the potential will be either $V^{\uparrow}(\boldsymbol{r})$ or $V^{\downarrow}(\boldsymbol{r})$.

2.6 Transport in Layered Systems

In this section, we shall generalize the semiclassical theory discussed in Sect. 2.4 primarily in terms of the FERPS model to more realistic electronic stuctures. The quantum theory of transport has also been successfully applied to realistic electronic structures as shown, for example, in references [2.18, 20–31]. The full quantum calculations become rather difficult as the number of atomic layers included in the calculation becomes large. More importantly, for our present purposes, the physics of the conduction process is less transparent than for the semiclassical theory. Fortunately, most of the important physics can be can be retained in a semiclassical calculation, at least approximately, if one uses realistic electronic structures to describe the layers. This can be accomplished by solving the Boltzmann transport equation for the electrons using realistic Fermi surfaces,and group velocities for the Bloch electrons at the Fermi energy, and using correct multiband transmission and reflection matrices at the interfaces for boundary conditions.

In this and the following sections we shall assume that the materials are homogeneous in the x and y directions but that they vary, (different materials, interfaces, boundaries, etc.) in the z direction. Because we have boundaries and interfaces, etc., the distribution function will vary with z and will satisfy (2.44). The new feature that we must address when we consider realistic electronic structures is that that they will vary from layer to layer if the layers consist of different metals or alloys. We will deal with this complication by solving the Boltzmann equation for each layer and applying the proper boundary conditions at the interfaces between the layers to obtain a solution valid for the entire film.

2.6.1 Boundary Conditions

As is well known, the solution to a differential equation is not uniquely determined until a proper set of boundary conditions is specified. The key to applying the boundary conditions for layered systems is to realize that electrons travelling in the $+z$ direction satisfy a different boundary condition from those travelling in the $-z$ direction. This was first worked out for single layer films by Fuchs [2.32] and the generalization to multilayers [2.33–36] is relatively straightforward.

The boundary conditions on[6] $h^{\pm,j}(z, \boldsymbol{k}_{\|})$ are obtained by requiring particle conservation at each of the interfaces. Since $h_i^{+,j}(z, \boldsymbol{k}_{\|})$, and $h_i^{-,j}(z, \boldsymbol{k}_{\|})$ represent the distribution functions in layer i for electrons travelling in the $+z$ and $-z$ directions

[6] The band index j is needed because in realistic electronic structures as opposed to free electron models, there may be more than one band for a given value of $\boldsymbol{k}_{\|}$.

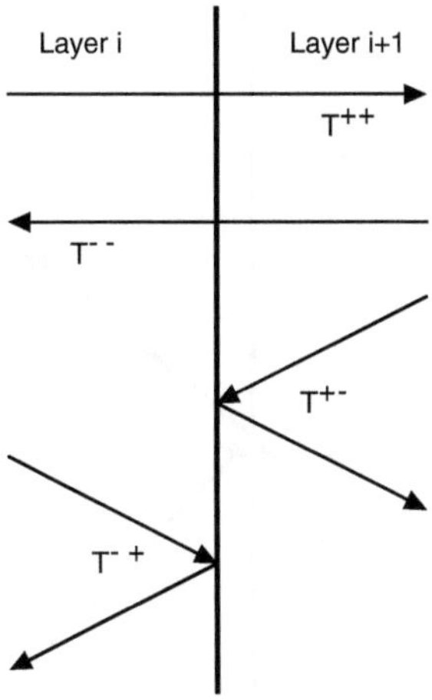

Fig. 2.9. Convention for the transmission and reflection probabilities

respectively, we can express the relationships between the distribution functions in layers i and $i+1$ (with interface at z_i) in terms of the transmission (T_i^{++}, T_i^{--}) and reflection (T_i^{+-}, T_i^{-+}) probabilities of the interfaces. We use a convention illustrated in Fig. 2.9 in which, for example, $T_i^{+-}(\boldsymbol{k}, \boldsymbol{k}')$ is the probability for a $-z$ going electron in Bloch state $\boldsymbol{k}'$ incident on interface i to leave the interface going in the $+z$ direction in Bloch state $\boldsymbol{k}$. Consider the flux of electrons leaving this interface travelling in the $+z$ direction (in layer $i+1$), $\sum_{j,\boldsymbol{k}_\parallel} h_{i+1}^{+,j}(z, \boldsymbol{k}_\parallel)$. This flux is the sum of the transmitted flux of $+z$ going electrons from layer i and the reflected flux from those electrons originally travelling in the $-z$ direction in layer $i+1$. A similar flux conservation argument relates the $-z$ going electron flux leaving the interface to the incoming fluxes in the two layers, thus,

$$
\begin{aligned}
h_{i+1}^{+,j}(z_i^+, \boldsymbol{k}_\parallel) &= \sum_{j',\boldsymbol{k}'_\parallel}^{N_\mathrm{R}} T_i^{+-}(j\boldsymbol{k}_\parallel, j'\boldsymbol{k}'_\parallel) h_{i+1}^{-,j'}(z_i^+, \boldsymbol{k}'_\parallel) \\
&\quad + \sum_{j',\boldsymbol{k}'_\parallel}^{N_\mathrm{L}} T_i^{++}(j\boldsymbol{k}_\parallel, j'\boldsymbol{k}'_\parallel) h_i^{+,j'}(z_i^-, \boldsymbol{k}'_\parallel) \\
h_i^{-,j}(z_i^-, \boldsymbol{k}_\parallel) &= \sum_{j',\boldsymbol{k}'_\parallel}^{N_\mathrm{L}} T_i^{-+}(j\boldsymbol{k}_\parallel, j'\boldsymbol{k}'_\parallel) h_i^{+,j'}(z_i^-, \boldsymbol{k}'_\parallel) \\
&\quad + \sum_{j',\boldsymbol{k}'_\parallel}^{N_\mathrm{R}} T_i^{--}(j\boldsymbol{k}_\parallel, j'\boldsymbol{k}'_\parallel) h_{i+1}^{-,j'}(z_i^+, \boldsymbol{k}'_\parallel) \,. \qquad (2.62)
\end{aligned}
$$

Here N_L and N_R denote the number of states on the left or right of the interface respectively for a given value of $\boldsymbol{k}'_\parallel$. If we assume that the layers have two dimen-

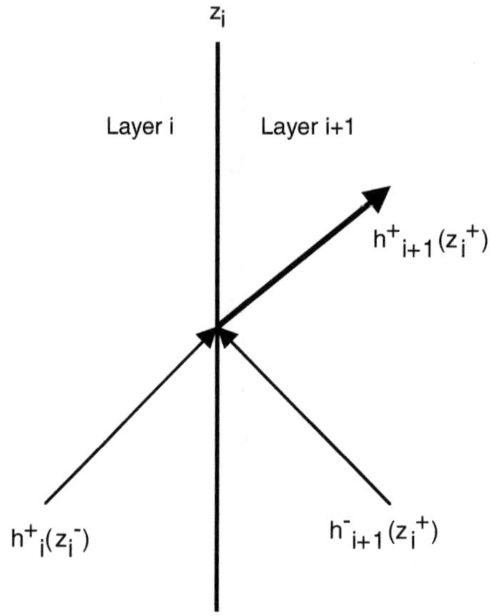

Fig. 2.10. The right-going beam in layer $i+1$, h^{+}_{i+1}, is the sum of the reflected part of the left going beam in that layer, h^{-}_{i+1} and the transmitted part of the right going beam in layer i, h^{+}_{i}

sional periodicity, so that the momentum parallel to the interface is conserved on transmission or reflection, the boundary conditions become,

$$h^{+,j}_{i+1}(z^{+}_{i},\boldsymbol{k}_{\|}) = \sum_{j'}^{N_{\mathrm{R}}} T^{+-}_{i}(j,j')h^{-,j'}_{i+1}(z^{+}_{i},\boldsymbol{k}_{\|}) + \sum_{j'}^{N_{\mathrm{L}}} T^{++}_{i}(j,j')h^{+,j'}_{i}(z^{-}_{i},\boldsymbol{k}_{\|})$$
$$h^{-,j}_{i}(z^{-}_{i},\boldsymbol{k}_{\|}) = \sum_{j'}^{N_{\mathrm{L}}} T^{-+}_{i}(j,j')h^{+,j'}_{i}(z^{-}_{i},\boldsymbol{k}_{\|}) + \sum_{j'}^{N_{\mathrm{R}}} T^{--}_{i}(j,j')h^{-,j'}_{i+1}(z^{+}_{i},\boldsymbol{k}_{\|})\,. \tag{2.63}$$

The first of these relations is shown pictorially in Figs. 2.10. The transmission and reflection matrices can be calculated from the underlying electronic structure of the layers and their interface [2.37]. Figures 2.11 and 2.12 show the transmission and reflection probabilities for Bloch waves in copper incident on cobalt. The transmission and reflection probabilities conserve electron flux. Thus, considering incident left and right-going waves of unit flux respectively, we can derive the following conservation rules,

$$\sum_{j}^{N_{\mathrm{R}}} T^{++}(j,j') + \sum_{j}^{N_{\mathrm{L}}} T^{-+}(j,j') = 1$$
$$\sum_{j}^{N_{\mathrm{L}}} T^{--}(j,j') + \sum_{j}^{N_{\mathrm{R}}} T^{+-}(j,j') = 1\,, \tag{2.64}$$

which state that the total probability for an electron to be transmitted or reflected is unity. Considering unit left and right-going fluxes leaving the interface we obtain,

$$\sum_{j'}^{N_L} T^{++}(j,j') + \sum_{j'}^{N_R} T^{+-}(j,j') = 1$$
$$\sum_{j'}^{N_R} T^{--}(j,j') + \sum_{j'}^{N_L} T^{-+}(j,j') = 1 , \tag{2.65}$$

which state that an electron leaving an interface must have been either transmitted or reflected.

Figures 2.11 and 2.12 show the transmission probabilities for the majority and minority channels for a Cu-Co interface. In the majority channel, the transmission is nearly unity except for values of $\boldsymbol{k}_\parallel$ near the neck where there are states in the cobalt layer, but not in the copper and for large values of $\boldsymbol{k}_\parallel$ for which there are states in

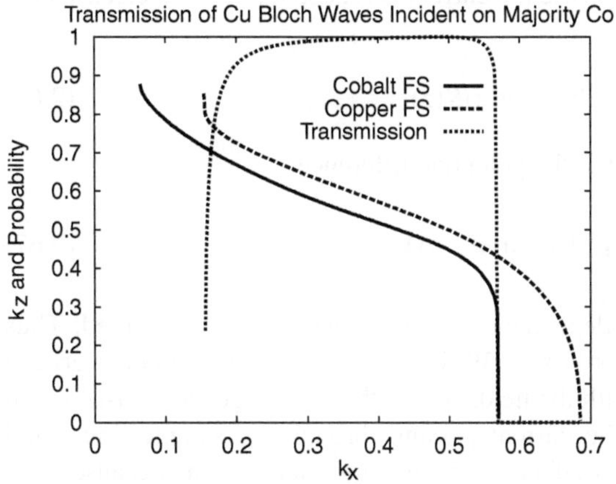

Fig. 2.11. Transmission probabilities of copper electrons incident on majority cobalt for a cut through the Fermi surface with $k_y = 0$. The Fermi surface of copper and majority cobalt are also shown.

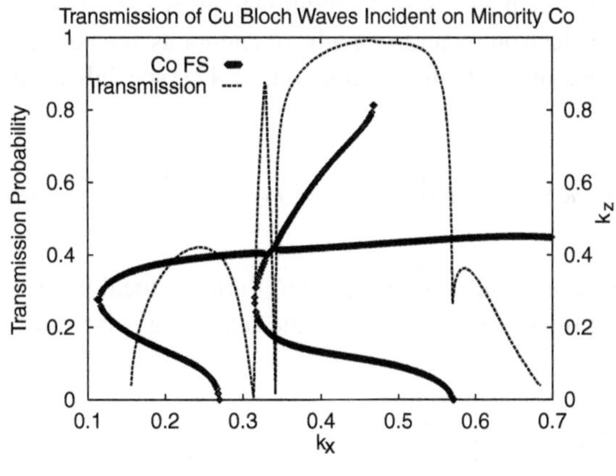

Fig. 2.12. Transmission probabilities of copper electrons incident on minority cobalt for a cut through the Fermi surface with $k_y = 0$. The Fermi surface of minority cobalt is also shown. Note that the transmission probability goes to zero if the velocity of the state receiving the electrons goes to zero. This is a consequence of flux conservation

copper, but not in cobalt. Electrons will be totally reflected by a perfect interface if there are no states across the interface with the same $\boldsymbol{k}_{\parallel}$.

In the minority channel, the total transmission probability consists of the sum of the probabilities of being transmitted from a copper Bloch state into any of the Bloch states of cobalt with the same value of $\boldsymbol{k}_{\parallel}$. Note that there are several values of $\boldsymbol{k}_{\parallel}$ for which the transmission drops dramatically. These are generally associated a dip in $v_z(\boldsymbol{k}_{\parallel})$ in the receiving band. Flux conservation implies that the transmission must vanish if the group velocity of the receiving state vanishes.

2.6.2 Boltzmann Equation for CPP

When the system is inhomogeneous in the direction in which the field is applied, it is necessary to deal with the accumulation of spin and charges. It is useful to think first about a simple system with a local but spatially varying conductivity. Let us assume that the system is homogeneous in the x and y directions and has a *local* conductivity that depends on z, the direction which sustains a current, J and a potential difference, ΔV. In this case the current is related to the local electric field through the conductivity,

$$J(z) = \sigma(z)\mathcal{E}(z) \ . \tag{2.66}$$

The integral of this local field gives the potential difference,

$$\int \mathrm{d}z\mathcal{E}(z) = -\Delta V \ . \tag{2.67}$$

In steady state, $J(z)$ must be independent of z if charge is to be conserved. Thus the local field, $\mathcal{E}(z)$, must vary as $1/\sigma(z)$. We can think of this local field as arising from an applied field together with the fields due to the inhomogeneous distribution of electrons that is set up by the current passing through the sample. We could ignore these charge accumulation effects in the previous subsections because there we assumed that the system was homogeneous in the direction of the applied field.

For the more general case in which the spatial inhomogeneities have a scale comparable to or smaller than the electron mean free path we cannot assume a local conductivity. We can, however, use the Boltzmann equation, (2.42), specialized to our geometry,

$$v_z(\boldsymbol{k})\frac{\partial f(z,\boldsymbol{k})}{\partial z} - \sum_{\boldsymbol{k}'} P_{\boldsymbol{k}\boldsymbol{k}'}\left[f(z,\boldsymbol{k}') - f(z,\boldsymbol{k})\right] = e v_z(\boldsymbol{k})\delta(\varepsilon_{\boldsymbol{k}} - \mu)\frac{\partial V}{\partial z} \ , \tag{2.68}$$

but we must still deal with charge accumulation effects and variable local electric fields. The analysis is greatly simplified if we assume that the scattering is isotropic. This assumption implies that

$$P_{\boldsymbol{k}\boldsymbol{k}'} = \frac{\delta(\varepsilon_{\boldsymbol{k}} - \varepsilon_{\boldsymbol{k}'})}{n\tau} \ , \tag{2.69}$$

where n is the Fermi energy DOS given by (2.58). For simplicity, we have temporarily suppressed the band index, j. It can be assumed that $\boldsymbol{k} = (\boldsymbol{k}, j)$.

Taking, $f(z, \boldsymbol{k}) = f_0(\varepsilon_{\boldsymbol{k}} - \mu) + \delta(\varepsilon_{\boldsymbol{k}} - \mu)h(z, \boldsymbol{k})$, and taking account of the fact that the distribution function, $f(z, \boldsymbol{k})$, may contain spatial charge inhomogeneities, we can write,

$$\sum_{\boldsymbol{k}} f(z, \boldsymbol{k}) = N + n\mu(z) , \tag{2.70}$$

where N is the number of electrons and n is the Fermi energy density of states. Thus the scattering terms of the Boltzmann equation are given by,

$$\sum_{\boldsymbol{k}'} P_{\boldsymbol{k}\boldsymbol{k}'} h(z, \boldsymbol{k})\delta(\varepsilon_{\boldsymbol{k}} - \mu) = h(z, \boldsymbol{k})\delta(\varepsilon_{\boldsymbol{k}} - \mu)/\tau \tag{2.71}$$

and

$$\sum_{\boldsymbol{k}'} P_{\boldsymbol{k}\boldsymbol{k}'} h(z, \boldsymbol{k}')\delta(\varepsilon_{\boldsymbol{k}'} - \mu) = \mu(z)\delta(\varepsilon_{\boldsymbol{k}} - \mu)/\tau . \tag{2.72}$$

Note that for CPP the scattering-in term cannot be neglected as it usually is for CIP.

Thus the Boltzmann equation for isotropic scattering can be written as

$$v_z(\boldsymbol{k})\frac{\partial h(z, \boldsymbol{k})}{\partial z} + \frac{h(z, \boldsymbol{k}) - \mu(z)}{\tau} = ev_z(\boldsymbol{k})\frac{\partial V}{\partial z} . \tag{2.73}$$

If we define the anisotropic part of the deviation function as, $h^A(z, \boldsymbol{k}) = h(z, \boldsymbol{k}) - \mu(z)$, we can write the Boltzmann equation in the form,

$$v_z(\boldsymbol{k})\frac{\partial h^A(z, \boldsymbol{k})}{\partial z} + \frac{h^A(z, \boldsymbol{k})}{\tau} = -v_z(\boldsymbol{k})\frac{\partial \bar{\mu}}{\partial z} \tag{2.74}$$

where $\bar{\mu} = \mu - eV$. At first glance this appears to be no more complicated than (2.44) with scattering-in term omitted which we used for the CIP case, but it is really much more complicated because $h^A(z, \boldsymbol{k})$ must be anisotropic for all z. In other words, it must satisfy

$$\sum_{\boldsymbol{k}} h^A(z, \boldsymbol{k})\delta(\varepsilon_{\boldsymbol{k}} - \mu) = \sum_{\boldsymbol{k}} [h(z, \boldsymbol{k}) - \mu(z)]\, \delta(\varepsilon_{\boldsymbol{k}} - \mu) = 0 . \tag{2.75}$$

When the electronic structure is different in successive layers, the solutions to the Boltzmann equation obtained for each layer need to be matched across the interfaces between the layers. The procedure for performing this matching is given by (2.62) and is relatively straightforward. However, it involves the full distribution function, $h(z, \boldsymbol{k})$, not just the anisotropic part, $h^A(z, \boldsymbol{k})$.

In order to properly match the solutions at the boundaries, it is necessary to utilize the general solution to the (2.74) and to admit exponentially varying solutions that could be omitted for the homogeneous case. Thus, it can be verified that the anisotropic distribution function for spin s in layer i can be written as

$$h_s^A(z,\boldsymbol{k}) = \frac{J_s}{\sigma_{si}} e v_z^{si}(\boldsymbol{k})\tau_{si} - F_{si}(\boldsymbol{k}) e^{(-z/v_z^{si}(\boldsymbol{k})\tau_{si})}$$
$$+ \left\langle \frac{F_{si}(\boldsymbol{k}')e^{(-z/v_z^{si}(\boldsymbol{k}')\tau_{si})}}{v_z^{si}(\boldsymbol{k}) - v_z^{si}(\boldsymbol{k}')} \frac{v_z^{si}(\boldsymbol{k})}{\left\langle \frac{v_z^{si}(\boldsymbol{k})}{v_z^{si}(\boldsymbol{k})-v_z^{si}(\boldsymbol{k}')} \right\rangle_{\boldsymbol{k}}} \right\rangle_{\boldsymbol{k}'} , \quad (2.76)$$

with the generalized chemical potential, $\overline{\mu}$ given by,

$$\bar{\mu}(z) = \alpha_{is} J_s - \frac{J_s}{\sigma_{si}} e z + \left\langle \frac{F_{si}(\boldsymbol{k}')e^{(-z/v_z^{si}(\boldsymbol{k}')\tau_{si})}}{\left\langle \frac{v_z^{si}(\boldsymbol{k})}{v_z^{si}(\boldsymbol{k})-v_z^{si}(\boldsymbol{k}')} \right\rangle_{\boldsymbol{k}}} \right\rangle_{\boldsymbol{k}'} . \quad (2.77)$$

Here, J_s is the current density for spin-channel s and σ_{si} is the bulk conductivity for spin s of the material in layer i,

$$\sigma_{si} = -\frac{e^2}{V} \sum_{\boldsymbol{k}} (v_z^{si}(\boldsymbol{k}))^2 \tau_{si} \delta(\varepsilon_{\boldsymbol{k}} - \mu) , \quad (2.78)$$

and α_{is} and $F_{si}(\boldsymbol{k})$ are parameters determined by the matching conditions at the boundaries, (2.62). The angular brackets indicate an average over the Fermi surface. Note that it is $\overline{\mu}$ rather than μ or eV that has physical significance.

If the distribution function, (2.76), is used to calculate the current density, the first term yields J_s and the contributions of the other two terms cancel. The second term would have been expected from the general solution to the CIP case. The third term is made necessary by the requirement that $h^A(z,\boldsymbol{k})$ be purely anisotropic. The constant α_{is} in the solution for $\overline{\mu}$ can vary between layers. Thus discontinuities in $\overline{\mu}$ can arise at interfaces from both the second and third terms of (2.77). These represent interfacial resistances caused by partial reflection of the electrons at the interfaces.

In order to avoid dealing with the effects of interfacial reflection and transmission, an alternative approach is often followed. If there were no interfaces, so that the electronic structure for both spins in all layers were the same, then, from (2.76) and (2.77)

$$h_s^A(z,\boldsymbol{k}) = \frac{J_s}{\sigma_{si}} e v_z^{si}(\boldsymbol{k})\tau_{si}$$
$$\bar{\mu}(z) = -\frac{J_s}{\sigma_{si}} e z . \quad (2.79)$$

This result can be obtained under slightly more general conditions from the requirement that the current be conserved and (2.74). Consider (2.74) but with the generalization that the lifetime (but not the band structure) is allowed to vary with z. Then the requirement that $\partial J/\partial z = 0$, together with the fact that (2.76) provides no mechanism to couple different values of $\boldsymbol{k}$ leads to $h^A(z,\boldsymbol{k}) = -v_z(\boldsymbol{k})\tau(z)(\partial\bar{\mu}/\partial z)$ and the requirement that $\tau(z)(\partial\bar{\mu}/\partial z)$ be independent of z. Thus

$$-\frac{\partial \bar{\mu}_s}{\partial z} = \frac{e J_s}{\sigma_s(z)} , \quad (2.80)$$

which leads immediately to

$$A\,R_s = \int \mathrm{d}z \rho_s(z)\;, \tag{2.81}$$

where A is the area and $\rho_s(z)$ is the local resistivity. Although the application of this result to magnetic multilayers such as spin valves requires the ridiculous assumption that the electronic structure does not depend on the material or the spin, the additional effects of the interfaces consist of discontinuous changes in the chemical potential plus additional perturbations that extend over the range of a mean free path on either side of an interface. If these are lumped into *ad hoc* "interfacial resistances" then (2.81) can be used in the form

$$A\,R_s = \int \mathrm{d}z \rho_s(z) + A \sum_i R_{i,s} \tag{2.82}$$

where i labels the interfaces. The accuracy of this type of model can be assessed from calculations presented in Sect. 2.7.2.

2.6.3 Effects of Diffuse Interfacial Scattering

If the interface is disordered, flux will still be conserved, but the conservation rules must be extended to include the scattering of electrons between different values of $\boldsymbol{k}_\parallel$, *i.e.* we must use (2.62) rather than (2.63). In the past, primarily because of lack of detailed knowledge of the diffuse scattering transmission and reflection probabilities, the diffuse scattering by interfaces and surfaces has been treated phenomenologically [2.38] by continuing to use (2.63), but adding a specularity parameter for each interface so that it becomes

$$h^{+,j}_{i+1}(z,\boldsymbol{k}_\parallel) = S_i \left[\sum_{j'}^{N_\mathrm{R}} T_i^{+-}(j,j') h^{-,j'}_{i+1}(z_i^+,\boldsymbol{k}_\parallel) + \sum_{j'}^{N_\mathrm{L}} T_i^{++}(j,j') h^{+,j'}_{i}(z_i^-,\boldsymbol{k}_\parallel) \right]$$
$$h^{-,j}_{i}(z_i^-,\boldsymbol{k}_\parallel) = S_i \left[\sum_{j'}^{N_\mathrm{L}} T_i^{-+}(j,j') h^{+,j'}_{i}(z_i^-,\boldsymbol{k}_\parallel) + \sum_{j'}^{N_\mathrm{R}} T_i^{--}(j,j') h^{-,j'}_{i+1}(z_i^+,\boldsymbol{k}'_\parallel) \right]\,. \tag{2.83}$$

Here $S_i = 1$ for purely specular scattering and $S_i = 0$ for purely diffuse scattering. Equation (2.83) with $S_i < 1$, is very much analgous the lifetime approximation, in that it neglects the diffusely scattered electrons and simply reduces the specular scattering probabilities to allow for these missing electrons. It can be shown that it is the vertex-corrections or the "scattering-in" term that is needed to calculate the diffuse scattering probabilities. One important drawback of this approach is that it does not conserve flux. Therefore, it can only be used for CIP when there is no net current flowing perpendicular to the interface. For CPP the diffuse or scattering-in flux must be included to maintain constant flux across the interface.

Even within the context of CIP transport, the constant specularity parameter approach to diffuse scattering is not very accurate. A model system in which there

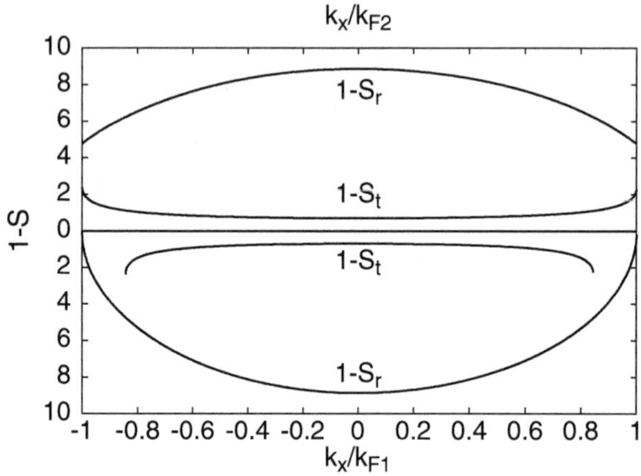

Fig. 2.13. $1 - S_t$ and $1 - S_r$ in the effective mass approximation. The Fermi momentum, k_{F1} corresponds to 0.5 electrons per spin channel (e.g. as in Cu) while k_{F2} corresponds to 0.3 electrons per spin channel (e.g. as in majority Co). The values of $1 - S_t$ and $1 - S_r$ are measured in terms of the dimensionless parameter $2m_e \gamma n_0 a/\hbar^2$. Where a is the lattice constant for an assumed fcc lattice

are random point scatterers at the interface between two free electron regions can be solved in closed form in the weak scattering limit [2.39, 40]. The result is a specularity function that is strongly dependent on $\boldsymbol{k}_\parallel$ and that is quite different for transmission and reflection. The specularity function for transmission in this model is,

$$S_t(\boldsymbol{k}_\parallel) = 1 - \frac{4\pi}{\hbar} \frac{\gamma n_0(0; E)}{v_z^L(\boldsymbol{k}_\parallel) + v_z^R(\boldsymbol{k}_\parallel)} , \tag{2.84}$$

where γ is a measure of the interface roughness[7], $n_0(0; E)$ is the DOS at the interface and v^L and v^R are the electron velocities on the right and left sides of the interface. Similarly, the specularity function for reflection is

$$S_r(\boldsymbol{k}_\parallel) = 1 - \frac{8\pi}{\hbar} \frac{v_z^L(\boldsymbol{k}_\parallel)\gamma n_0(0; E)}{\left(v_z^L(\boldsymbol{k}_\parallel)\right)^2 - \left(v_z^R(\boldsymbol{k}_\parallel)\right)^2} . \tag{2.85}$$

Note that within this model, interfacial disorder can only decrease the specular transmission, but it can either decrease or enhance the specular reflection depending on the velocities on either side of the interface. Specular transmission remains symmetric in the sense that that transmission from left to right remains the same as from right to left. Specular reflection, however, is no longer symmetric in the presence of a disordered interface.

The specularity parameters are plotted both as functions of $k_\parallel/k_{F1}$ and $k_\parallel/k_{F2}$ in Fig. 2.13. Note that within this model the diffuse scattering vanishes as $k_\parallel \rightarrow$

[7] γ is defined by the correlation function of the random interfacial potential, $\langle V(\boldsymbol{r})V(\boldsymbol{r}')\rangle = \gamma\delta(\boldsymbol{r} - \boldsymbol{r}')\delta(z)$.

k_{F1}. Generally, however, the model predicts the effects of diffuse scattering to be significantly greater for the reflected beam than for the transmitted beam.

A more realistic and quantitative model for the effect of diffuse scattering on the specular reflection and transmission can be obtained by using the coherent potential approximation to describe the average Green function in the presence of interfacial disorder. In this approach, a finite number of atomic layers near the interface are modeled as a random alloy with defined, layer-dependent concentrations. Figure 2.14 shows the transmission and reflection probabilities for the majority cobalt-copper interface. In this case the interface is perpendicular to the (100) direction.

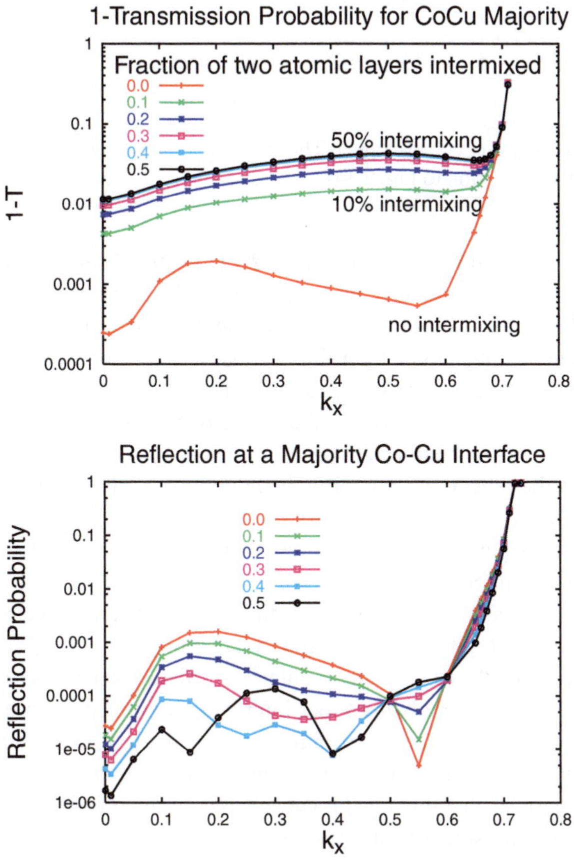

Fig. 2.14. Majority channel specular transmission (subtracted from unity) and specular reflection probabilities for an interface between fcc-cobalt and copper calculated in the coherent potential approximation. There are two intermixed atomic layers, one with Cu_xCo_{1-x} on the cobalt side of the interface and one with Co_xCu_{1-x} on the copper side where $x = 0.1, ..., 0.5$. The interface is assumed to be perpendicular to the (100) direction

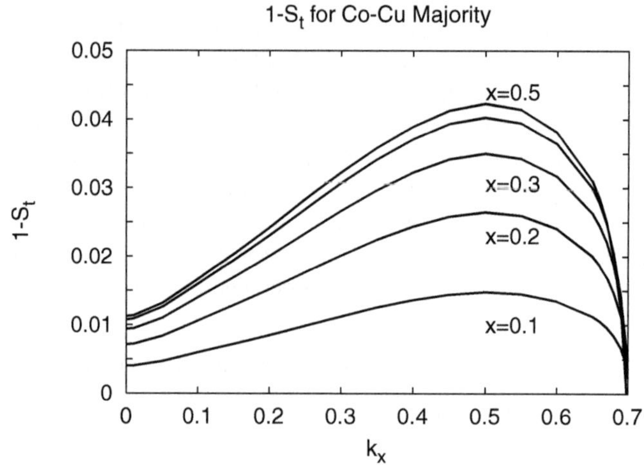

Fig. 2.15. 1-S$_t$ for majority cobalt-copper. The fraction of intermixing of two interfacial layers is indicated by x (see caption to Fig. 2.14)

Two features are surprising about Fig. 2.14. In the absence of intermixing the transmission probability is nearly unity (and the reflection probability nearly zero[8]) over most of the range of $k_{\|}$. The increase in reflection for large $k_{\|}$ is due to the Fermi surface of copper being larger than that of majority cobalt so that total reflection occurs when for a given value of $k_{\|}$ there is no cobalt state for the electron to be refracted into. The second surprising feature is that substantial intermixing of two atomic layers has only a small effect on the transmission and reflection probabilities. The structure in the reflection probability is thought to result from interference within the disordered layers. Figure 2.15 shows the transmission specularity parameter for majority cobalt-copper which may be compared with that for the free electrons with random point interfacial scatterers model shown in Fig. 2.13. One important difference is the vanishing of the effect of diffuse scattering for electrons incident on the interface with the highest values of transverse momentum, i.e. grazing incidence.

Figure 2.16 shows the calculated specular transmission and reflection probabilities for minority electrons at a cobalt-copper interface. The reflection probability can be viewed as the probability of an electron in copper being reflected off of the interface. Because the minority spin channel for cobalt has multiple bands for most values of $k_{\|}$, the transmission probability can be viewed as the total probability that an electron in copper will be transmitted into any of the bands with the same value of $k_{\|}$. It is interesting that these calculations predict that disorder can actually increase the specular transmission for a few values of $k_{\|}$ for which the transmission of the ordered interface is particularly low and can increase the specular reflection for values of $k_{\|}$ where it is especially low.

[8] In the absence of intermixing, $R = 1 - T$. The small differences between $1 - T$ and R seen in Fig. 2.14 are due to numerical inaccuracies.

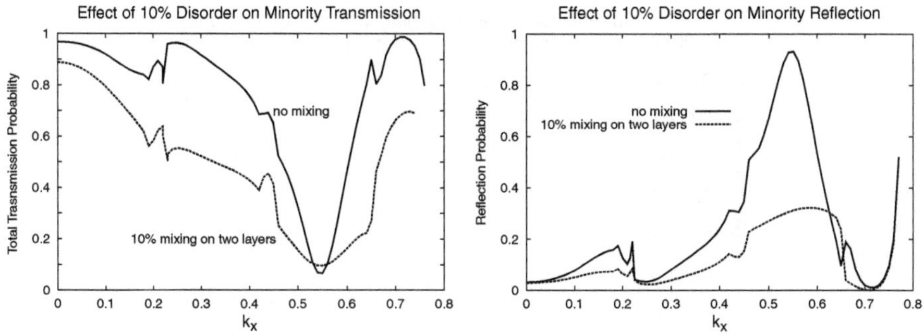

Fig. 2.16. Transmission and reflection probabilities for minority cobalt-copper interface

One of the important results of Figs. 2.14 and 2.16 is that for most values of $k_{\|}$ the transmission probability is much higher in the majority channel than for the minority channel. In addition the effects of disorder are not nearly so dramatic on the majority transmission as on the minority. This difference is one of the major contributors to the GMR effect and is due to the relatively good matching of the electronic structures of cobalt and copper in the majority channel and the relative poor match in the minority channel.

2.7 Giant Magnetoresistance

GMR in magnetic multilayers is observed in important geometries that must be treated quite differently both experimentally and theoretically. GMR was first observed the "Current In the Plane" or CIP geometry [2.1]. As implied by its name; in this geometry the current flows in the plane of the layers and the resistance of the multilayer film is lower if the magnetic layers have their moments aligned rather than anti-aligned. This was also the first geometry to be exploited commercially. GMR can also be observed when the current flows perpendicular to the planes [2.41], a geometry referred to as "CPP".

2.7.1 GMR for Current In the Planes

GMR for CIP is somewhat more subtle than for CPP. CIP GMR arises from the non-local nature of electrical conduction. A necessary requirement for CIP GMR is that the electron mean free path be at least comparable to the thicknesses of the layers. One contribution to CIP GMR can be thought of as an effect similar to the effect of a boundary decreasing the conductivity. Consider a three layer system, e.g. a layer of copper sandwiched between two layers of cobalt. We shall assume that the (outer) boundaries of the film are sufficiently rough that they produce no specular scattering, but that the transmission and reflection at the internal interfaces is purely specular.

Suppose that copper and cobalt matched perfectly in the majority channel. Then the majority electrons, when the moments of the two cobalt layers are aligned, would effectively see a film thickness equal to the sum of the thicknesses of the three layers

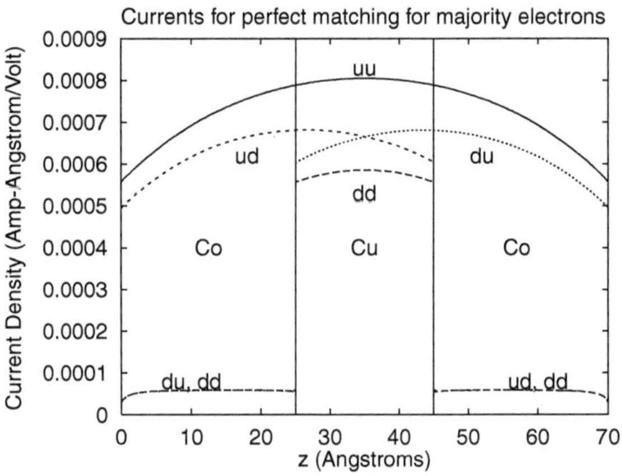

Fig. 2.17. Currents for spin channels for an idealized Co-Cu spin valve with perfect matching in the majority channel. uu and dd refer to the majority and minority channels respectively for aligned Co moments. ud and du refer to the channels that are locally majority on the left and on the right respectively for the case of antiparallel alignment of the Co moments

while the minority electrons would tend to be confined within the individual layers because of the changes in electronic structure at the interfaces. When the moments are anti-aligned, however, both of the spin channels would see effectively two layers. This situation is shown in Fig. 2.17. It is also assumed in this example that the mean free path is much longer for the majority cobalt channel than for the minority.

The sum of the "uu" and "dd" currents gives the total current for parallel alignment and the sum of the "ud" and "du" currents give the total current for anti-parallel alignment. These are shown in Fig. 2.18 together with the difference which yields the giant magnetoconcuctance.

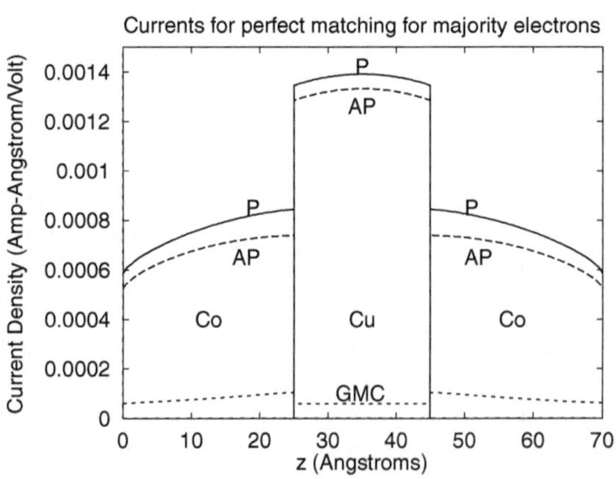

Fig. 2.18. Currents for Parallel(P) and Anti-Parallel (AP) alignments of the cobalt moments. The difference is the giant magnetoconductance, (GMC)

In fact, of course, as indicated in Figs. 2.6, 2.7, and 2.11, there is a difference between cobalt and copper in the majority channel. The copper majority Fermi surface is larger than that of cobalt. It holds 0.5 electrons while that of cobalt holds only 0.3. A cut through the Fermi surfaces of copper and majority cobalt is shown in Fig. 2.11. The z direction (perpendicular to the layers) is towards the top of the figure. The directions perpendicular to this direction are in the plane of the layers. If the interfaces are smooth on an atomic scale then the component of the momentum parallel to the interface ($\boldsymbol{k}_{\parallel}$) does not change on reflection or refraction at an interface. Thus from Figure 2.11 it is clear that there are values of $\boldsymbol{k}_{\parallel}$ for which states exist in the copper but not in the cobalt (i.e. $0.57 < k_x < 0.69$). This means that these states cannot refract into the cobalt, they must reflect back into the copper. This can lead to a significant contribution to the GMR if the interface is sufficiently smooth because some of the majority electrons can be "trapped" inside the copper where the resistance is significantly lower for both spin channels than for cobalt. This "trapping" of the electrons inside the copper layer is analogous to the trapping of light waves within a waveguide [2.25]. Note from Figs. 2.13 and 2.14 that interfacial disorder is relatively ineffective in reducing the specular reflection for values of $k_{\parallel}$ where total reflection occurs.

A calculation for the current density in a cobalt|copper|cobalt spin valve using realistic electronic structures is shown in Figs. 2.19 and 2.20. Figure 2.19 shows the majority and minority currents for both parallel (uu,dd) and anti-parallel (ud,du) alignment. In this example, the scattering rate in the copper is chosen to give the copper a resistivity of 3 μΩ cm, a typical value for sputter deposited copper films at room temperature. The scattering rates for cobalt were chosen to give it a resistance of 15 μΩ cm which is also typical of sputtered films. A much higher scattering rate was chosen for the minority than for the majority cobalt. It can also be seen that the current density is significantly higher in the copper than in the cobalt. It can be seen from

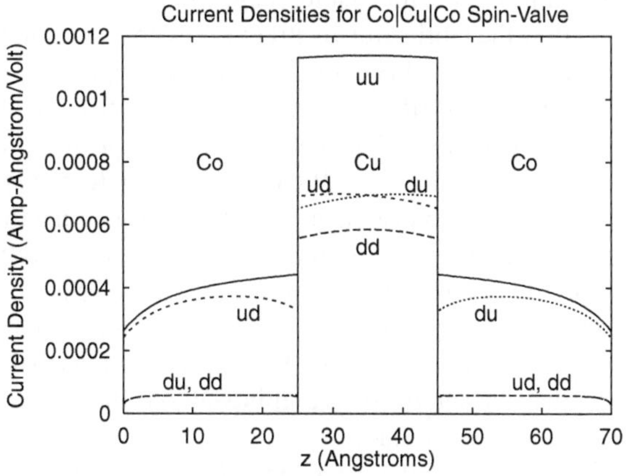

Fig. 2.19. Current Densities for Co|Cu|Co spin valves. The scattering rates correspond to bulk resistivities of 3 μΩcm for copper and 15 μΩcm for cobalt with the majority lifetime ten times longer than the minority lifetime for cobalt

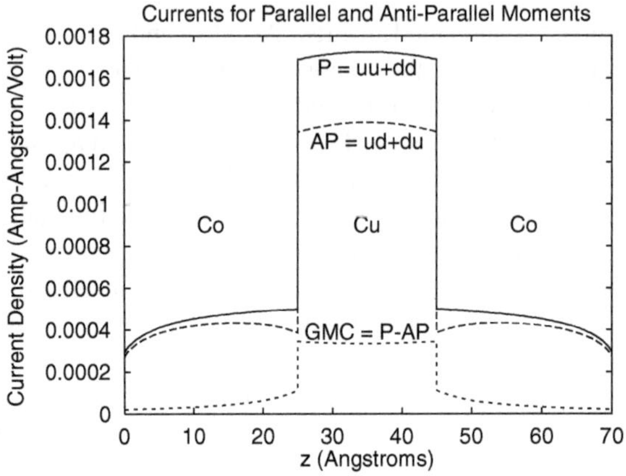

Fig. 2.20. Current densities for parallel alignment, anti-parallel alignment and difference or giant magnetoconductance

Fig. 2.20 that the largest contributions to the giant magnetoconductance arise from the copper spacer layer indicating the importance of the channeling effect. Comparison of calculated and measured values of GMR for a series of cobalt-copper spin valves with different thicknesses of the cobalt layers indicated the existence of important contributions to GMR from both the channeling effect and from the differences in bulk scattering rates for the majority and minority channels of cobalt [2.22].

2.7.2 Current Perpendicular to the Planes

Since magnetic multilayer films generally have a huge lateral extent compared to their thickness, measuring the CPP resistance not to mention GMR can be a challenge. However, the increasing demand for smaller magnetic sensors coupled with advances in lithography that allow the fabrication of structures with much smaller lateral extent make it likely that CPP GMR will become increasingly important.

Although the origin of GMR is easier to conceptualize for the CPP geometry, it is actually much harder to treat theoretically except in the limit of uniform electronic structure. Similarly to the CIP case, the Layer Korringa Kohn Rostoker Approach [2.37] can be used to calculate the self-consistent electronic structure of interfaces and to evaluate the transmission (T^{++}, T^{--}) and reflection probabilities (T^{+-} and T^{-+}) for Bloch electrons impinging on the interfaces. The Boltzmann equation, (2.74) and (2.75), including the boundary matching equations, (2.63), can then be solved using an iterative procedure. For the calculations that we present here, it will be assumed that the interfaces are epitaxial and that there is no additional disorder in the vicinity of the interface.

Figures 2.21 and 2.22 show the calculated electrochemical potential, $\bar{\mu}$ for the majority and minority channels in the vicinity of a copper-cobalt interface. The scattering rates are similar to those used for the CIP calculations, the bulk resistivity of the copper is approximately 3 $\mu\Omega$ cm and that of the cobalt is approximately

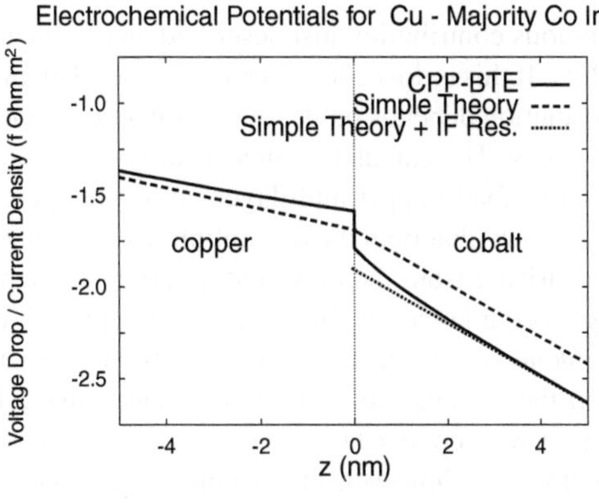

Fig. 2.21. Chemical Potential divided by current density for copper-cobalt majority interface. Dashed line is simple theory, dotted line is simple theory plus interfacial resistance, and solid line is calculated voltage drop using Boltzmann equation

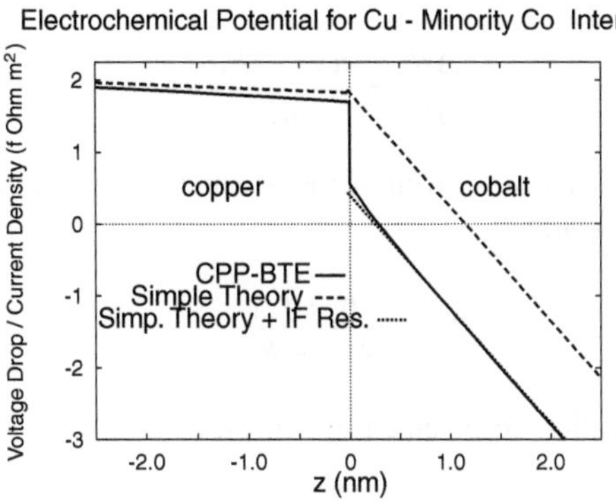

Fig. 2.22. Chemical Potential divided by current density for copper-cobalt minority interface. See caption for Fig. 2.21

15 μΩ cm with a scattering rate about 15 times higher for minority cobalt than for majority cobalt. The electrochemical potential has been divided by the current density so that the plots yield $\int dz \rho_s(z)$. For the simple theory represented by (2.81), this is simply a straight line for each layer with the slope of the line for each layer being its resistivity. The actual solution to the Boltzmann equation differs in two ways: (1) There is a discontinuity in the chemical potential at the interface which is equivalent to an interfacial resistance. This interfacial resistance is not due to intermixing or additional scattering at the interface (although this effect can be included in the model if desired) but to the mismatch of the bands across the interface which causes some of the electrons incident on the interface to be reflected. (2) There are exponential terms in the electrochemical potential in the vicinity of the interface that decay at a rate comparable to the component of the mean free path perpendicular to the layers. The effect of these terms can be included as an additional interfacial resistance that

is added to the discontinuous contribution just described as is indicated in the dotted lines of Figs. 2.21 and 2.22. If this is done, however, it must be taken into consideration that this additional contribution depends on the environment of the interface, e.g. the proximity of other interfaces. The calculated interfacial resistances are comparable to those observed [2.41] in cobalt-copper multilayers at low temperature.

Figures 2.23 and 2.24 show that both the discontinuous interfacial resistance and the interfacial resistance arising from the exponential terms can depend on the type and thicknesses of neighboring layers. If Fig. 2.23, for example, it can be seen that the discontinuous interfacial resistance on both sides of the Co-Cu-Co (majority) spin valve increase with the thickness of the copper spacer layer. The exponential contributions also increase with copper layer thickness. For the case of antiparallel alignment shown in Figure 2.24, however, the discontinuous contribution increases slightly with spacer layer thickness on the minority side but *decreases* with thickness on the majority side.

GMR in the "additive" approximation described by (2.82) is relatively simple. The GMR ratio will be given by

$$\frac{\Delta R}{R} = \frac{R_{\mathrm{AP}} - R_{\mathrm{P}}}{R_{\mathrm{P}}} \tag{2.86}$$

where the "parallel" resistance is given by the resistance of the majority and minority channels conducting in parallel,

$$R_{\mathrm{P}} = \frac{R_{\uparrow\uparrow} R_{\downarrow\downarrow}}{R_{\uparrow\uparrow} + R_{\downarrow\downarrow}} . \tag{2.87}$$

Similarly, the "anti-parallel" resistance is given by,

$$R_{\mathrm{AP}} = \frac{R_{\uparrow\downarrow} R_{\downarrow\uparrow}}{R_{\uparrow\downarrow} + R_{\downarrow\uparrow}} . \tag{2.88}$$

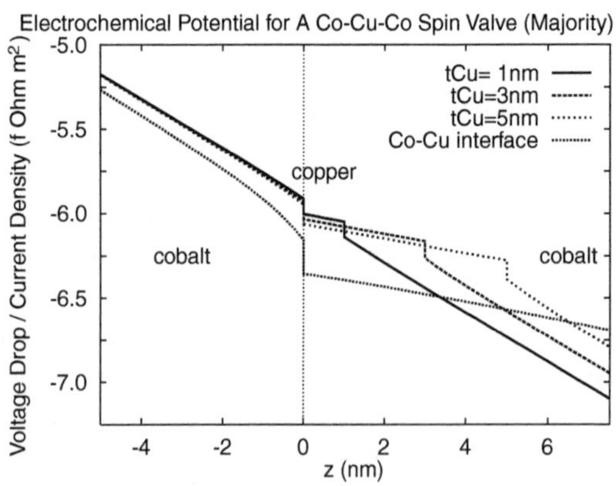

Fig. 2.23. Chemical potential divided by current density for the majority channel of a cobalt-copper-cobalt spin valve for various thickness of the copper layer

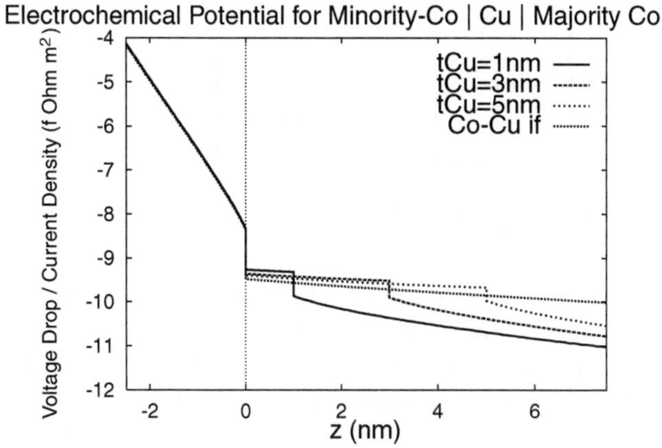

Fig. 2.24. Chemical potential divided by current density for a cobalt-copper spin valve with cobalt moments aligned anti-parallel

Here $R_{\uparrow\uparrow}$, $R_{\downarrow\downarrow}$, $R_{\uparrow\downarrow}$, $R_{\downarrow\uparrow}$ represent the series resistance in each particular channel, e.g. for a spin valve consisting of two ferromagnetic layers separated by a non-magnetic spacer layer, one has

$$\begin{aligned} AR_{\uparrow\uparrow} &= (\rho_{F1\uparrow} t_{F1} + r_{F1\uparrow,N}) + \rho_N t_N + (r_{N,F2\uparrow} + \rho_{F2\uparrow} t_{F2}) \\ &= A(R_{1\uparrow} + R_N + R_{2\uparrow}) \\ AR_{\uparrow\downarrow} &= (\rho_{F1\uparrow} t_{F1} + r_{F1\uparrow,N}) + \rho_N t_N + (r_{N,F2\downarrow} + \rho_{F2\downarrow} t_{F2}) \, , \\ &= A(R_{1\uparrow} + R_N + R_{2\downarrow}) \end{aligned} \tag{2.89}$$

where t is the thickness of a layer and r is an interfacial resistance. Because the resistances are assumed to be additive, $R_{\uparrow\uparrow} + R_{\downarrow\downarrow} = R\uparrow\downarrow + R_{\downarrow\uparrow}$ and the GMR ratio is

$$\frac{\Delta R}{R} = \frac{(R_{1\uparrow} - R_{1\downarrow})(R_{2\uparrow} - R_{2\downarrow})}{R_{\uparrow\uparrow} R_{\downarrow\downarrow}} \, . \tag{2.90}$$

2.8 Landauer Approach to Ballistic Transport

There is another approach to CPP conduction that can be applied in the limit in which the transport is ballistic rather than diffusive. Usually transport in metallic systems is diffusive in nature; between the time that an electron enters the sample from one lead and an electron leaves the sample through another lead many scattering events occur. Devices (usually on the nanometer scale) can be constructed however, in which very few scattering events take place between the two leads. In this limit, sometimes called the ballistic limit, there is a very simple expression due to Landauer [2.42] which relates the conductance to the probability of an electron being transmitted through the sample.

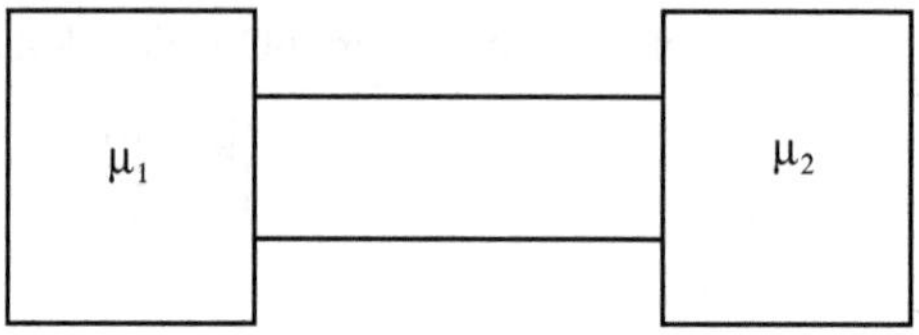

Fig. 2.25. Two electron reservoirs connected by a sample

To understand the Landauer conductance formula it is helpful to consider two reservoirs of electrons connected by a sample as shown in Fig. 2.25. If we imagine the left reservoir, with chemical potential μ_1, to be an emitter of right going electrons, we can write the current density of those electrons that leave the reservoir on the left and enter the reservoir on the right as

$$J^+ = \frac{e}{(2\pi)^3} \int \mathrm{d}^3k v_z^+(\boldsymbol{k}) f_0^+(\mu_1) T^+(\boldsymbol{k}) \tag{2.91}$$

where a "+" superscript indicates an electron with $v_z > 0$, z being the direction from reservoir 1 to reservoir 2 and where

$$T^+(\boldsymbol{k}) \equiv \sum_{k'} T^{++}(\boldsymbol{k}, \boldsymbol{k}') . \tag{2.92}$$

We can perform the integral over k_z,

$$J^+ = \frac{e}{A} \sum_{k_\parallel, j} \frac{1}{2\pi} \int \mathrm{d}k_z \frac{1}{\hbar} \frac{\partial \varepsilon}{\partial k_z} f_0(\mu_1) T^+(\boldsymbol{k}) , \tag{2.93}$$

which yields an expression for the current,

$$I^+ = \frac{e}{h} \int^{\mu_1} \mathrm{d}\varepsilon \sum_{k_\parallel, j} T^+(k_\parallel, j) . \tag{2.94}$$

A similar line of reasoning leads to an expression for the current of electrons emitted in the $-z$ direction by the reservoir on the right which enter the reservoir on the left,

$$I^- = \frac{e}{h} \int^{\mu_2} \mathrm{d}\varepsilon \sum_{k_\parallel, j} T^-(k_\parallel, j) . \tag{2.95}$$

Assuming time reversal invariance, we can equate T^+ and T^-. This allows us to write the net current as

$$I = I^+ - I^- = \frac{e^2}{h} \sum_{k_\parallel, j} T^+(k_\parallel, j) \frac{\mu_1 - \mu_2}{e} , \tag{2.96}$$

which yields the Landauer conductance formula,

$$G = \frac{e^2}{h} \sum_{k_\parallel, j} T^+(k_\parallel, j) . \tag{2.97}$$

The original Landauer formula has the ratio of transmission probability divided by reflection probability (T/R) where we have only the transmission probability in (2.97). It is argued that this additional factor of $1/R$ arises from the reflected electrons changing the chemical potential of the reservoirs. It is now usually accepted that this additional factor of $1/R$ is present or not depending on exactly how the measurement is performed, that is, on whether or not one measures current and voltage using the same leads, as is assumed in the derivation here, or whether a separate set of probes is used to determine the voltage across the sample. In the application in the next section, the difference between the two formulas will usually be negligible because tunneling transmission probabilities are very small and the reflection probabilities are near unity.

2.9 Spin-Dependent Tunnelling

One important application of the Landauer conductance formula is to the calculation of tunneling conductance. It is assumed that the two electrodes act as electron reservoirs and that the electrons traverse the barrier region in a single tunneling event. Recently, it has been observed that the tunneling conductance between two ferromagnetic electrodes depends on the relative orientation of the magnetic moments in the two electrodes [2.3, 4]. Generally, the tunneling rate is higher when the moments of the two electrodes are parallel.

The first observation of spin-dependent tunneling as well as the first theory was due to Julliere [2.2]. Julliere's theory was based on the not unreasonable assumption that the tunneling rate was proportional to the products of the Fermi energy densities of states of the two electrodes. Thus the tunneling current for parallel and anti-parallel alignment of the moments in the ferromagnetic electrodes would be

$$\begin{aligned} I_{\mathrm{P}} &\propto n_{\mathrm{L}}^{\uparrow} n_{\mathrm{R}}^{\uparrow} + n_{\mathrm{L}}^{\downarrow} n_{\mathrm{R}}^{\downarrow} \\ I_{\mathrm{A}} &\propto n_{\mathrm{L}}^{\uparrow} n_{\mathrm{R}}^{\downarrow} + n_{\mathrm{L}}^{\downarrow} n_{\mathrm{R}}^{\uparrow} \end{aligned} \tag{2.98}$$

where $n_{\mathrm{L,R}}^{\uparrow,\downarrow}$ are the majority and minority Fermi energy densities of states for the left and right electrodes respectively. Defining the polarization of the Fermi energy density of states for the two electrodes as,

$$P_{\mathrm{L}} = \frac{n_{\mathrm{L}}^{\uparrow} - n_{\mathrm{L}}^{\downarrow}}{n_{\mathrm{L}}^{\uparrow} + n_{\mathrm{L}}^{\downarrow}} \qquad P_{\mathrm{R}} = \frac{n_{\mathrm{R}}^{\uparrow} - n_{\mathrm{R}}^{\downarrow}}{n_{\mathrm{R}}^{\uparrow} + n_{\mathrm{R}}^{\downarrow}} , \tag{2.99}$$

we can write the magnetoresistance as

$$\frac{I_{\mathrm{P}} - I_{\mathrm{A}}}{I_{\mathrm{A}}} = \frac{2P_{\mathrm{L}}P_{\mathrm{R}}}{1 - P_{\mathrm{L}}P_{\mathrm{R}}} . \tag{2.100}$$

Unfortunately, the assumption that the tunneling rate is proportional to the density of states is not supported by experimental observations of tunneling from ferromagnets into superconductors [2.43]. These observations indicate that it is almost always the majority channel conductance which dominates the tunneling from ferromagnets, despite the fact that the minority Fermi energy density of states is often many times larger than the minority, e.g. in cobalt and nickel. Despite this rather serious problem this approach is still widely used to analyze and rationalize experimental data. The argument is sometimes made that the densities of states and polarizations refer to "those electrons that participate in the tunneling" and that these "effective" polarizations can be obtained from superconducting tunneling experiments.

Recently, the Landauer approach was used to evaluate the the spin-dependent tunneling conductance for some epitaxial systems of the form Fe|S|Fe where S represents an insulator or semiconductor. The same techniques that were used to calculate the transmission and reflection probabilities in metals can be used to calculate the transmission probabilities for spin-dependent tunneling. The semiconductors Ge, GaAs, and ZnSe and the insulator MgO, which have lattices that match almost perfectly to Fe using (100) interfaces for both electrode and barrier materials [2.44–47] were studied and it was found that these systems show a remarkably large magnetoresistance the microscopic origins of which can be analyzed in detail.

A number of surprising results have emerged from these studies. Interfacial resonance states can be very effective in enhancing the tunneling conductance for certain values of $\boldsymbol{k}_{\parallel}$, if the barrier is sufficiently thin. These interfacial resonance states seem to occur primarily in the minority channel for Fe(100). Another unexpected result was quantum interference between evanescent states in the barrier. This interference leads to an oscillatory dependence of the transmission probability on $\boldsymbol{k}_{\parallel}$ and a damped oscillatory dependence on the thickness for certain fixed values of $\boldsymbol{k}_{\parallel}$.

As the semiconducting or insulating barrier layer is made thicker, it is primarily the states near $\boldsymbol{k}_{\parallel} = 0$ that contribute to the tunneling conductance, i.e. those electrons whose momentum in the iron is perpendicular to the interface with the barrier. Figure 2.26 shows the density of states (DOS) for each of the bands in Fe at the Fermi energy for $k_{\parallel} = 0$. It also shows how the DOS decays within the barrier which in this case is MgO. This density of states is calculated for a particular incident Bloch state on the left, all possible reflected Bloch states on the left and all possible transmitted Bloch states on the right. It can be seen that the incident Bloch states differ in how well they are *injected* into the semiconductor, their *rate of decay* within the semiconductor and how well they are *extracted* from the semiconductor to form the transmitted wave on the right.

For both majority and minority Fe for $k_{\parallel} = (0, 0)$ at E_{F} there are 4 bands. A doubly degenerate Δ_5 band (compatible with p and d DOS), a $\Delta_{2'}$ band (compatible with d DOS) are found for both majority and minority. In addition, there is a majority Δ_1

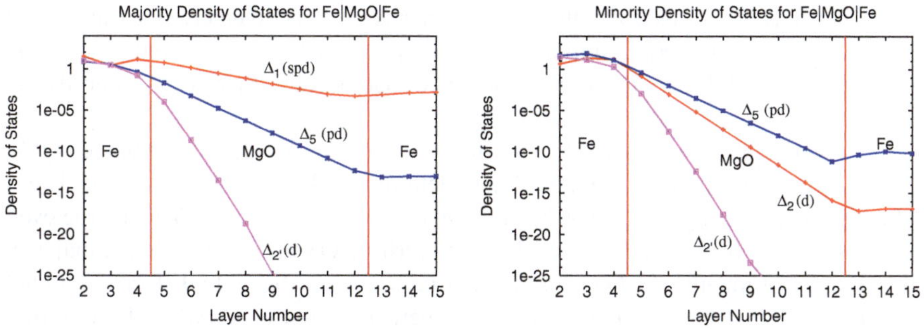

Fig. 2.26. Majority (*left panel*) and minority (*right panel*) density of states for each of the Fe Fermi energy Bloch states for $k_{\parallel} = 0$. The magnetic moments on the two Fe layers are aligned parallel

band (compatible with s, p and d DOS) and a minority Δ_2 band (compatible with d DOS).

Four decay rates can be discerned within the MgO barrier, a relatively slow decay associated with the Δ_1 Bloch state, a somewhat faster decay associated with the doubly degerate Δ_5 state, a still-faster decay associated with the Δ_2 state and a very fast decay associated with the $\Delta_{2'}$ state. The angular momentum composition of the DOS for each of these symmetries is noted on the figure for each of the bands. For parallel alignment, it is clear that the conductance is dominated by majority electrons because of the presence of a Δ_1 symmetry Bloch state at the Fermi energy for the majority that is absent for the minority.

Figure 2.27 shows the DOS for the case in which the Fe moments are aligned anti-parallel. The left panel shows the case in which the incident Bloch states (on the left) are majority. The right panel shows the case in which the incident Bloch states (on the left) are minority. The left panel is similar to the left panel of Fig. 2.26 except that the Δ_1 states continue to decay into the minority Fe electrode on the right hand

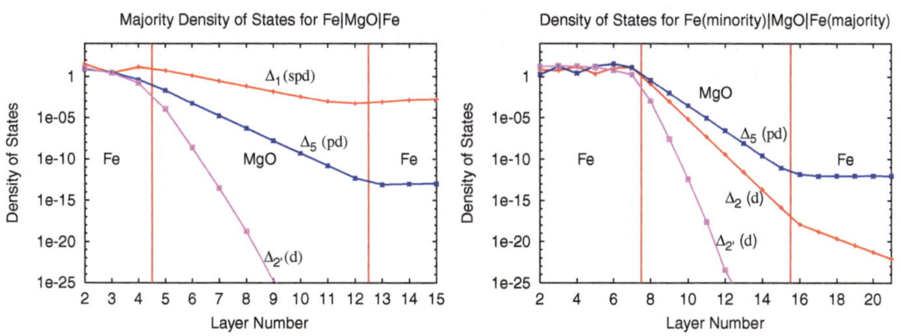

Fig. 2.27. DOS for each of the Fe(100) Fermi energy Bloch states for $k_{\parallel} = 0$. The magnetic moments on the two Fe layers are aligned anti-parallel. In the *left panel*, the Fe majority Bloch states are incident from the left. In the *right panel*, minority Bloch states are incident

side. The reason for this is that there is no minority Δ_1 state to receive the transmitted electron. The right panel is similar to the right panel of Fig. 2.26 with the exception that the Δ_2 band decays in the right side electrode because there is no minority Δ_2 Bloch state.

These results give insight into a mystery concerning spin-dependent tunneling. In every instance that it has been possible to determine which of the spin channels has the highest tunneling rate, it has been the majority channel [2.43]. This is true even for systems such as nickel for which the minority DOS at E_F exceeds the majority by an order of magnitude. One possible reason for this is that not all electrons tunnel equally well. For the particular states shown here it is those bands which have some *s* character that seem to be most efficient at tunneling. Many of the magnetic systems, e.g. nickel and cobalt have filled majority *d*-bands. These "strong" magnets will clearly have more states with *s* character in the majority channel than in the minority channel. Fe does not have a filled majority *d*-band, but, at least in the 100 direction, the band with *s*-character dominates.

It has been emphasized [2.48] that the fundamental difference between the Δ_1 state[9] and the other states is the absence of lateral oscillations in the wave functions of the former. Because the wave functions in the electrode must join smoothly to the wave functions in the barrier layer, they must induce lateral oscillations of the wave function in the barrier. These oscillations require energy that effectively increases the barrier height for tunneling.

These considerations should be valid more generally than the class of epitaxial systems considered in the calculations described here. In systems with disordered interfaces and barrier layers, however, there is the opportunity for non-Δ_1 Bloch states to be scattered into the slowly decaying state within the barrier that does not have the lateral oscillations. This effect can significantly reduce the magnetoresistance.

Acknowledgement. It is a pleasure to acknowledge colleagues and collaborators who have worked with us on transport in magnetic multilayers, especially Thomas C. Schulthess, James M. MacLaren, Derek A. Stewart and Viktor Los. This work was supported by NSF grant-DMR 0213985 and by DARPA/ONR through grant N00014-02-1-0590. Work at Oak Ridge was supported by the Office of Basic Energy Sciences Division of Materials Sciences of the U.S. Department of Energy. Oak Ridge National Laboratory is operated by UT-Battelle, LLC, for the U.S. Department of Energy under contract DE-AC05-00OR22725.

[9] We usually associate Δ_1 symmetry with *s* character, but it should be remembered that one of the three *p* states and one of the 5 *d* states is also compatible with this symmetry.

References

2.1. M.N. Baibich, J.M. Broto, A. Fert, F. Nguyen Van Dau, F. Petroff, P. Eitenne, G. Creuzet, A. Friederich, and J. Chazelas, Phys. Rev. Lett. **61**, 2472 (1988).
2.2. M. Julliere, Phys. Lett. 54A, 225 (1975).
2.3. J.S. Moodera, L.R. Kinder, T.M. Wong, and R. Meservey Phys. Rev. Lett. **74**, 3273 (1995).
2.4. T. Miyazaki, N. Tezuka, J. Mag. Magn. Mat. **151**, 403 (1995).
2.5. J.C. Slonczewski, J. Magn. Magn. Mat. **159**, L1 (1996).
2.6. [Berger 96] L. Berger, Phys. Rev. B **54**, 9353 (1996).
2.7. H. Ohno, J. Magn. Magn. Mater. **200**, 110 (1999) and references therein.
2.8. E.Y. Tsymbal and D.G. Pettifor, Solid State Physics **56**, 113 (2001).
2.9. R. Kubo, J. Phys. Soc. Japan **12**, 570 (1957).
2.10. D.A. Greenwood, Proc. Phys. Soc. London **71**, 585 (1958).
2.11. H.U. Barranger and A.D. Stone, Phys. Rev. B **40**, 8169 (1989).
2.12. X.-G. Zhang and W.H. Butler, Phys. Rev. B **51**, 10085 (1995).
2.13. N. Trivedi and N.W. Ashcroft, Phys. Rev. B **38**, 12298 (1988).
2.14. H.E. Camblong, Phys. Rev. B **51**, 1855 (1995). H.E. Camblong and P.M. Levy, Phys. Rev. Lett. **69**, 2835 (1992); H.E. Camblong and P.M. Levy, J. Appl. Phys. **73**, 5533 (1993).
2.15. See, for example, *Density Functional Theory*, by R.M. Dreizler and E.K.U. Gross (Springer-Verlag, Berlin, 1990).
2.16. *Solid State Physics*, by N.W. Ashcroft and N.D. Mermin (Holt, Rinehart and Winston, New York, NY 1976).
2.17. See, for example, Appendix E in [2.16].
2.18. W.H. Butler, X.-G. Zhang, D.M.C. Nicholson, and J.M. MacLaren, Phys. Rev. B **52** 13399 (1995).
2.19. See, for example, *Quantum Mechanics*, by A. Messiah, (North Holland, New York, 1961).
2.20. W.H. Butler, X.-G. Zhang, D.M.C. Nicholson, and J.M. MacLaren, J. Appl. Phys. **76** 6808 (1994).
2.21. W.H. Butler, X.-G. Zhang, D.M.C. Nicholson, T.C. Schulthess, and J.M. MacLaren, J. Appl. Phys. **79**, 5282 (1996).
2.22. W.H. Butler, X.-G. Zhang, T.C. Schulthess, D.M.C. Nicholson, J.M. MacLaren, V.S. Speriosu, and B.A. Gurney, Phys. Rev. B, **56**, 14574 (1997).
2.23. P. Weinberger, P.M. Levy, J. Barnhart, L. Szunyogh, B. Ujafalussy J. Phys. Cond. Matt. **8** 7677 (1996).
2.24. W.H. Butler and X.-G. Zhang, in "Magnetic Interactions and Spin Transport" Yves Idzerda, ed. Kluwer, New York, (2002).
2.25. W.H. Butler, X.-G. Zhang, D.M.C. Nicholson, T.C. Schulthess, and J.M. MacLaren, Phys. Rev. Lett., **76**, 3216 (1996).
2.26. C. Blaas, L. Szunyogh, P. Weinberger, C. Sommers, P.M. Levy, and J. Shi, Phys. Rev. B **65**, 134427 (2002).
2.27. P. Zahn, J. Binder, I. Mertig, R. Zeller, and P.H. Dederichs, Phys. Rev. Lett. **80**, 4309 (1998).
2.28. J. Binder, P. Zahn, and I. Mertig, J. Appl. Phys. **87**, 5182 (2000).
2.29. P. Zahn and I. Mertig, Phys. Rev. B **63**, 104412 (2001).
2.30. F. Erler, P. Zahn, and I. Mertig, ibid. **64**, 094408 (2001).
2.31. J. Binder, P. Zahn, and I. Mertig, J. Appl. Phys. 89, 7107 (2001).

2.32. K. Fuchs, Proc. Philos. Camb. Soc. **34**, 100 (1938). E.H. Sondheimer, Adv. Phys. **1**, 1 (1952).
2.33. P.F. Carcia and A. Suna, J. Appl. Phys., **54** 2000 (1983).
2.34. R.E. Camely and J. Barnaś, Phys. Rev. Lett. **63**, 664 (1989).
2.35. R.Q. Hood and L.M. Falicov, Phys. Rev. B **46**, 8287 (1992). L.M. Falicov and R.Q. Hood, J. Appl. Phys. **76**, 6595 (1994).
2.36. W.H. Butler, X.-G. Zhang, J.M. MacLaren, IEEE Trans. Magnetics **34**, 927 (1998).
2.37. J.M. MacLaren, X.-G. Zhang, and W.H. Butler, Phys. Rev. B **59**, 5470 (1999).
2.38. R.Q. Hood, L.M. Falicov, and D.R. Penn, Phys. Rev. B **49**, 368–377 (1994).
2.39. V.K. Dugaev, V.I. Litvinov, and P.P. Petrov, Phys. Rev. B **52**, 5306 (1995).
2.40. D.A. Stewart, V. Los, X.-G. Zhang, and W.H. Butler (unpublished).
2.41. W.P. Pratt Jr. et al., J. Magn. Magn. Mater. **126**, 406 (1993); Q. Yang et al., Phys. Rev. B **51**, 3226 (1995).
2.42. R. Landauer, IBM J. Res. Develop. **1**, 223 (1957).
2.43. R. Meservey and P.M. Tedrow, Physics Reports **238** 173 (1994).
2.44. W.H. Butler, X.-G. Zhang, X.-D. Wang, J. van Ek, and J.M. MacLaren J. of Appl. Phys. **81**, 5518–5520 (1997).
2.45. J.M. MacLaren, W.H. Butler, and X.G. Zhang, J. Appl. Phys. **83** 6521 (1998).
2.46. W.H. Butler, X.-G. Zhang, T.C. Schulthess, and J.M. MacLaren, Phys. Rev. B **63**, 054416 (2001).
2.47. J. Mathon and A. Umerski, Phys. Rev. B **63**, 220403 (2001).
2.48. W.H. Butler, X.-G. Zhang, T.C. Schulthess, and J.M. MacLaren, Phys. Rev. B **63**,092402 (2001).

3

Spin Polarized Electron Tunneling

P. LeClair, J.S. Moodera and H.J.M. Swagten

The quantum mechanical tunnel effect is one of the oldest quantum phenomena, and yet continues to enrich our understanding of many fields of physics, as well as creating sub-fields of its own. The generality of the tunneling phenomenon is such that virtually any introductory textbook on quantum mechanics treats tunneling through a potential barrier, and the possibility of tunneling in solid state structures was already recognized in the early 1930's [3.1]. Despite this long history and much experimental work, electron tunneling continues to be an actively used tool in many areas of physics. Spin dependent tunneling between two ferromagnets, first proposed [3.2] and observed [3.3] in 1975, has only in the last few years been reliably demonstrated [3.4]. Since the first observation of large magnetoresistance at room temperature, shown in Fig. 3.1, there has been an enormous increase of research in this field. The large Tunnel Magnetoresistance (TMR) effects possible in ferromagnet-insulator-ferromagnet tunnel junctions (magnetic tunnel junctions, or MTJs) have garnered much attention for magnetic tunnel junctions, or MTJs possible application in non-volatile Magnetic Random Access Memories (MRAMs; see the chapter by S.S.P. Parkin in this volume) and next-generation magnetic field sensors (e.g., in hard disks). However, the fundamental physics behind these devices is only beginning to be understood.

It is the purpose of this review to present a tutorial overview of the fundamental physics behind spin dependent tunneling in magnetic junctions. Much attention will be paid in the first sections to the crucial conceptual developments which led to spin polarized tunneling, and are ultimately necessary for understanding MTJs. Once the proper background has been introduced, we will discuss the basic features and behavior of MTJs, followed by a brief presentation of the earliest experiments and a short discussion of fabrication and characterization. In the main section of this review, we present recent experiments which are in our view crucial to understanding the fundamental physics behind MTJs. Finally, we will discuss a few related topics and the future outlook for this new and growing field of research.

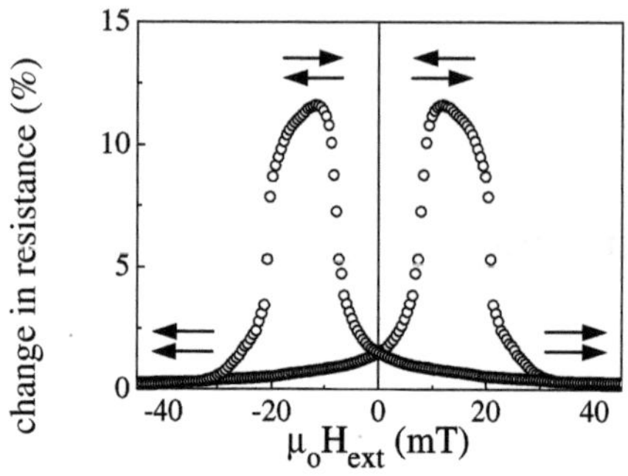

Fig. 3.1. First observation of large room temperature magnetoresistance in a $CoFe/Al_2O_3/Co$ magnetic tunnel junction. The arrows indicate the relative magnetization orientation of the CoFe and Co layers. From [3.4]

3.1 Tunneling Between Two Free-Electron Metals

Though a complete historical and conceptual background for electron tunneling and all its nuances is beyond the scope of this review, and has been discussed thoroughly in [3.5, 6], we will briefly introduce the tunnel effect in solid state structures as is relevant to MTJs. Electron tunneling is a quantum phenomena by which an electric current may flow from one electrode, through an insulating barrier, into another electrode. A simple way to understand how tunneling is possible is by considering an electron wave which encounters a potential step, see Fig. 3.2. Though most of the intensity is reflected at the potential step, a portion decays exponentially through the barrier. For sufficiently thin barriers (typically a few nm thick), some intensity remains on the other side of the potential step, and therefore, the electron will have a finite probability of being found on the other side of the potential barrier. The most straightforward realization of this structure is in a metal-insulator-metal (M-I-M) trilayer structure, commonly called a tunnel junction, with the insulator typically provided by a metal oxide (e.g., Al_2O_3).

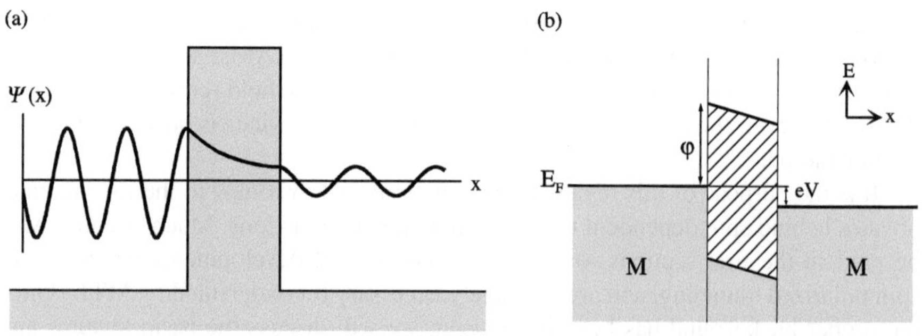

Fig. 3.2. Tunneling in metal-insulator-metal structures. **(a)** Electron wave function decays exponentially in the barrier region, and for thin barriers, some intensity remains in the right side. **(b)** Potential diagram for a M-I-M structure with applied bias eV. Shaded areas represent filled states, open areas are empty states, and the hatched area represents the forbidden gap in the insulator

In most studies, tunneling phenomenon in M-I-M structures are studied by observing the current (or its derivative) as a function of applied voltage across the junction. As an illustration, we consider phenomenologically an idealized M-I-M structure, with the electrode-tunnel barrier system modeled as a step potential (Fig. 3.2b). Without a voltage across the junction, the two metals will equilibrate, and the Fermi levels will be at the same energies for the two electrodes. When a bias V is applied across the junction, one Fermi level will shift by eV with respect to the other, where e is the electron charge (Fig. 3.2b). The number of electrons tunneling from one electrode to the other is given by the product of the density of states at a given energy in the left electrode, $\varrho_{\mathrm{l}}(E)$, and the density of states at the same energy in the right electrode, $\varrho_{\mathrm{r}}(E)$, multiplied by the square of a matrix element $|M|^2$, which is essentially the probability of transmission through the barrier. One must also then multiply by the probabilities that the states in the left electrode are occupied, $f(E)$, and that the states in the right electrode are empty, $1-f(E-eV)$, where $f(E)$ is the Fermi-Dirac function. This product is an expression of the requirement that electrons on one side of the barrier must have empty states to tunnel into on the other side of the barrier. For the general case [3.7], the tunnel current (I) from the left electrode (l) to the right electrode (r) is given by:

$$I_{\mathrm{l}\to\mathrm{r}}(V) = \int_{-\infty}^{+\infty} \varrho_{\mathrm{l}}(E)\cdot \varrho_{\mathrm{r}}(E+eV)\,|M|^2 f(E)\,[1-f(E+eV)]\;\mathrm{d}E \tag{3.1}$$

where the subscript l(r) refers to the left (right) electrode. The total tunnel current is then given by $I_{\mathrm{l}\to\mathrm{r}} - I_{\mathrm{r}\to\mathrm{l}}$. Simmons [3.8] used the WKB approximation to obtain the matrix elements $|M|^2$ for an arbitrary barrier of *average* height $\overline{\varphi}$ above the common Fermi level E_{F}. He then calculated the tunnel current from (3.1), using a free electron relation for $\varrho_{\mathrm{l(r)}}$, and approximating the Fermi-Dirac functions as step functions (i.e., $T=0$). His well-known result for a trapezoidal barrier (as shown in Fig. 3.2) is

$$\begin{aligned} J(V) &= \frac{J_{\mathrm{o}}}{d^2}\left(\overline{\varphi}-\frac{eV}{2}\right)\exp\left[-A\,d\sqrt{\overline{\varphi}-\frac{eV}{2}}\right] \\ &\quad -\frac{J_o}{d^2}\left(\overline{\varphi}+\frac{eV}{2}\right)\exp\left[-A\,d\sqrt{\overline{\varphi}+\frac{eV}{2}}\right] \end{aligned} \tag{3.2}$$

where J is the tunnel current density, $A = 4\pi\sqrt{2m_{\mathrm{e}}^*}/h$ and $J_{\mathrm{o}} = e/2\pi h$ are constants, m_{e}^* is the electron effective mass, d is the barrier thickness, $\overline{\varphi}$ is the average barrier height above the Fermi level, and V is the applied bias. If we take the barrier thickness in Ångströms (10^{-10} m), the barrier height in electron Volts, and the bias in Volts, then $A = 1.025\,\mathrm{eV}^{-0.5}\text{Å}^{-1}$ and $J_{\mathrm{o}} = 6.2\times10^{10}\,\mathrm{eV}^{-1}\text{Å}^{2}$, with the resulting current density J in A/cm^2. This equation, or its variants [3.9], are often used to fit experimental $J(V)$ characteristics to obtain *effective* barrier heights and thicknesses. For $V \ll \overline{\varphi}$, it is easily seen that the $J(V)$ is linear, while for larger voltages it becomes rapidly nonlinear. At moderate voltages, Simmons showed that $J \sim \alpha V + \beta V^3$, which leads to one of the hallmark characteristics of tunneling: a parabolic dependence of conductance ($G \equiv \mathrm{d}I/\mathrm{d}V$) on voltage, which is often observed experimentally for tunnel junctions with non-magnetic electrodes. Further, the expected exponential dependence of the tunnel current on barrier thickness and the square root of the barrier height are correctly recovered. However, any dependence on the electronic density of states (DOS) in the electrodes is suspiciously absent [3.10], which is a direct result of the over-simplified model used [3.11, 12]. As we shall scc presently, this over-simplification has far-reaching consequences.

3.2 Role of the Density of States in Tunneling

In this section, we present an overview of the subtle role of the density of states in solid state tunneling. The early models of tunneling, as well as contemporary experimental studies, all interpreted tunneling within a single-particle picture [3.5, 10] as discussed above, in which the tunnel current is independent of the electronic density of states in the metal electrodes. However, in 1960 Giaever [3.13, 14], and subsequently Shapiro et al. [3.15], performed tunneling experiments with superconducting electrodes which appeared to be the measurement of a many-body effect, viz. the superconducting energy gap of the Pb electrode in a Al/Al_2O_3/Pb junction, which had no adequate theoretical explanation at the time. In fact, several of Giaever's coworkers were initially skeptical of his "naive" experiment to measure directly the superconducting energy gap [3.16]. This "naive" experiment later led to a shared Nobel prize for Giaever in 1973.

3.2.1 Early Experiments of Giaever

In his original experiments, Giaever [3.13] investigated current-voltage and conductance-voltage characteristics, $I(V)$ and $\mathrm{d}I/\mathrm{d}V(V)$, of Al/$Al_2O_3$/Pb normal metal-insulator-superconductor (N-I-S) tunnel junctions well below the superconducting transition temperature of the Pb electrode. The Pb electrode could also be driven into the normal state by the application of a magnetic field larger than the critical field (H_C) to study transport characteristics between the two normal metals. When the Pb electrode was in the normal state, $I(V)$ was linear at low voltages (see inset to Fig. 3.3), as expected from the discussion of Sect. 3.1. However, when the Pb was in the superconducting state, the tunnel current was very much reduced at low voltages, independent of current polarity, as seen in the inset of Fig. 3.3.

More detailed information on the current transport can be obtained from the conductance curve, $\mathrm{d}I/\mathrm{d}V(V)$, shown in Fig. 3.3. Most interestingly, the conductance curve in this case very closely resembled the BCS quasiparticle "density of states" [3.6, 13] :

$$\varrho_s(E) = \varrho_n(E)\frac{|E|}{(E^2 - \Delta^2)^{\frac{1}{2}}} \tag{3.3}$$

where ϱ_n is the density of states in the normal state, and Δ is the energy gap in the quasiparticle excitation spectrum. Giaever interpreted this result [3.13], correctly as it

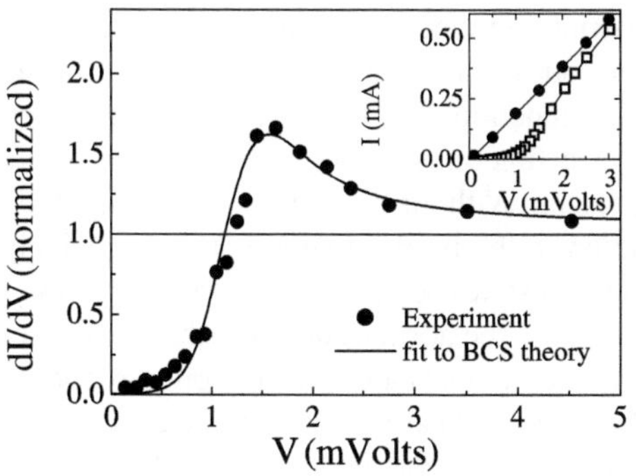

Fig. 3.3. First observation of the superconducting energy gap by tunneling. Conductance-voltage curve $dI/dV(V)$ for an $Al/Al_2O_3/Pb$ junction at $T = 1.6\,K$, $\mu_0 H = 0$, along with a fit to the BCS theory. Inset: current-voltage characteristics $I(V)$ at $T = 1.6\,K$, $\mu_0 H = 0$, Pb superconducting (*squares*); and $T = 4.2\,K$, $\mu_0 H = 0.27\,T$, Pb normal (*circles*). After Giaever [3.13]

turned out, as an indication that tunneling conductance is proportional to the density of states in the superconducting electrode. Subsequent measurements of Giaever et al. [3.17, 18] and Shapiro et al. [3.15] confirmed these results shortly thereafter.

3.2.2 Theoretical Explanation

Tunneling into superconductors is now so ubiquitous that a phenomenological explanation, closely following Giaever's original treatment, is presented in nearly every introductory solid state physics text, e.g., Kittel [3.19]; we will not discuss it further here. However, a proper theoretical justification of Giaever's ideas is subtle, with important consequences for MTJs, and we will now discuss one interpretation briefly. Despite the fact that a simple calculation starting with (3.1) seems to include the density of states directly, it has been shown that [3.5, 10] in a simple independent electron model the matrix elements $|M|^2$ are *inversely proportional to the density of states* and thus they exactly cancel the density of states factors in (3.1). Within this model, therefore, the density of states does *not* directly enter the expression for the tunnel current [3.5, 10] (see previous section). However, shortly after Giaever's experiments, Bardeen [3.11], and later Cohen et al. [3.20] and Schrieffer et al. [3.21], clarified this apparent controversy.

In fact, the failure of the independent particle models was not in the use of (3.1), but in the failure to capture the true many-body nature of tunneling in solid state systems. The presence of many-body interactions in the metal electrodes – e.g., the interactions responsible for superconductivity or ferromagnets – subtly changes the content of $|M|^2$ and ϱ in (3.1). Bardeen [3.11], appealing to many-body arguments, showed that the densities of states are indeed those of the metal electrodes, but that the matrix elements are a reflection of what is happening within the insulating barrier. The key difference with this method is in the approach to the tunneling process itself.

Within the many body approach, tunneling occurs when the insulating barrier is sufficiently thin such that the waves decaying into the barrier from each electrode overlap within the insulating barrier. The number of electrons tunneling, and hence the tunnel current, is given by the amount of overlap within the barrier. When evaluating the matrix elements and densities of states in (3.1), one must consider the different character of the wavefunctions in the barrier and electrode regions. The density of states are indeed those of the metallic electrodes near the interface (see the following section), ϱ_s and ϱ_n, respectively for a S-I-N junction. The matrix elements, however, represent the overlap of the electrode wave functions *in the barrier region* where many-body interactions are no longer present. Without the presence of many-body interactions in the barrier, the wavefunction in the barrier region is not appreciably different in the superconducting or the normal state – both are decaying functions –, and the $|M|^2$ are still essentially characteristic of the normal state. This has the consequence, according to Bardeen, that the matrix elements are inversely proportional to the *normal state* density of states, and the ϱ's in (3.1) are no longer canceled, so that the tunnel conductance is indeed proportional to ϱ_s, exactly as observed by Giaever. The crucial point is that the dependence of the tunnel conductance on the density of states does not come as simply as one might expect from (3.1), but results only from carefully considering many body effects when evaluating (3.1). For MTJs, this means that although an explanation of the TMR effect (Julliere's model, Sect. 3.5.1) seems to come quite easily from (3.1), the proper justification of such an explanation is in fact subtle.

3.2.3 Theoretical Refinements and Interface Sensitivity

Several extensions of this model were subsequently put forward [3.20–23], which essentially recover the result of Bardeen in the limit of a simple weak-coupling BCS superconductor. More sophisticated treatments by Appelbaum and Brinkman [3.22, 23] and Zawadowski [3.24], as well as more recent refinements [3.12, 25–30], put the role of the electrode density of states on firmer ground. They were able to show that tunneling is specifically sensitive to the *local density of states at the electrode-barrier interface* in superconducting as well as normal metal junctions, i.e., only the DOS near the interface is important for tunneling properties. However, this interfacial sensitivity has very different consequences for junctions with superconducting electrodes and those with normal metal electrodes.

The difference arises from the scale over which the density of states is sampled in the tunneling process – the range of the many body interactions in the superconducting or normal metal electrodes [3.22, 23]. In a normal metal, this is only on the order of a few Fermi wavelengths. However, the wavefunction in the normal metal is strongly perturbed by the presence of the metal-insulator interface in this region near the interface. This leads to the complication that in normal metal tunneling structures, one can only measure the density of states in the normal metal *in this strongly perturbed region* rather than measuring bulk-like behavior. Put slightly differently, although one may probe non-superconducting many body effects by tunneling (e.g., ferromagnetism), these effects are probed only within a few Fermi wavelengths of the electrode-barrier interface, where the density of states may differ greatly from that in the bulk. One may imagine that because of this interfacial sensitivity, tunneling characteristics will be a property of both the electrode and barrier *together*. In fact, junctions with the same electrodes but different barriers *do* behave differently, a phenomena which will be discussed in Sect. 3.6.1. Generally speaking, the interfacial sensitivity of tunneling in non-superconducting junctions has extremely important consequences for spin-dependent tunneling [3.31, 32], particularly in MTJs [3.32, 33] (see Sects. 3.6.1, 3.6.4).

For superconducting electrodes, however, the range of the many body interactions – the coherence length – is extremely large. The density of states is still sampled within a few coherence lengths of the electrode-barrier interface, but now this is a much larger length scale than the scale on which the electrode-barrier interface modifies the superconducting wave function [3.23]. This has the consequence that where the distance that the density of states is sampled, the superconducting wavefunction is essentially unchanged by the interface. Thus, for superconductor-insulator based junctions, the density of states *as measured by tunneling,*, and therefore the tunnel conductance, are essentially "bulk-like" properties, while for normal metal based junctions both are a property of the metal-insulator interface.

3.3 The Beginnings of Spin Dependent Tunneling

In order to properly discuss spin polarized transport in magnetic junctions, one must necessarily discuss the spin-polarized tunneling technique (SPT), developed by Meservey and Tedrow [3.31] in 1970, which created the field of spin-dependent tunneling. These pioneering experiments are the fundamental basis for the magnetoresistance effect in magnetic tunnel junctions, as well as many other spin polarized tunneling phenomenon. Though the full details of SPT are beyond the scope of this review (and have been excellently reviewed elsewhere, see [3.31]), we will briefly describe the concepts behind this powerful technique.

3.3.1 The Spin Polarized Tunneling Technique

Essentially, SPT utilizes a superconducting electrode in a N-I-S tunnel junction to probe the tunneling spin polarization of electrons tunneling electrons within a few 100's of μeV of the Fermi level (i.e., on the order of Δ). If the superconductor is sufficiently thin ($\sim$ 4–5 nm), such that the orbital screening currents in the superconductor are largely suppressed, the effects of the electron spin interaction with the magnetic field may be observed, resulting in Zeeman splitting of the quasiparticle states in the superconductor. However, not only must the critical field of the superconductor be large enough to have observable Zeeman splitting (Zeeman splitting larger than the thermal smearing, i.e., $3.5\,\mathrm{k_B}T < \mu_\mathrm{B} H_\mathrm{ext}$), elements with low atomic number must be used to avoid significant spin scattering via the spin-orbit effect (spin-orbit scattering).

If Zeeman splitting is achieved, the density of the states in the superconductor is then a superposition of spin-up and spin-down densities of states, as in (3.3), each of which shifted in energy by $\mp\mu_\mathrm{B} H_\mathrm{ext}$ from the zero-field curve, as shown in Fig. 3.4a. Assuming that spin is conserved in the tunneling process, which will be addressed shortly, the total conductance is simply the sum over the spin up and spin down channels. For a normal unpolarized metal, this results in a conductance curve as in

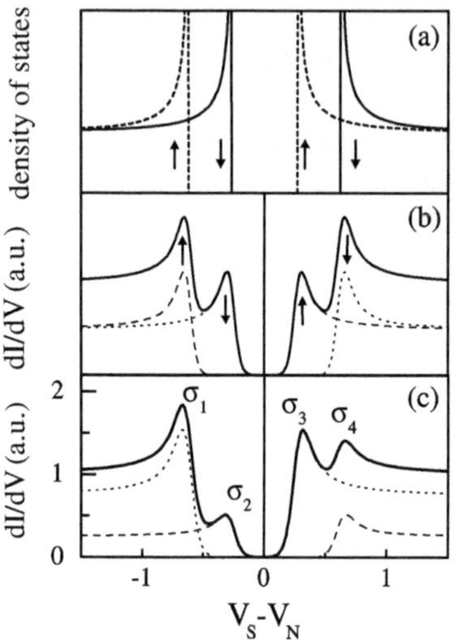

Fig. 3.4. **(a)** Magnetic field splitting of the quasi-particle states into spin-up and spin-down densities of states. **(b)** Spin resolved conductances (*dashed and dotted lines*) and resulting total conductance (*solid line*) of a S-I-N junction with $P_N = 0$, **(c)** and spin resolved (*dashed and dotted lines*) and total conductance (*solid line*) for $P_N = +0.50$

Fig. 3.4b, exhibiting 4 peaks, relating to the 2 BCS density of states peaks for each spin channel as indicated by the spin resolved conductances (dashed and dotted lines). However, when the normal metal has an unequal spin population *at the Fermi level*, $\varrho_\uparrow(E_F) \neq \varrho_\downarrow(E_F)$, each of the separate spin conductances is weighted by the relative density of states for that spin channel, as shown in Fig. 3.4c for $\varrho_\uparrow(E_F) = 3\varrho_\downarrow(E_F)$. Assuming a simple proportionality between conductance and $\varrho_l\varrho_r$, the tunneling spin polarization may be *estimated* by the relative heights of the 4 conductance peaks, σ_{1-4}:

$$P(E_F) \approx \frac{(\sigma_4 - \sigma_2) - (\sigma_1 - \sigma_3)}{(\sigma_4 - \sigma_2) + (\sigma_1 - \sigma_3)}. \tag{3.4}$$

Thus, as defined above, the full curve in Fig. 3.4c has tunneling spin polarization of $P = 50\%$. The difference in conductance peak heights is only an estimate of P, and an accurate determination must account for spin-orbit scattering in the superconductor [3.27, 31, 34] (using (3.4) tends to overestimate P, see the caption to Fig. 3.5). Further, we have not yet precisely defined what the "tunneling spin polarization" is – this will be addressed in Sect. 3.3.2. For comparison, Fig. 3.5 shows the first SPT measurement by Meservey and Tedrow [3.35] using $Al/Al_2O_3/Ni$ junctions. While the zero-field curve is symmetric, the curves for $\mu_0 H = 2.26\,T$ and $\mu_0 H = 3.37\,T$ show an increasing asymmetry, consistent with Fig. 3.4c. Though the spin polarization is small in this case, $P_{Ni} = 8.5 \pm 0.3\%$ is obtained from fitting to a model incorporating spin-orbit scattering and orbital depairing in the Al [3.31, 36], the presence of a finite and *positive* tunneling spin polarization in Ni is clear. Parenthetically, we note that the recent value of tunneling spin polarization for (polycrystalline) Ni (see Table 3.1) is more than a factor of 4 higher, almost certainly due to three decades of improved deposition techniques resulting in cleaner junctions with better interfaces. Table 3.1 lists recently obtained tunneling spin polarization values obtained for several ferromagnetic metals, after correction for spin-orbit scattering. Of particular intersest is

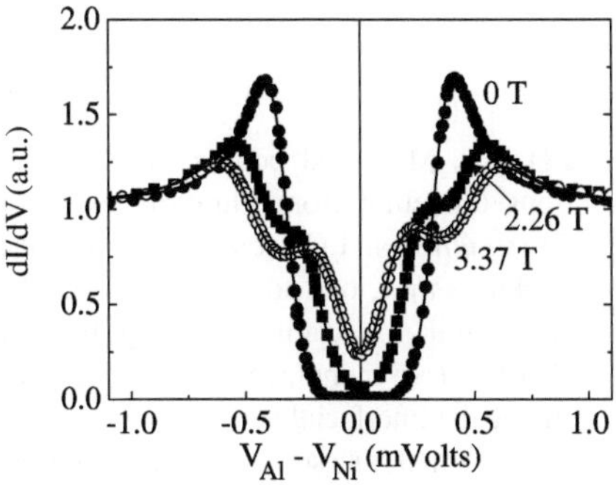

Fig. 3.5. First observation of spin polarized tunneling using Ni. The curves in a magnetic field show an obvious asymmetry, as in Fig. 3.4c. Fitting the curves (*lines*) with a model accounting for spin-orbit scattering and orbital depairing [3.31, 36] gives a polarization of 8.6±0.3%, while using (3.4) gives only ≈7.5% All curves are measured at ~0.45 K. After [3.35]

Table 3.1. Tunneling spin polarization values obtained from FM/Al_2O_3/Al junctions, after correction for spin orbit scattering.

Material	P (%)	Ref.	Material	P (%)	Ref.
Ni	31–33	[3.27, 34]	$Ni_{95}Fe_5$	34	[3.34]
Co	42	[3.27]	$Co_{40}Fe_{60}$	51	[3.34]
Fe	44–45	[3.27]	$Co_{50}Fe_{50}$	50–55	[3.27, 37]
$Ni_{40}Fe_{60}$	55	[3.34]	$Co_{60}Fe_{40}$	50	[3.34]
$Ni_{80}Fe_{20}$	45–48	[3.27, 37]	$Co_{84}Fe_{16}$	52–55	[3.27, 34]
$Ni_{90}Fe_{10}$	36	[3.34]	$La_{0.67}Sr_{0.33}MnO_3$	78	[3.38]
$SrRuO_3$	–9.5	[3.39]			

that the sign of the tunneling spin polarization for the 3*d* metals and alloys is in all cases positive. In fact, the only negative polarization measured with SPT to date is for $SrRuO_4$.

In the preceding analysis, the determination of spin polarization with the SPT technique relied on spin being conserved during the tunneling process. In fact, the Zeeman splitting of the quasiparticle density of states in a magnetic field may be also used directly to *prove* that spin flipping does not take place during tunneling. In an elegant experiment [3.40], Meservey and Tedrow studied spin polarized tunneling in Al/Al_2O_3/Al (S-I-S) junctions with *two* Zeeman split superconductors, and showed that no spin flipping takes place during the tunneling process, a crucial result for the development of spin polarized tunneling and magnetic tunnel junctions.

In another application of this powerful technique, spin polarization measurements on ultra-thin ferromagnetic films [3.31, 41] first demonstrated the anticipated interfacial sensitivity [3.22–24, 26] (see Sect. 3.2.3) in tunnel junctions with a normal metal electrode. In this case, ultrathin (0–30 Å) films of Fe or Co were deposited on a substrate normal metal to form the bottom electrode of the N-F-I-S junctions, on which the Al_2O_3 tunnel barrier and superconducting Al electrode were grown (Fig. 3.6). The spin polarization of electrons tunneling from these ultrathin layers was then measured as a function of average layer thickness, and compared with the polariza-

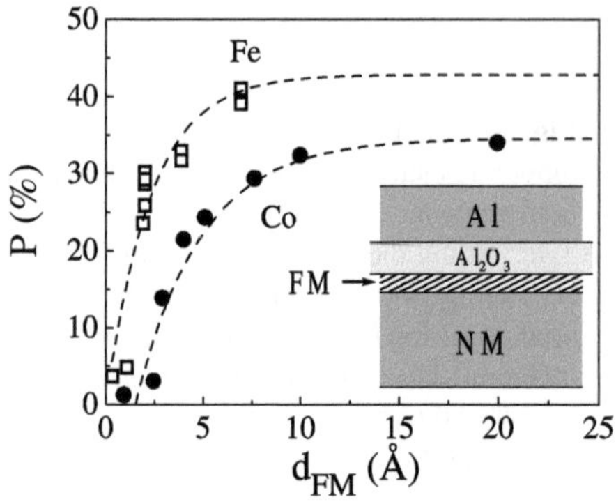

Fig. 3.6. Measured polarization of electrons tunneling from ultrathin layers of Fe (*squares*) and Co (*circles*), of average thickness d_{FM}, on (normal-state) Al. The nearly full polarization for only 10 Å of Co or Fe (∼5 monolayers) clearly demonstrates the interfacial sensitivity of tunneling in normal state structures. After Tedrow and Meservey [3.41]

tion values for thick Fe or Co layers. For Fe and Co backed with (normal-state) Al, Fig. 3.6, the polarization increased rapidly with increasing thickness, showing clear spin polarization (and hence, ferromagnetism) for even a single monolayer. The most striking result is that the "bulk" tunneling spin polarization measured on thick layers is nearly reached for only 7–10 Å approximately 3–5 monolayers. Not only does this experiment clearly indicate that the onset of ferromagnetism extremely rapid, it definitively demonstrates that tunneling in normal-state structures is highly interface sensitive, and apparently, the outermost 3–5 monolayers dominate the transport properties. This is in excellent agreement with theoretical expectations [3.22–24, 26], and of crucial importance for the development and understanding of magnetic tunnel junctions. Further, this result implies that the tunneling characteristics, such as spin polarization, will be highly susceptible to the interface quality, which has great ramifications for the growth of high-quality magnetic tunnel junctions.

3.3.2 What is Tunneling Spin Polarization?

The most striking feature of these results initially was the fact that the tunneling spin polarization of 3*d* ferrmagnetic metals was *positive* (i.e., spin up majority) in all cases, seemingly at odds with the bulk band structures [3.42] of these materials, which show a dominant *minority* contribution to the density of states at the Fermi level. Though the details of exactly what the tunneling spin polarization is are rather complex and still the subject of controversy, we will attempt here to present a physically transparent description of what in our interpretation determines the tunneling spin polarization. For more in-depth discussions, we refer the reader to the references.

Typically, for SPT experiments $|M|^2$ and $\rho_{1,r}$ in (3.1) are taken as constants, and for vanishing external bias this results in the tunneling spin polarization being simply the (normalized) difference between spin up and spin down densities of states at the Fermi level [3.31]. However, this is generally not justified, even for a vanishing bias. *In reality, the density of states measured via tunneling is never the "raw" density of states, but always weighted by* $|M|^2$. As we have seen in Sect. 3.2.2, a proper consideration of the matrix elements $|M|^2$ is crucial in explaining the role of the density of states in tunneling, and as discussed below, even more crucial in understanding P. A slightly more precise definition of tunneling spin polarization P (considering transmission independent of $k_{||}$), may be given for vanishing bias by [3.43]

$$P|_{V=0} = \frac{\varrho_\uparrow |M_\uparrow|^2 - \varrho_\downarrow |M_\downarrow|^2}{\varrho_\uparrow |M_\uparrow|^2 + \varrho_\downarrow |M_\downarrow|^2}. \tag{3.5}$$

For non-vanishing bias, the energy dependence of $|M|^2$ and $\varrho_{\mathrm{l,r}}$ must be taken into account as well when integrating over all energies in (3.1), although only the energy range where there are filled states in one electrode and empty states in the other (the applied bias) will contribute. For SPT experiments, P is still essentially evaluated at E_F due to the small bias involved (> 1 mVolt), but for MTJs the energy dependence of P is significantly more important. In addition, P *gains several rather subtle dependencies via the matrix elements – such as the crucial dependence of the tunneling spin polarization on the insulating barrier used* – some of which will be discussed below. From the above definition it is also clear that due to the presence of the matrix elements, the tunneling spin polarization is *not* directly related to the magnetization or even the spin polarization measured by other techniques (e.g., spin polarized photoemission or Andreev reflection). Here we will briefly discuss a few approaches which illustrate how considering the ϱ and $|M|^2$ in realistic systems can resolve the apparent discrepancy between band structure and tunneling spin polarization.

One early attempt to explain the sign of the spin polarization measured by SPT was given by Stearns [3.44]. She recognized that within the independent electron model of tunneling [3.5, 10], the transmission probability depends on the electron effective mass m_e^*, which is different for different bands. More generally, one may say that the $|M|^2$ in (3.1) are different for states of different symmetry [3.29, 45–49]. In the $3d$ ferromagnets studied by Meservey and Tedrow [3.31], the relatively localized d electrons carry most of the total magnetic moment, while the mobile s electrons contribute little. However, the d electrons have a large m_e^* and decay extremely rapidly into the barrier region, while the mobile s-hybridized electrons decay slowly into the barrier region. One would expect, then, that for reasonably thick tunnel barriers ($\sim$ several monolayers or more) that the s-hybridized electrons would come to dominate the tunnel current, given that decay rate is exponential in $\sqrt{m_\mathrm{e}^*}$ but only proportional to the total DOS. In $3d$ ferromagnets, the s bands are spin polarized through hybridization with the d bands, and tend to have a spin polarization opposite in sign to the $3d$ bands. In this simplistic model, Stearns could explain the *positive* tunneling spin polarization and the *negative* spin polarization of the total DOS. In fact, recent calculations based on a tight binding model through a vacuum barrier [3.27] essentially reproduce this simple analysis (as do different treatments by Mazin [3.43] and Butler et al. [3.46]). For very thin barriers (e.g., a few monolayers), the large d DOS dominates, despite the lower probability of d electrons tunneling, giving a large negative P. For thicker barriers (more than a few monolayers), the slower decay rate of s electrons compensates their smaller DOS, and the s electrons with positive spin polarization begin to dominate the tunnel current, with a positive spin polarization resulting.

For a simple vacuum barrier, this view is reasonable, but this picture neglects the role of the insulating barrier and the ferromagnet-insulator bonding. Based on the previous discussions (Sect. 3.2.3), one expects that the tunnel current is sensitive to the *interfacial* DOS at the ferromagnet-insulator barrier, which is strongly modified by the presence of the barrier and bonding at the ferromagnet-insulator interface. Tsymbal and Pettifor [3.25] first addressed the influence of the ferromagnet-insulator bonding, finding that the tunnel conductance strongly depends on the type of covalent bonding at the interface. They showed that only $ss\sigma$ bonding between Co or Fe and the insulating barrier could explain a positive spin polarization measured by SPT. In other words, Stearns' conjecture [3.44] that the positively polarized s-hybridized electrons have a lower decay rate into the barrier is valid for real insulators only in the case of interfacial $ss\sigma$ bonding. Tsymbal and Pettifor further predicted that if, e.g., $sd\sigma$ bonding were to dominate, the Co or Fe d electrons would actually couple through the barrier more effectively, and their large negative polarization should dominate. For relatively simple insulators of s and p band elements, such as Al_2O_3, $ss\sigma$ bonding is dominant and positive polarization is expected, consistent with the observed $P > 0$. Indeed, an interfacial $sp-d$ hybridization consistent with this idea has been experimentally observed at Co/Al_2O_3 interfaces [3.50]. However, as will be seen in Sect. 3.6.1, if more complex insulators containing transition metals are used, dominant $sd\sigma$ bonding may be realized, and a negative polarization may be found [3.51–53]. Thus, in realistic systems, the tunneling spin polarization is not simply a property of the electrode alone, but is (at least) a property of the *electrode-barrier interface* – again a reflection of the fact that the $|M|^2$ in (3.1,3.5) can generally *not* be neglected. Further, this once again essentially eliminates in most cases any relation between P and M, or tunneling spin polarization and spin polarization measured by other techniques.

For ordered systems, such as fully epitaxial tunnel junctions, where strict $k_{||}$ conservation must be considered [3.54] the situation is more complex. As recent *ab initio* calculations [3.28, 29, 45–49, 55–58] have demonstrated, the idea of a "tunneling spin polarization" in terms of e.g. (3.5) then tends to lose meaning, with the transport characteristics becoming a rather complex function of k, barrier thickness, and the symmetry of the Bloch states involved. The resulting spin polarization of the tunnel current depends in detail on the entire quantum system. For systems with non-epitaxial electrodes and amorphous barriers, as typically studied, k conservation rules are broken [3.54], and any analysis of tunneling based on strict k conservation is generally invalid. However, these discussions are beyond the scope of this review. For most of the experiments discussed here, these considerations play a negligible role, and are referred to only briefly in Sect. 3.6.3.

3.3.3 Spin Filter Tunneling

Up until now, we have focused on a spin polarized density of states as the source of tunneling spin polarization. However, as alluded to by (3.5), a spin dependent tunneling probability, e.g., $|M_{\uparrow}|^2 > |M_{\downarrow}|^2$, also results in a spin polarized tunnel current, since in this example spin up electrons have a larger tunneling probability than spin down electrons. Clearly, this is the case even with nonmagnetic electrodes. The question remains, though, of how to physically realize this potential source of spin polarization. Given the form of (3.1) and (3.2), the most obvious and physically realistic way to realize a spin dependent tunneling probability is via a spin dependent tunnel barrier height.

Indeed, a spin dependent barrier height may be achieved by utilizing a (ferro-) magnetic tunnel barrier. As an example, consider the ferromagnetic semiconductor, EuS, a model Heisenberg system with $T_C = 16.7\,K$ [3.60, 61]. Above T_C the barrier is nonmagnetic, and tunneling electrons see a single spin independent barrier height φ_o determined by the bottom of the EuS conduction band, Fig. 3.2b. Below T_C, however, the EuS conduction band is exchange split by an amount $\Delta\varphi_{exch}$, Fig. 3.7a, and tunneling electrons now see a *spin dependent* barrier height of $\varphi_{\uparrow(\downarrow)} = \varphi_o \mp \Delta\varphi_{exch}/2$. For EuS, typical tunnel barrier heights are $\sim 1.5-2\,eV$ [3.59, 62], with a conduction band exchange splitting of $\approx 0.36\,eV$ [3.60, 61]. Given the exponential dependence of tunnel current on $\sqrt{\varphi}$, it is easy to see that a highly spin polarized current may result. Further, since most electrons utilize this lower tunnel barrier, the overall junction resistance decreases below T_C [3.59, 62, 63], a signature of spin filtering.

The principle of spin filtering has been experimentally demonstrated in field emission experiments [3.61, 63], and most dramatically, in SPT experiments [3.62]. In the latter case, Moodera et al. [3.59, 62] performed SPT experiments using Al/EuS/M junctions, where M was Ag, Au, or Al. Fig. 3.7b shows a SPT curve for a Al/EuS/Au junction [3.59], showing a tunneling spin polarization of approximately 80%. Further,

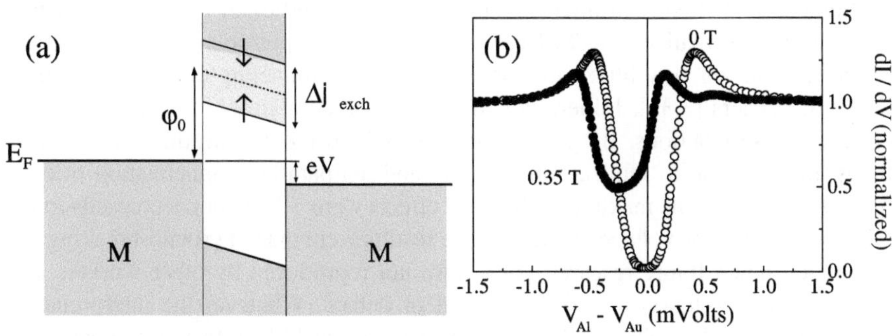

Fig. 3.7. The principle of spin filter tunneling with magnetic barriers. Above the T_C of the barrier, there is no conduction band exchange splitting, and both spin up and spin down electrons see a single barrier height φ_0 (see Fig. 3.2). **(a)** Below T_C, the bottom of the conduction band is exchange split by an amount $\Delta\varphi_{exch}$. Since spin up electrons see a significantly lower barrier height, they tunnel preferentially and large tunneling spin polarizations (approaching 100%) may be achieved. **(b)** Conductance vs. voltage at $\mu_o H = 0\,T$ and $\mu_o H = 0.35\,T$ for a Al/EuS/Au junction at 0.4 K demonstrating the spin filter effect. For 0.35 T almost no minority spin conductance is observed; a model fit gives a polarization of $\approx 80\%$. In this case, the Zeeman splitting in the Al electrode is enhanced via an exchange interaction with the EuS. From [3.59]

using a related Eu-chalcogenide EuSe, Moodera et al. were able to demonstrate essentially 100% spin polarization. Given that the largest spin polarization observed with ferromagnetic electrodes thusfar is 78%, using the low T_C material $La_{0.67}Sr_{0.33}MnO_3$, spin filtering is an attractive route for the generation and manipulation of highly spin polarized currents, e.g., for spin injection into semiconductors [3.64]. Finally, by *combining* both a spin dependent density of states (i.e., ferromagnetic electrodes) and a spin dependent tunneling probability (i.e., spin filtering), additional hybrid devices and novel effects may be imagined. One recent demonstration of these ideas is discussed in Sect. 3.6.5.

3.3.4 Early MTJ Experiments

Given the previous discussions of the role of the density of states in tunnel junctions, and the conservation of spin during tunneling, it is reasonable to anticipate that spin-polarized tunneling effects may be observed in *non-superconducting* junctions. In 1975, Julliere [3.3] and Slonczewski [3.2] had exactly this idea in mind: tunneling between two ferromagnets. In the SPT technique, a tunnel current is passed between a spin-dependent density of states in a metallic ferromagnet, and a spin-split BCS density of states in a superconductor. In tunneling between ferromagnets, the basic idea is this: rather than use a spin-split superconductor as a spin detector, the spin-dependent density of states in a second ferromagnetic electrode may also be used. In this case, the tunnel current is expected to depend on the relative magnetization orientation of the two ferromagnetic electrodes (this will be discussed in more depth in Sect. 3.5), giving rise to a magnetoresistance effect – the TMR effect.

Julliere [3.3] first measured the conductance properties of Fe/Ge/Co junctions in an attempt to realize the TMR effect. Though later spin-polarized tunneling experiments showed very little spin conservation when tunneling through amorphous Ge or Si barriers [3.31], Julliere nonetheless observed a sizable magnetoresistance effect in these junctions. By growing the Fe and Co layers with different coercive fields, he was able to realize both parallel and antiparallel magnetization orientations. At zero bias, the maximum observed effects were ~14%, but decreased rapidly with increasing bias voltage. Though these results were indeed ground-breaking and stimulated much future research, they were not reproduced by other workers, and their true interpretation is still the subject of debate. Whatever the interpretation, Slonczewski [3.2] quickly realized the potential of the TMR effect, and others soon followed this research.

In a related experiment, Helman and Abeles [3.65] studied transport in thin films composed of Ni granules in a SiO_2 matrix. In this case, the magnetizations of the Ni grains were presumably aligned randomly in zero field, while in a high magnetic field the grains were forced to align with parallel magnetizations. Up to 1.5% magnetoresistance was observed at low temperatures and in high fields. Though much less ambiguous than the results of Julliere [3.3], the effects were still disappointingly far from those expected. Maekawa and Gäfvert [3.66] successfully and unambiguously reproduced Julliere's results in 1982, using Ni/NiO/Ni and

Ni/NiO/Co junctions. For Ni/NiO/Co junctions at 4.2 K, up to $\approx 2\%$ magnetoresistance was observed, which was, for the first time, clearly correlated with measured hysteresis behavior of the magnetic electrodes. The effects were still small compared to those anticipated, but this work was the first to establish that the magnetoresistance effects were due to the relative magnetization alignment between the two electrodes.

Over the next decade, several workers (see [3.67–70] for example) attempted to realize the anticipated large TMR effects, with little success. In most cases, the observed effects were $\sim 1\%$ or less, and present only at relatively low temperatures. In retrospect, this lack of success was in most cases probably due to the magnetic oxide barriers (e.g., NiO, Gd_2O_3) used, which are likely to introduce significant spin-flip scattering during tunneling [3.71–73], and hence, largely negate the TMR effect (see Sect. 3.4.3 on barrier doping [3.74, 75] and Sect. 3.6.4 on interface dusting [3.32, 33]). Further, given the interfacial sensitivity of tunneling in normal state structures, and the extremely short range of ferromagnetic interactions, extremely clean ferromagnet-insulator interfaces are required, which would have to wait for improved deposition techniques to become widely used.

In 1995, nearly 20 years after the original "discovery" of the TMR effect, Moodera et al. [3.4] and Miyazaki and Tezuka [3.76] solved most of the earlier problems with MTJs, and independently demonstrated $>10\%$ TMR at room temperature (shown in Fig. 3.1). The latter case, however, was shown to suffer from a measurement geometry artifact [3.77] which inflated the measured TMR values. Nonetheless, these first demonstrations quickly garnered a great deal of attention, and catalyzed many groups to investigate MTJs. The first demonstration by Moodera et al. [3.4] already established many of the basic features associated with MTJs, most notably, the bias and temperature dependence of the TMR, which are discussed in Sects. 3.5.4 and 3.5.5.

3.4 Fabrication and Characterization of FM-Al_2O_3-FM Junctions

Though the possibility of large magnetoresistance in MTJs had been recognized since 1975 [3.3, 78], the experiments which actually *demonstrated* large, reproducible magnetoresistance at room temperature came only in 1995 [3.4]. To some degree, this breakthrough can certainly be ascribed to the tremendous advances in thin film growth technology since 1975. The extreme sensitivity of MTJs to interface contamination, surface and interface roughness, and barrier quality clearly point to a need for extremely high quality growth. Further, the growth of ultra-thin oxidic films which are flat and pinhole-free by itself presents a host of difficulites. Though these topics are somewhat beyond the scope of this review, we will briefly discuss a few topics related to fabrication and characterization of MTJs. For a more complete discussion of the challenges in preparing MTJs, we refer to [3.27].

3.4.1 A Fabrication Recipe

Given the recent explosion of research focused on MTJs, a wide variety of methods have been developed for fabricating MTJs. Here, we outline one used in the authors' laboratories which has proven successful. Ferromagnetic tunnel junctions have been prepared in our laboratory by UHV dc/rf magnetron sputtering, utilizing metal contact masks to create a cross-geometry junction structure, with junction areas of 300 μm × 300 μm to 500 μm × 500 μm and 24 junctions per sample. The base pressure is typically $<5\times10^{-10}$ mbar. Si(100) or glass substrates are typically used, which are *in-situ* cleaned in an O_2 plasma to remove remaining carbon and water contamination and also provide an insulating layer to prevent leakage conduction through the Si substrates. Our standard junction structures consist of Si/SiO_2/Ta 3.5 nm/Co 8 nm/FeMn 10 nm/Co 5 nm/Al_2O_3/Co 15 nm/Ta 2 nm, with the Al_2O_3 layers formed by plasma oxidation of 2.2 nm Al in 10^{-1} mbar O_2 for 200 seconds. Post-growth annealing in magnetic field of ≈ 20 mT for 30 minutes at 200 °C is used to promote a uniform exchange biasing direction (see Sect. 3.5). This routinely provides junctions with magnetoresistance of $\Delta R/R_p = 25-30\%$ and resistance-area products of 10^7–10^8 $\Omega \cdot \mu m^2$ at 295 K.

3.4.2 A Few Characterization Techniques

There are a myriad of possible techniques to characterize MTJs either during the growth process or after fabrication. We will not go into any detail, but, rather, only outline several common techniques that have been used successfully and provide references for further inquiry. Some in situ methods for chemical characterization are X-ray photoelectron spectroscopy (XPS), Ultraviolet Photoelectron Spectroscopy (UPS), and Auger electron spectroscopy (AES). UPS can give information on valence band structure [3.79], while XPS [3.80–84], as well as other ex situ techniques like Electron Energy Loss Spectroscopy (EELS) [3.50], Fourier-transform infrared spectroscopy (FTIR) [3.85], and X-ray absorption edge spectroscopy (XAS) [3.86], may be used to determine the chemical state of the barrier (e.g., degree of oxidation) and electrodes. AES also has been used to study the growth modes of ultrathin interfacial layers in MTJs [3.33]. Other possibilities for characterizing oxidation are in situ differential ellipsometry [3.80, 87] and in situ resistivity measurements [3.88], where the oxidation of the thin Al layers can be monitored real-time, allowing true process control. Structural characterization can also be realized with in situ scanning tunneling microscopy (STM) [3.79, 89, 90] or atomic force microscopy (AFM) [3.91–93]. Surface structure and roughness can be analyzed at each stage of growth using AFM or STM, and local transport characteristics of Al_2O_3 barriers have also been recently reported [3.89, 91–93] using STM and conducting AFM. More complex STM-related techniques, like Ballistic Electron Emission Microscopy (BEEM) [3.94, 95] may also be used to study buried interfaces and barrier transmission characteristics. Once fabrication is completed, many ex situ techniques are also possible. Rutherford backscattering (RBS) [3.96, 97] and Transmission Electron Microscopy (TEM) [3.50, 98, 99] can be used to gain structural and chemical information on the sub-nanometer scale.

Often overlooked techniques for local structural characterization on an atomic scale are ^{59}Co NMR [3.100–103] and ^{27}Al NMR [3.81], which are also capable of studying interface structure and buried layers in completed structures. Magnetic characterization can be performed using the magneto-optical kerr effect (MOKE), superconducting quantum interference device (SQUID) magnetometry, or vibrating sample magnetometry (VSM). Most important, perhaps, are transport measurements. Measurements of junction resistance or conductance ($\mathrm{d}I/\mathrm{d}V$) as a function of field, temperature, and applied bias are at the heart of MTJ characterization, and reveal the "hallmark" features associated with MTJs, which will be discussed in Sect. 3.5.

3.4.3 Sensitivity of MTJs to Barrier Impurities and Annealing

The extreme sensitivity of MTJs to *barrier* impurities was demonstrated by Moodera and Jansen [3.74, 75, 104]. In a series of experiments, an ultrathin "δ" doping layer was inserted in the middle of the Al layer which was subsequently oxidized to form the tunnel barrier. As Auger electron spectroscopy (AES) showed [3.104], the result was an Al_2O_3 barrier with *oxidized* metallic impurities at the midpoint of the barrier. The magnetoresistance decreased extremely rapidly as a function of δ dopant layer thickness for most dopants, shown in Fig. 3.8. Impurities that had no magnetic moment in the oxidic state, such as Co, showed only a modest decrease for ~1 monolayer of dopant, while those impurities which possessed a moment in the oxidic state, such as Ni or Cu, nearly quenched the magnetoresistance by ~1 monolayer. Further analysis showed that the slow decrease for non-magnetic impurities was due to an excess impurity-assisted conductance contribution (hopping conductance through defect levels within the barrier), which merely dilutes the magnetoresistance, while the rapid decrease for magnetic impurities was due to spin-dependent scattering of the tunneling electrons [3.71–73, 105] by the magnetic impurities, which affects the magnetoresistance much more severely. Curiously, for Fe doping the magnetoresistance actually exhibited a relative *increase*, a result which has not yet been adequately explained [3.104], but certainly outlines the complexity of transport in MTJs. These results clearly highlight the fact that not only ultra-clean *interfaces* are required, as anticipated in Sects. 3.2.2 and 3.2.3, but ultra-clean *barriers* are crucial as well,

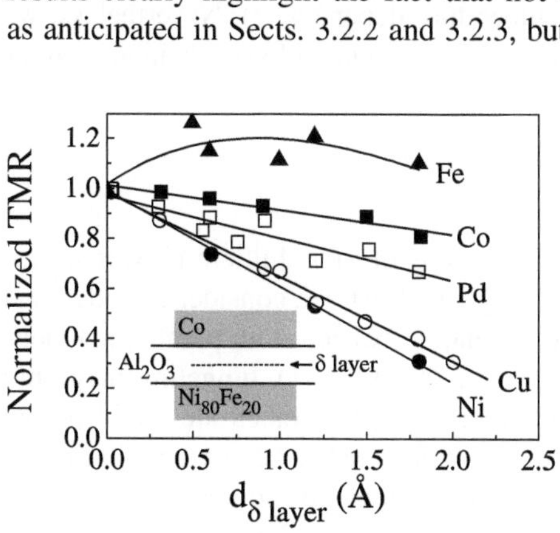

Fig. 3.8. Normalized $\Delta R/R_{\mathrm{ap}}$ as a function of barrier δ doping layer thickness at 77 K for Fe, Co, Pd, Cu, and Ni dopants. Line for Fe is a guide to the eye, the rest are linear fits. From [3.74, 75, 104]

since any spin-independent conductance contribution or spin scattering mechanism will necessarily decrease the magnetoresistance. Spin scattering by magnetic oxides formed by over-oxidation in the fabrication process has been shown to be particularly effective in this regard [3.86, 106], which as mentioned in Sect. 3.3.4, perhaps explains the lack of success in earlier work. [3.66–70] Further, a similar sensitivity to impurities at the ferromagnet-insulator *interface* has also been demonstrated [3.32].

Along similar lines, one may expect that MTJs are sensitive to structural changes within the barrier, e.g., induced by annealing. Despite the extreme sensitivity of MTJs to barrier impurities and non-ideal interfaces, current state-of-the-art MTJs have in fact proven to be extremely robust. Though it is perhaps surprising that Al_2O_3 layers of only 10 Å or less are stable during thermal processing, recent experiments by Sousa et al. [3.107] have shown that not only are MTJs thermally stable beyond 300 °C, annealing can actually *improve* the MTJ properties. Crucial to improved thermal stability of MTJs is the control of the interface oxidation state, again related to the interface sensitivity of tunneling (see Sect. 3.2.3). Zhang et al. [3.108, 109] further showed that by using interfacial FeO_x layers adjacent to the Al_2O_3 barriers, the interface oxidation state could be manipulated at elevated temperatures. In this way, MTJs with 40% TMR at up to 380 °C annealing temperatures could be prepared. The improvement of MTJ thermal stability is of crucial importance for the application of MTJs, and this remains an area of intensive research. These results further point to the tremendous advances in thin film growth technology over the last few decades, and clearly many more can be expected.

3.5 Hallmark Features of MTJs

Here we will discuss the basis for the TMR effect and the hallmark features of MTJs, viz., the dependence of the TMR and conductance on applied bias, temperature, and magnetization orientation. Though some of these features may be simply understood from the previous sections (for instance, the magnitude of the TMR and the angular dependence), others are more subtle and their origins have not been unambiguously determined (for instance, the bias and temperature dependence). Nevertheless, we will outline the main features of MTJs and the fundamental physics behind them as they are now understood.

3.5.1 Basis for the TMR Effect

Following directly Meservey and Tedrow's analysis of SPT, Julliere [3.3] first proposed a model for tunneling between ferromagnets. First, we consider two identical ferromagnetic electrodes, separated by an insulating barrier, with parallel magnetization orientations. Assuming spin conservation (Sect. 3.3.1), tunneling may only occur between bands of the same spin orientation in either electrode, i.e., from an up spin band to an up spin band, and vice versa. One further key assumption in the Julliere model is that the matrix elements in (3.1) can be considered constant. This is physically equivalent to saying that the tunneling *probability* is independent of spin or magnetization orientation, the spin reservoirs on either side of the barrier are coupled in the same way for each spin orientation (recall the discussion of Sect. 3.3.2). Using these two ideas (and further assuming a small bias voltage such that the ferromagnet DOS is essentially constant), along with a proportionality between tunnel conductance and the density of states in the electrodes (recall Sect. 3.2.2), we may determine the conductance for parallel magnetizations in a manner similar to SPT analysis for the zero temperature case:

$$G_{\mathrm{p}} = G_{\uparrow} + G_{\downarrow} \propto \varrho_{\mathrm{maj}}^2 + \varrho_{\mathrm{min}}^2, \tag{3.6}$$

where $G_{\uparrow(\downarrow)}$ is the conductance in the up- (down-) spin channel and $\varrho_{\mathrm{maj(min)}}$ is the majority (minority) spin density of states at E_{DF} (keeping in mind interface sensitivity and the complex nature of P, see Sects. 3.2.3 and 3.3.2). When changing the magnetization orientation of one ferromagnetic electrode (relative to that of the other ferromagnetic electrode), the axis of spin quantization is also changed in that electrode. Tunneling between like spin orientations now means tunneling from a majority to a *minority* band, and vice versa. The conductance for antiparallel alignments is then simply:

$$G_{\mathrm{ap}} = G_{\uparrow} + G_{\downarrow} = \varrho_{\mathrm{maj}}\, \varrho_{\mathrm{min}} + \varrho_{\mathrm{min}}\, \varrho_{\mathrm{maj}} = 2\varrho_{\mathrm{maj}}\, \varrho_{\mathrm{min}}. \tag{3.7}$$

Thus the conductances are different for parallel or antiparallel magnetizations, i.e., ferromagnet-ferromagnet tunnel junctions display a magnetoresistance. This tunnel magnetoresistance (TMR), which we define following the majority of workers as *the difference in conductance between parallel and antiparallel magnetizations, normalized by the antiparallel conductance*, can be calculated from (3.5–3.7):

$$TMR|_{V=0} \equiv \frac{G_{\mathrm{p}} - G_{\mathrm{ap}}}{G_{\mathrm{ap}}} = \frac{R_{\mathrm{ap}} - R_{\mathrm{p}}}{R_{\mathrm{p}}} = \frac{2P_{\mathrm{l}}P_{\mathrm{r}}}{1 - P_{\mathrm{l}}P_{\mathrm{r}}}, \tag{3.8}$$

where $P_{\mathrm{l(r)}}$ is the tunneling spin polarization in the left (right) ferromagnetic electrode, keeping in mind that even when the matrix elements in (3.5) are taken as constants, P is not simply the difference in total density states at the Fermi level, but is still determined by, e.g., the effective masses of different band electrons, interface bonding, and even k selection in some cases (see Sects. 3.2.3, 3.3.2). Using the measured tunneling spin polarization for Co with Al_2O_3 tunnel barriers (see Table 3.1), we may expect a TMR effect of more than 40% for Co/Al_2O_3/Co MTJs, only slightly above the observed (low temperature) value (see Table 3.2). In general, the most recent spin polarization values with Al_2O_3 barriers [3.27, 34] obtained via the SPT technique agree well with the maximum TMR values reported with Al_2O_3 barriers [3.34, 110]. Table 3.2 compares the expected TMR values based on Julliere's model, using values of P obtained from SPT experiments, with measured TMR values *using the same barriers in both cases*. However, we caution that the Julliere model is only a phenomenological guide to estimate the magnitude of the TMR effect *when tunneling spin polarizations are known*.

Obviously, one expects the largest TMR values for materials with the largest tunneling spin polarization. This explains a great deal of the recent interest in so-called

Table 3.2. Comparison of TMR expected from Julliere's model with measured low-temperature TMR values, for Al_2O_3 barriers (except where noted).

Junction	$TMR_{Expt.}$ (%)	$TMR_{Julliere}$ (%)	Ref.
LSMO/$SrTiO_3$/LSMO*	310	400	[3.38, 111, 112]
$Co_{75}Fe_{25}/Al_2O_3/Co_{75}Fe_{25}$	67-74	69	[3.113, 114]
$Co/Al_2O_3/Co$	38-40	37	[3.115]
$Ni/Al_2O_3/Ni$	25	24	[3.116]

*$La_{0.67}Sr_{0.33}MnO_3^*$.

"half-metallic" ferromagnets, materials for which only one spin band is occupied at the Fermi level, resulting in perfect 100% spin polarization [3.117]. Many compounds have been predicted to be half metallic, such as the half- and full-Heusler alloys NiMnSb [3.118] and Co_2MnSi [3.119]; the oxides CrO_2 [3.120], Fe_3O_4 [3.121, 122], and $La_{0.67}Sr_{0.33}MnO_3$ [3.123, 124]; and the sulfide $Co_xFe_{1-x}S_2$ [3.125]. However, only $La_{0.67}Sr_{0.33}MnO_3$ (LSMO) [3.126], NiMnSb [3.127], and CrO_2 [3.128] have any experimental evidence in favor of half metallic behavior, and little progress has been made in the fabrication of MTJs with half-metallic electrodes. Thusfar, no material has been shown to be half-metallic by a SPT experiment. Still, there has been much success with $La_{0.67}Sr_{0.33}MnO_3$ (LSMO) by Lu et al. [3.111] and Viret et al. [3.112], who observed TMR effects of more than 400% at low temperature utilizing $SrTiO_3$, $PrBaCu_{2.8}Ga_{0.2}O_7$, or CeO_2 barriers. Using (3.8), this implies a spin polarization of more than 80%, in agreement with SPT experiments [3.38] (see Table 3.1). Similarly, Jo et al. [3.129, 130] have used another mixed-valence manganite, $La_{0.7}Ca_{0.3}MnO_3$ (LCMO) and investigated LCMO/$NdGaO_3$/LCMO and LCMO/$NdGaO_3$/LSMO MTJs, also observing more than 400% TMR. Thus, although still not truly half metallic, LSMO and LCMO have shown the highest tunneling spin polarizations yet observed in MTJs, and byfar the largest magnetoresistances. MTJs with epitaxial NiMnSb electrodes have been successfully studied, but a spin polarization of only 25–28% was deduced from TMR values, in agreement with SPT experiments [3.131]. Clearly, much work is still needed in this very promising area.

3.5.2 Resistance vs. Field

Julliere's simple model for the TMR effect was discussed previously in this section, with the result that the conductance (resistance) of a MTJ *with spin polarizations of the same sign for both electrodes* is higher (lower) when the magnetizations of the two ferromagnets are parallel. Once said, one must realize experimentally both a parallel and antiparallel magnetization alignment. Perhaps the simplest way to realize this is to use two ferromagnets with different coercive fields [3.132, 133], for example by using two different ferromagnets like Co ("hard") and $Ni_{80}Fe_{20}$ ("soft"). In this case, the soft $Ni_{80}Fe_{20}$ electrode switches at relatively low fields (~0.5 mT) while the hard Co electrode switches at higher fields (~1.5 mT). TMR vs. magnetic field behavior for a $Ni_{80}Fe_{20}/Al_2O_3/Co$ junction (at $V = 0$) is shown in Fig. 3.9a. When the field is swept through zero and reaches values between the $Ni_{80}Fe_{20}$ and Co coercive fields (sweeping from P_1 to AP_1 or P_2 to AP_2 in Fig. 3.9a), an antiparallel magnetization alignment (AP_1 and AP_2) is reached. This state is maintained even if the field is removed, allowing a MTJ to potentially be used as a non-volatile memory element [3.134, 135] (see also the chapter by S.S.P. Parkin in this volume).

More technologically relevant are exchange biased MTJs. In this case, one of the magnetic electrodes is in direct contact with an antiferromagnetic (AFM) material (e.g., FeMn, IrMn, NiO). The presence of an exchange anisotropy at the FM/AFM interface [3.136, 137] shifts the entire magnetization-field loop of the ferromagnet away from zero field, such that it is centered at a finite magnetic field, the exchange bias effect. Typical TMR vs. magnetic field behavior for an exchange biased system (at $V = 0$) is shown in Fig. 3.9. As the magnetic field is swept from negative values through zero, the second magnetic electrode switches as in the previous example, giving the high resistance antiparallel state (AP). As the field is swept further, the exchange biased layer switches, returning the system to the low resistance parallel (P_2) state. As the field is reversed, this time the exchange biased layer reverses *first*, returning the system to the antiparallel state (AP), again at positive fields. The antiparallel state is retained until the field is swept through zero and the coercive field of the second electrode is reached for negative fields. The exchange biased MTJ is the most commonly studied configuration by far. Technologically, exchange biasing is advantageous because the resistance transition takes place near zero magnetic field, and it generally results in greater magnetic stability [3.138], while from a fundamental point of view, it allows one to study MTJs with nominally identical electrodes, leading to (hopefully) simplified analysis.

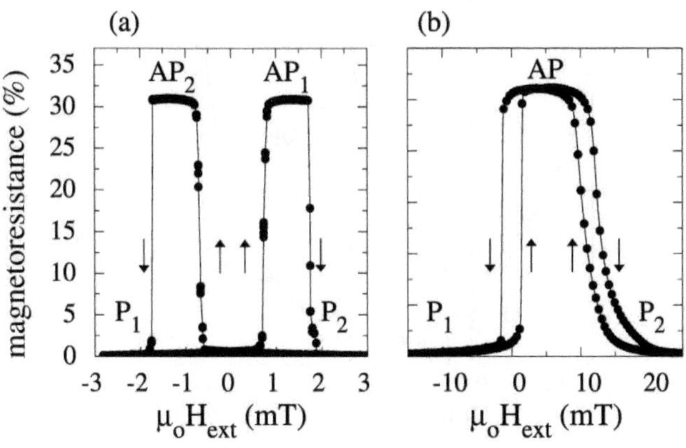

Fig. 3.9. Magnetoresistance vs. magnetic field for **(a)** a hard-soft MTJ and **(b)** an exchange biased MTJ, both at 10 K. P and AP refer to parallel and antiparallel magnetization orientations, respectively, and vertical arrows refer to sweep direction. Both curves are taken at $V = 0$. From [3.115, 116]

3.5.3 Conductance vs. Voltage

As we have seen previously in Sect. 3.1, tunneling between two non-magnetic free-electron metals should give rise to a conductance (dI/dV) which is quadratic in applied bias (V). Indeed, for non-magnetic junctions such as $Al/Al_2O_3/Al$, this behavior is often observed [3.6]. However, in MTJs (or even junctions with one magnetic electrode), the conductance behavior deviates significantly from the expected parabolic dependence, as shown in Fig. 3.10a for a $Co/Al_2O_3/Co$ junction (with 2 *polycrystalline* electrodes, see Sect. 3.6.2), most noticeably at low voltages. Though the conductance is approximately symmetric with respect to voltage, as expected for (nominally) identical electrodes, there is a pronounced, sharp decrease below $\sim$150 mVolts, where the conductance is approximately linear in voltage. This sharp dip in conductance about zero bias is similar to so-called "zero-bias anomalies" observed by many groups [3.6], which were attributed to (magnetic) impurities in the tunnel barrier or near the electrode-barrier interfaces. However, in contrast to those zero-bias anomalies, the conductance dip has no strong temperature dependence or magnetic field dependence, persisting even in the cleanest junctions. Further, the energy scale of the anomaly is far greater than usually observed. Other mechanisms, such as the presence of metal particles in the barrier (Giaever-Zeller anomalies [3.6]) are similarly inconsistent. Further, the characteristic (nearly) linear conductance contribution near zero bias occurs *only* when at least one electrode is magnetic [3.139–141].

Moodera et al. [3.139, 140] first suggested that magnon excitations may be at least partly responsible for the conductance anomalies as well as the TMR bias depen-

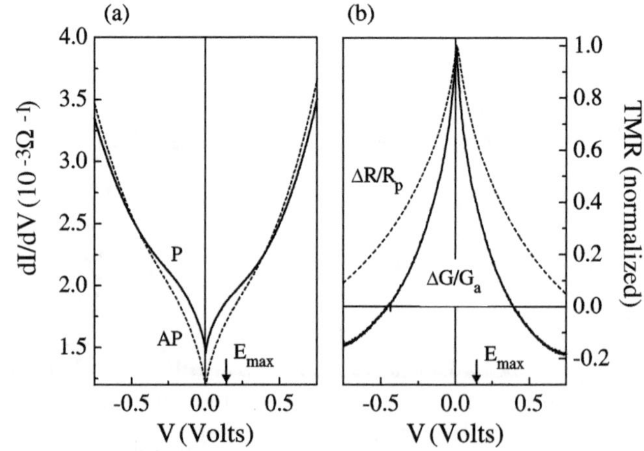

Fig. 3.10. **(a)** Conductance vs. applied bias, $dI/dV(V)$, at 5 K for parallel (*solid*) and antiparallel (*dashed*) magnetizations of a $Co/Al_2O_3/Co$ MTJ. Vertical arrow indicates the maximum magnon energy (E_{max}) for bulk Co in a mean field approximation. **(b)** Bias dependence of the TMR ($\Delta R/R_p$, *dashed*) and differential TMR ($\Delta G/G_{ap}$, *solid*) for a $Co/Al_2O_3/Co$ junction at 5 K. Note that the normal ($\Delta R/R_p$) and "differential" ($\Delta G/G_{ap}$) magnetoresistances are only equivalent at *zero bias*. From [3.100, 115]

dence. A theoretical explanation based on this idea was given by Zhang et al. [3.142] and Bratkovsky [3.143] in terms of magnon excitations localized at the FM-Al_2O_3 interface. Electrons at the Fermi level of one ferromagnet tunnel across the junction, reaching the second ferromagnet with an energy equal to the junction bias eV above the Fermi level in the second electrode (assuming no other inelastic tunneling processes). These "hot" electrons may then lose energy by emitting a magnon of energy $\hbar\omega \leq eV$ (a similar process holds for magnon absorption), and as one might expect (see following section) this leads to a decrease of the TMR. Zhang et al. [3.142] and Bratkovsky [3.143] found that these processes also influence the conductance characteristics, giving an additional inelastic conductance contribution $G_{\text{magnon}} \propto V$, simply reflecting the fact that a larger bias allows more magnons to be excited. The slope of the linear contribution is larger in the antiparallel case, with the difference between parallel and antiparallel slopes simply given by (3.6,3.7). The linearity is preserved up to a maximum voltage determined by the maximum magnon energy in the ferromagnetic electrode, which, within a mean-field approximation, corresponds to $E_{\max} = 3k_B T_C/(S+1)$, where k_B is Boltzman's constant, and T_C is the Curie temperature of the ferromagnet with spin S. Considering two identical Co electrodes with the bulk T_C, $E_{\max} \sim 144\,\text{meV}$. Given that the interface T_C is expected to be significantly lower, this is in agreeement with width of the conductance dip shown in Fig. 3.10a. For a detailed application of this model, we refer to Han et al. [3.110], who have performed a careful analysis of the conductance and magnetoresistance as a function of voltage and temperature for $Co_{75}Fe_{25}/Al_2O_3/Co_{75}Fe_{25}$ tunnel junctions using an extension of the aforementioned model. The primary conclusions of this study were that by *combining* different MTJ characteristics (such as $\mathrm{d}I/\mathrm{d}V(V,T)$, $\Delta R/R_p(V,T)$, see Sects. 3.5.4, 3.5.5), a consistent set of all model parameters could be obtained, allowing a much more stringent test of the magnon excitation model. Excellent agreement between experimental data and model calculations were demonstrated, suggesting that indeed the additional conductance contribution consistently observed in MTJs (the zero-bias "dip") can be largely, if not completely, explained by interfacial magnons excited by hot tunneling electrons.

One further question with regard to conductance-voltage behavior that frequently arises is the role of the density of states in the tunnel conductance. Given the form of (3.1), which explicitly contains the density of states as a function of energy in each electrode, it is rather surprising that clear observations of density of states and band structure effects have *not* been reported until recently. The reasons for this will be discussed in more detail in Sect. 3.6.2.

3.5.4 TMR vs. Voltage

Experimental Observations

Perhaps the most surprising feature of MTJs initially was the dependence of the TMR effect on applied dc bias. First observed by Julliere [3.3], and confirmed by Moodera et al. [3.4], it was at first unclear if this was an intrinsic effect, or simply due to inelastic tunneling through a non-ideal interface and barrier (e.g., impurity-assisted tunneling). However, subsequent measurements on clean junctions, as well as the observation of this effect by many other groups in the years that followed, produced the conclusion that the bias dependence of the TMR is an *intrinsic* effect [3.139], although its magnitude may vary considerably. A customary figure-of-merit is the voltage at which the TMR ($\Delta R/R_p$) is reduced by a factor of two. Moodera et al. in their initial observation [3.4] found a "half voltage" (usually denoted $V_{1/2}$) of $\sim$ 200 mVolts, while recently several groups have improved this figure to > 500 mVolts. Moodera et al. [3.139] also showed that the bias dependence is relatively temperature independent.

Figure 3.10b shows a representative TMR ($\Delta R/R_p$) vs. bias behavior for Co/Al_2O_3/Co junctions at 5 K [3.115], showing both $\Delta R/R_p$ and $\Delta G/G_{ap}$. Note that these two definitions of the TMR are *not* identical at finite bias, since, as seen in (3.2) the tunnel current is a non-linear function of voltage [3.144]. To avoid confusion, we will refer to $\Delta R/R_p$ as simply TMR, following the majority of workers, and to $\Delta G/G_{ap}$ as the *differential* TMR. The differential TMR is typically observed to be approximately linear in bias up to $\sim$ 0.5 Volts [3.145], becoming negative at higher biases and then tending to zero. The normal TMR shows a roughly parabolic voltage dependence at intermediate biases ($\sim$0.5 Volts), and tends smoothly to zero at higher biases, with a half-voltage of typically $\sim 0.3-0.5$ Volts. One insight into its possible origin is the fact that the differential TMR has exactly the same voltage dependence as the magnon-assisted conductance contribution [3.142, 143] discussed in the previous section.

Theoretical Explanations

The aforementioned models of Zhang et al. [3.142] and Bratkovsky [3.143] were in fact originally proposed to explain the bias dependence of the TMR. We have seen that the magnon-assisted conductance contribution is linear in voltage, but with a differing slope for parallel and antiparallel magnetization orientations. Thus, the conductance change ΔG from parallel to antiparallel orientations will also be approximately linear in voltage up to $eV = E_{max}$, roughly as observed in $\Delta G/G_{ap}$ (after further analysis to remove other conductance contributions). The fact that both the conductance-voltage and magnetoresistance-voltage agree favorably with model calculations, and further, yield realistic parameters [3.110], emphasizes that magnon excitations may play a dominant role in the bias dependence. However, several other possible origins have also been proposed and must be considered.

Davis and MacLaren [3.146] have proposed that the bias dependence also has an *intrinsic* component resulting from the underlying electronic structure. They considered free electron tunneling through a square potential barrier, using simple parabolic bands modeled on the itinerant electron bands in Fe determined from *ab initio* calculations. Two primary electronic structure effects were considered, viz., shifting of the Fermi level of one electrode relative to the other, and the altered barrier shape at finite bias. The shifting of the chemical potential allows new states to be accessed for tunneling, which essentially makes the spin polarization a voltage-dependent quantity, whereas the altered barrier shape allows higher-energy states to tunnel more easily.

This makes the matrix elements in (3.1) *spin and bias dependent.* A strong decrease of the TMR with bias is also found within their model, and also gives good agreement with experimental data with realistic parameter choices. Further, early work by Moodera et al. [3.4, 134, 139] as well as recent experiments utilizing composite insulating barriers [3.51–53] and ultra-thin non-magnetic layers at the FM/Al_2O_3 interface [3.101] have pointed out the importance of a bias-dependent spin polarization, as will be discussed in Sects. 3.6.1 and 3.6.4. In particular, at large biases ($eV > E_{max}$) one must consider alternative mechanisms to a purely magnon-assisted bias dependence. As one example, the effects of phonon [3.143] and impurity [3.147, 148] assisted tunneling on the bias dependence of the TMR have also been considered theoretically. Experimental support exists for both magnon-assisted and intrinsic band-structure mechanisms of bias dependence as well as for an impurity-assisted contribution [3.74, 75, 104]. The currently accepted viewpoint is that all of these mechanisms play a key role to some degree.

3.5.5 TMR Temperature Dependence

It was first noticed by Shang et al. [3.149] that the temperature dependence of the tunnel resistance for magnetic tunnel junctions greatly exceeds that for non-magnetic junctions with nominally identical barriers. Typically, standard Al/Al_2O_3/Al junctions showed only a 5–10% [3.149] change in resistance from 4.2 to 300 K, while magnetic tunnel junctions always exhibited a 15–25% change in resistance, as shown in Fig. 3.11 for a Co/Al_2O_3/Co junction. Further, the TMR ($\Delta R/R_p$) can change as much as 25% or more from 4.2 to 300 K depending on the magnetic electrodes (also shown in Fig. 3.11). Shang et al. explained these results within a simple phenomenological model, in which it was assumed that the tunneling spin polarization P decreases with increasing temperature due to spin-wave excitations, as does the surface magnetization. They thus assumed that both the tunneling spin polarization and the interface magnetization followed the same temperature dependence, the famous Bloch $T^{3/2}$ law [3.19], i.e., $M(T) = M(0)(1 - \alpha T^{3/2})$ for low T/T_C. This temperature dependence holds for surfaces as well as the bulk, though the former has a larger decay constant α [3.150, 151]. Given the interfacial sensitivity of tunneling (see Sect. 3.2.3), Shang et al. [3.149] assumed that α was also larger than the bulk value in MTJs, and thus gave a satisfactory explanation for the temperature dependence of the TMR. MacDonald et al. [3.152] provided a more rigorous theoretical justification of these ideas, essentially reproducing the proportionality between $M(T)$ and $P(T)$, though the microscopic origin was slightly different than that considered by Shang et al. The fact that a microscopic model is able to justify the phenomenological approach of Shang et al. lends support to the idea that the temperature dependence of the (surface) magnetization and spin polarization may be intimately connected. However, we again caution that this does not imply any general relation between M and P [3.153]. That two quantities have the same *temperature dependence* is only a reflectaion of the fact that the same physical process causes both M and P to decrease with temperature, viz., magnetic disorder due to thermal excitation of spin waves. In our interpretation,

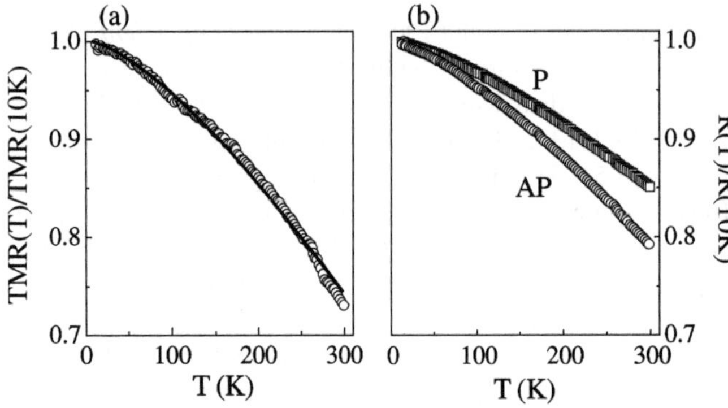

Fig. 3.11. **(a)** Temperature dependence of the TMR, $\Delta R/R_p$ (*circles*) for a Co/Al_2O_3/Co MTJ, along with a fit to the model of Shang et al. [3.149] (*line*). **(b)** Temperature dependence of the normalized tunnel resistance for parallel (P, *squares*) and antiparallel (AP, *circles*) magnetization orientations, R_p, for the same junction. All curves taken at $V=0$. Data from [3.115, 154]

the temperature dependence of the TMR should be regarded as a special coincidence rather than the signature of any more fundamental relationship.

Davis et al. [3.154] also attempted to put the model of Shang et al. on firmer theoretical ground, in a slightly different manner, by assuming that the temperature dependence of the polarization arises from a Stoner-like collapse of the exchange splitting. As with their model of the bias dependence [3.146], they used free electron-like bands, modeled after the *ab initio* band structure of Co. The model itinerant bands were exchange-split parabolic bands, with a spin dependent effective mass. Recent photoemission data [3.155, 156] indicate that these itinerant bands are Stoner-like, i.e., the exchange splitting depends on temperature, and collapses at T_C. Further, citing the fact that the exchange splitting is nearly proportional to $M(T)$ [3.157], they used the known $M(T)$ behavior of Co to obtain the temperature dependent exchange splitting: $\Delta E_{ex} = \beta M(T)$. Using the bulk $M(T)$ behavior of Co and TMR data of LeClair et al. [3.115, 154], Davis et al. were able to explain only about a third of the experimental drop in TMR from 0 to 300 K. This is not surprising, given that the interface T_C is expected to be significantly lower [3.150, 151]. Strikingly, however, they calculated a 18% change in TMR despite the fact that the magnetization changes by only 1.5% over the same temperature range, once again indicating that in general P and M are not simply related. The TMR temperature dependence over this range could be well described by assuming a lower interface ordering temperature for Co, viz., $T_C \approx 982$ K, though this is perhaps unphysically low. As with their work on the TMR bias dependence, the main message is that purely *intrinsic* band effects are able to explain much of the TMR temperature dependence, without resorting to inelastic processes.

One further attempt to explain the TMR temperature dependence comes again from the model of Zhang et al. [3.142]. In the case of temperature dependence, it is the *thermal* excitation of spin waves which leads to a decrease in TMR. We note parenthetically that this is *not* inconsistent with the above model of Davis et al. [3.154], as the $M(T)$ behavior is a result of spin wave excitations. In contrast to the TMR voltage dependence obtained by Zhang et al., it is no longer the maximum magnon energy, E_{max}, which sets the relevant energy scale, but rather the lower wavelength cutoff in the spectrum, E_C. This cutoff results physically either from anisotropy, always present for the spins near the FM-I interface, or from a finite coherence length, due to, e.g., a finite grain size. In this model, the zero-bias conductance behaves with temperature roughly as $G(T) \approx T \ln(k_B T/E_c)$ dependence. For a detailed comparison between this model and experimental data, we again refer to Han et al. [3.110]. In fact, all three models presented [3.142, 149, 154] reproduce experimental data reasonably well and yield physically realistic parameters. One problem with all studies thus far the Curie temperature of the ferromagnetic electrodes used is well beyond temperatures at which MTJs can be studied, so that the true critical behavior of the TMR, which may possibly distinguish between different models, cannot be studied. Careful studies of low T_C materials, such as rare earth materials or transition metal alloys, are required – only GaMnAs-based junctions (see Sect. 3.6.3) have thusfar been measured close to T_C. Finally, as with the bias dependence, the role of impurity-assisted tunneling must also be considered [3.74, 75, 104, 147, 148]. The current opinion is generally that both intrinsic (e.g., band effects) and extrinsic (inelastic processes) contributions are of importance.

3.6 Recent Magnetic Tunnel Junction Experiments

In the following sections, we will attempt a more detailed description of MTJs by describing several experiments which have crucially expanded our understanding of the fundamental physics involved. Section 3.6.1 emphasizes the importance of interface bonding and the importance of the insulating barrier in determining the properties of MTJs. Section 3.6.2 illustrates a clear relation between the electronic and physical structure of MTJ electrodes and the spin-polarized transport properties, as well as a successful application of a theoretical model. Section 3.6.3 covers recent developments of semi- or fully-epitaxial MTJs, and the dependence of spin polarization on crystallographic orientation. Section 3.6.4 covers ultra-thin interfacial layers in MTJs, highlighting the importance of interfacial sensitivity of MTJs and novel properties which may be realized by so-called "interface engineering." Finally, Sect. 3.6.5 discusses one new hybrid magnetoresistive device based on the *combination* of spin filtering and spin-dependent tunneling. Though necessarily not complete, these experiments highlight some of the more subtle concepts touched on previously.

3.6.1 Composite Barriers and the Role of Interface Bonding

As discussed previously in Sect. 3.3.2, the sign and magnitude of the tunneling spin polarization are expected to depend on not only the ferromagnetic electrode, but the barrier and the electrode-barrier interface bonding as well. In 1999, two different groups [3.51–53] independently observed this prediction [3.25]. Later theories confirmed this prediction [3.27, 47] and demonstrated the possibility of creating novel

MTJ properties by not only varying the magnetic electrode, but by varying the insulating barrier as well.

Recently, de Teresa et al. [3.51, 52] also observed that the tunneling spin polarization depends explicitly on the insulating barrier used. In this case, half metallic $La_{0.7}Sr_{0.3}MnO_3$ (LSMO) was used as one of the electrodes with barriers of Al_2O_3, $SrTiO_3$ (STO), $Ce_{0.69}La_{0.31}O_{1.845}$ (CLO), or a composite $Al_2O_3/SrTiO_3$ barrier. Since it is known that LSMO has a only majority states at E_F, the tunneling spin polarization must be positive and close to 100% [3.38, 126] (see Sect. 3.5.1), regardless of the insulating barrier used. Indeed, SPT experiments have measured a tunneling spin polarization of $P = +78\%$ (see Table 3.1). Thus, LSMO electrodes serve as a spin analyzer for the polarization of the second electrode, in the spirit of the SPT technique of Meservey and Tedrow (see Sect. 3.3.1). Moreover, a half metallic analyzer has the added advantage that to some extent the energy dependence of the polarization may be studied.

As expected, de Teresa et al. found that Co/Al_2O_3/LSMO MTJs gave a positive TMR for all biases, not surprising since both LSMO and the Co/Al_2O_3 interface are known to have positive polarizations. On the other hand, $Co/SrTiO_3$/LSMO junctions showed *negative* TMR values at zero bias, and further displayed a strong bias dependence as shown in Fig. 3.12b. In this case the polarization of the $Co/SrTiO_3$ interface must be *negative*, opposite that of Co/Al_2O_3 interfaces, in agreement with the d electron polarization of Co at E_F (see Sect. 3.3.2). In order to show this more conclusively, de Teresa et al. investigated $Co/Al_2O_3/SrTiO_3$/LSMO junctions, with the expectation that the positive LSMO and Co/Al_2O_3 polarizations would yield a normal positive TMR for all biases. As shown in the inset to Fig. 3.12b, a normal positive TMR is observed for all biases, with a bias dependence that is essentially identical to standard $Co/Al_2O_3/Co$ junctions (see Sect. 3.5.4).

The sign change of the Co tunneling spin polarization can be understood in terms of interface bonding, as discussed in Sect. 3.3.2. For Co/Al_2O_3 interfaces, $ss\sigma$ bonding across the Co/Al_2O_3 interface dominates, preferentially selecting the s-partial density of states in Co which is *positively* polarized. For $Co/SrTiO_3$ interfaces, however, the bonding may be predominantly of a $d-d$ nature between Co and Ti at the interface [3.25, 158], preferentially selecting the d partial density of states which is *negatively* polarized. Indeed, this conjecture is borne out by recent *ab inito* calculations [3.159] for $Co/SrTiO_3/Co$ MTJs. Further experimental proof is evidenced in the bias dependence of $Co/SrTiO_3$/LSMO junctions, shown in Fig. 3.12, which correlates with the d spin polarization of Co. At zero bias, the Co d DOS is minority dominated, as shown in Fig. 3.12a (for the Co(100) surface), and exhibits a large negative tunneling spin polarization. For negative bias (electrons tunneling *from* the LSMO), the LSMO detector predominantly scans the DOS (convoluted with the energy-dependent tunneling probability) above the Fermi level in the Co electrode. As the bias is decreased to -0.4 Volts, in the antiparallel state the Fermi level of LSMO is at the same energy as the peak of the Co minority d DOS, giving a larger tunneling spin polarization and, hence, a large (negative) TMR. As the bias becomes more negative, the minority d DOS progressively decreases, and the tunneling spin polarization decreases.

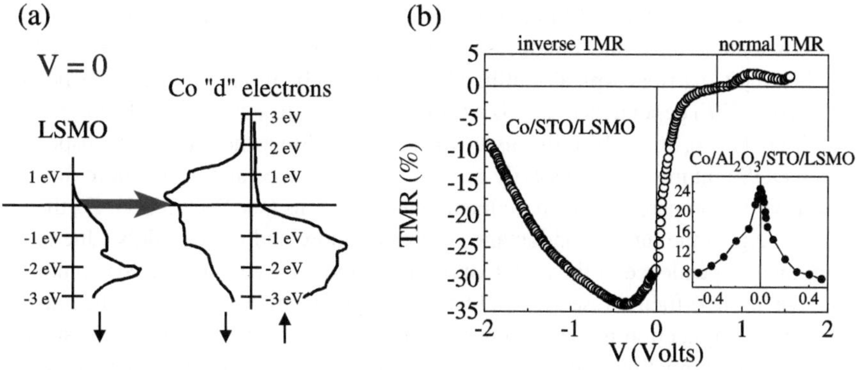

Fig. 3.12. (a) Schematic of the spin polarized densities of states of LSMO (as derived from photoemission) and the Co(100) surface (calculated). **(b)** TMR ratio vs. applied bias for a Co/$SrTiO_3$/LSMO junction at 5 K. Inverse TMR is observed for $V < 0.8$ Volts, while normal TMR is observed for $V > 0.8$ Volts, indicating that the Co/$SrTiO_3$ spin polarization is negative for $V < 0.8$ Volts. Inset: TMR ratio vs. applied bias for a Co/Al_2O_3/$SrTiO_3$/LSMO junction. In this case, the polarization of Co/Al_2O_3 and LSMO are both positive, and a normal positive TMR is seen. From [3.51, 52]

For positive bias (electrons tunneling *from* Co), the LSMO detector predominantly scans the DOS *below* the Fermi level in the Co electrode. As the bias is increased, the majority *d* DOS steadily increases, until at approximately 0.8 Volts the majority and minority densities of states are equal, resulting in a zero crossing of the TMR. For still higher bias, the majority *d* DOS of Co continues to increase (while the minority DOS steadily decreases), resulting in a positive tunneling spin polarization and TMR. At approximately 1.15 Volts, the Fermi level of the LSMO is at the same energy as the *majority d* DOS peak in Co, and the positive TMR is maximal. Thus, all of the features in the bias dependence of Co/$SrTiO_3$/LSMO junctions can be attributed to the structure of the spin polarized Co *d* DOS, in contrast to recent experiments with Co/Al_2O_3/Co junctions (see below, Sect. 3.6.2) which revealed structure in the Co *s* DOS. More recent results by Sugiyama et al. [3.160] and Sun et al. [3.161] essentially corroborate these results.

Finally, recent experiments by Sharma et al. [3.53] utilizing composite Ta_2O_5/Al_2O_3 barriers have indicated that the sign of the spin polarization at Ta_2O_5 interfaces is also negative, and have proposed an explanation similar to that of de Teresa et al. However, these results have not yet been independently verified. Still, these results illustrate the rich physics behind spin polarized tunneling in MTJs, as well as the intriguing possibility of "engineering" MTJs with tailored properties.

3.6.2 Role of Electrode Electronic and Physical Structure

The absence of density of states features in most MTJ experiments is apparently at odds with (3.1), which indicates a direct proportionality of the tunnel conductance with the density of states. This discrepancy is in part due to several competing factors.

One limitation is that one can only hope to see band or DOS features for those bands that contribute to the tunnel conductance. These bands and the DOS they generate are (for the Al_2O_3 barriers typically utilized) believed to be limited to highly dispersive (i.e., strong $E(k)$ dependence) bands whose states at finite k are s-hybridized [3.6, 12, 25, 27, 31, 44–46, 56, 162], as discussed in Sect. 3.3.2. Since they are highly dispersive bands, DOS features are necessarily much less pronounced and hence more difficult to observe as compared to the more localized d electrons (see Sect. 3.6.1). Further, at higher energies the tunnel conductance increases more and more rapidly (due to the strong bias dependence of the effective barrier height at high biases [3.8]), making it more and more difficult to see subtle features on the strong background conductance. A second complication is the presence of many conductance contributions, such as normal elastic tunneling, see (3.1,3.2), and inelastic excitations (phonons, magnons; see Sects. 3.5.3 and 3.5.4), which may obscure electronic structure features, but which may be essentially eliminated with careful analysis [3.6] as discussed below. A third complication is the extreme difficulty in theoretical analysis of these structures, particularly MTJs. A last, and perhaps most fatal complication is the absence of tunneling electrodes with a *known physical and electronic structure which can be modulated* in order to convincingly compare theory and experiment.

In a recent series of experiments, LeClair et al. [3.100] utilized Co/Al_2O_3/Co MTJs with a view to observe features of the Co s-hybridized DOS. Using ^{59}Co NMR combined with X-ray diffraction, it was shown that two different types of Co bottom electrodes could be grown, viz., highly-textured fcc(111)-Co (denoted fcc-Co) and polycrystalline, polyphase Co (denoted poly-Co). This enabled both fcc-Co(111)/Al_2O_3/poly-Co and poly-Co/Al_2O_3/poly-Co MTJs to be studied in order to determine the influence of the fcc-Co structure on MTJ properties. Figure 3.13a shows the $\mathrm{d}I/\mathrm{d}V(V)$ characteristics for fcc(111)-Co 35 Å, 50 Å/Al_2O_3/poly-Co junctions, and a poly-Co/Al_2O_3/poly-Co junction for parallel magnetizations at 5 K (the fcc-Co electrode is biased positively for $V > 0$). The poly-Co/Al_2O_3/poly-Co junctions show almost perfectly symmetric behavior for both magnetization orientations, with a parabolic background and a low-voltage linear contribution, consistent with normal MTJ behavior as discussed in Sect. 3.5.3. The fcc(111)-Co/Al_2O_3/poly-Co junctions, however, show an obvious conductance asymmetry, with most notably a local minimum at approximately −0.25 Volts and a slight "shoulder" at the same positive voltage, as also observed by others [3.139, 142, 163, 164]. Similar conductance minima have been observed in Ni/Al_2O_3/Ni junctions by Jansen et al. [3.165], where in that case the origin was clearly the electronic structure of the Ni. This suggests that the band structure and spin-dependent DOS of the fcc-Co structure must be responsible.

If the fcc-Co DOS is responsible for the conductance features, one expects the TMR to be also affected. Although the TMR is the most commonly used criterion to gauge MTJs, the TMR magnitude can be very susceptible to slight differences in preparation conditions [3.27] (see Sect. 3.4), and is not generally well suited to detailed comparison with theory. The normalized *differential* TMR vs. voltage characteristics, $\Delta G/G_{ap}(V)$, though generally nearly identical for many junctions [3.145], suffer from a strong contribution by inelastic excitations [3.142], as discussed below, which may mask the underlying electronic structure effects. Further, these inelas-

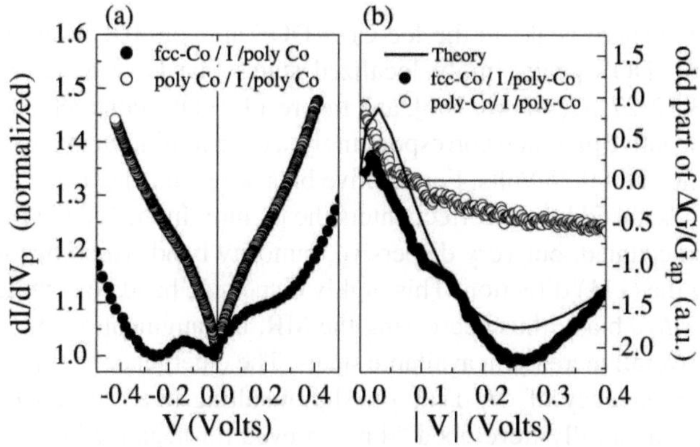

Fig. 3.13. (a) Conductance-voltage $\mathrm{d}I/\mathrm{d}V(V)$ characteristics for fcc(111)-Co/Al_2O_3/poly-Co junction (*closed circles*), and a poly-Co/Al_2O_3/poly-Co junction (*open circles*), for parallel magnetization alignments at 5 K. **(b)** Odd part of $\Delta G/G_{\mathrm{ap}}$ (see text) as a function of voltage for the same junctions as **(a)**, and a theoretical calculation for a fcc-Co/I/hcp-Co junction (*line*). From [3.100]

tic excitations depend on extrinsic factors (such as interface quality [3.139, 149]) and are extremely difficult to model quantitatively. However, to a reasonable approximation most inelastic excitations give a contribution *symmetric* in applied bias (particularly for physically symmetric junctions), and thus by plotting the *odd* part of the differential TMR-voltage behavior, one is primarily sensitive to only electronic structure effects [3.6]. On the other hand, by plotting the *even* part of $\Delta G/G_{\mathrm{ap}}$, one may more selectively view inelastic excitations [3.6]. In this manner, the DOS features may be more clearly evidenced, and more easily compared with theory, effectively eliminating those contributions which are difficult to incorporate theoretically.

Figure 3.13b shows the odd portion of the differential magnetoresistance, i.e., $\Delta G/G_{\mathrm{ap}}(V > 0) - \Delta G/G_{\mathrm{ap}}(V < 0)$, as a function of voltage for both types of junctions studied. As noted above, this quantity is a measure of the asymmetry in the bias dependence of the *differential* TMR. For poly-Co/Al_2O_3/poly-Co junctions, only a small asymmetry is present, as expected for nominally identical electrodes. However, for fcc(111)-Co/Al_2O_3/poly-Co junctions, there is a clear strong minimum, corresponding to the same voltage where the local minimum and shoulder features were seen in the $\mathrm{d}I/\mathrm{d}V(V)$ curves, suggesting again that the fcc-Co DOS and band structure may play a key role.

Utilizing a ballistic model of tunneling [3.146, 154, 166], LeClair et al. [3.100] were able to explain these results in terms of the s-hybridized densities of states of hcp-Co and fcc-Co (see Sect. 3.3.2). Figure 3.13b shows the odd part of the differential TMR, *calculated* for a fcc-Co/I/hcp-Co MTJ using the model described in [3.146, 154]. The hcp-Co DOS was used to model the poly-Co DOS, since both are essentially featureless. The underlying tendency for $\mathrm{d}I/\mathrm{d}V(V)$ to dip in fcc(111)-Co/Al_2O_3/poly-Co structures at approximately ± 0.25 Volts can be explained by the presence of two sharp peaks in the fcc-Co s-DOS at about 0.4 eV above *and* below E_F [3.25, 166]. DOS peaks imply localized states that have a *reduced* tunneling probability [3.12, 27, 31, 44–46, 166], and this results in the conductance dips at both positive and negative bias, and correspondingly, a deep minimum in the odd part of the bias dependence at ≈0.25 Volts. For positive bias *only*, i.e., electrons tunneling from hcp to fcc bands, an additional effect enters the picture. In the fcc-Co band structure, there is an unoccupied, but very dispersive, minority band which begins just above E_F [3.166] in the (111) direction. This highly dispersive band contributes spin down states (for *positive* bias), thus decreasing the MR, but augmenting $dI/dV(V)$, since it leads to an overall increase in available states. The emergence of this band partially suppresses the tendency of $dI/dV(V)$ to dip, resulting in a "shoulder" for positive biases, rather than a fully-developed "dip" obseved for negative biases, and leads to the small peak at low biases and subsequent decrease in the odd part of the differential TMR. The calculated bias dependence for hcp-Co/I/hcp-Co junctions is completely symmetric, as expected, and the odd part of the differential TMR is zero. These results represent a clear experimental demonstration of density of states effects evidenced in tunnel conductance and TMR, as well as a convincing theoretical explanation, of key importance for a fundamental understanding of MTJs.

3.6.3 Epitaxial Junctions

One of the greatest problems in analyzing MTJs is the use of amorphous tunnel barriers like Al_2O_3, which are extremely difficult to model theoretically. Much theoretical work has been devoted to studying fully-epitaxial systems, such as Fe/ZnSe/Fe [3.47, 48, 58] or Fe/MgO/Fe(100) [3.45, 46, 56]. In addition to the fact that these systems are theoretically tractable, novel effects are anticipated in all-epitaxial structures, e.g., much larger TMR values due to resonant transmission in a single spin channel, quantum confinement effects in thin interfacial layers, and novel thickness and bias dependencies. On the other hand, fully-epitaxial metal-insulator-metal systems are not easy to prepare experimentally. MTJs with a single epitaxial NiMnSb electrode were prepared by Moodera et al. [3.131] (see Sect. 3.5.1), though the barrier was still amorphous Al_2O_3. As discussed above, de Teresa et al. [3.51, 52] prepared LSMO/$SrTiO_3$/Co junctions, with epitaxial LSMO electrodes and an epitaxial $SrTiO_3$ barrier, but with polycrystalline Co electrodes (see Sect. 3.6.1). Probably the first truly epitaxial MTJs used LSMO electrodes, and have been fabricated by several groups, as discussed briefly in Sect. 3.5.1. However, as yet no studies have been performed detailing the dependence on, e.g., crystallographic orientation for these systems. Further, given the complexity of LSMO itself, there is far less theoretical work on these structures.

Epitaxial Fe Electrodes

Yuasa et al. [3.167] have recently prepared *semi-epitaxial* Fe(100,110,211)/Al_2O_3/CoFe MTJs (i.e., *only* the bottom Fe layer is epitaxial), with a view to study the effect of the Fermi surface anisotropy on transport properties. Since the barrier is still

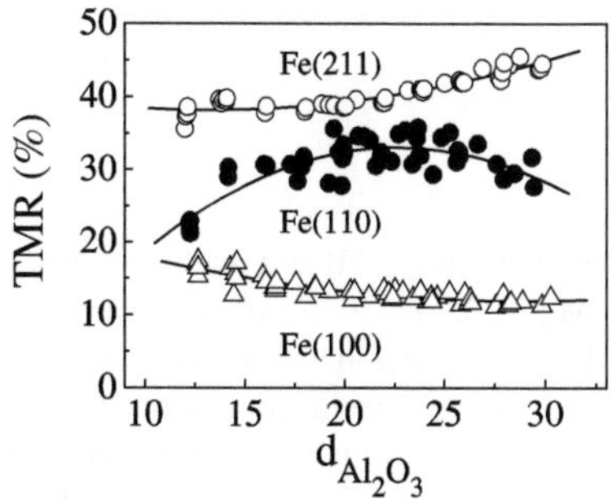

Fig. 3.14. TMR at 2 K as a function of Al_2O_3 thickness for Fe(211), Fe(110), and Fe(100) epitaxial electrodes in Fe(*hkl*)/Al_2O_3/CoFe junctions. Lines are only a guide to the eye. From [3.167]

amorphous, however, many of the novel features of fully-epitaxial systems (such as spin-dependent resonant transmission) are not expected to be observed. They have observed a strong dependence of the TMR on crystallographic orientation, as shown in Fig. 3.14, where the TMR is plotted as a function of (amorphous) Al_2O_3 thickness for different Fe crystallographic orientations [3.168]. They found the largest TMR for Fe(211) electrodes, while the TMR for Fe(100) is nearly a factor of 4 lower. Fe(110) electrodes yield a TMR value between that of the other directions, but with a much larger dependence on Al_2O_3 thickness. The reason for this thickness dependence (for any direction) is not yet well understood, particularly the fact that it is non-monotonic and distinctly different for each Fe orientation. One possible reason could be a slightly different growth mode [3.168] of the amorphous Al_2O_3 on the different crystalline facets of Fe, giving rise to a slightly different barrier quality for each electrode orientation and barrier thickness. However, there is no direct evidence for this mechanism. One other possibility is momentum filtering [3.6], i.e., the fact that the tunneling probability as a function of the electron incidence angle depends on barrier thickness. The latter explanation is perhaps unlikely, given that the barriers are amorphous and strict $k_{||}$ conservation may no longer apply.

The fact that the TMR varies so strongly with crystallographic orientation clearly points to the details of the Fe band structure [3.42, 169] and momentum filtering. Naively looking at the most dispersive s-like bands near E_F in Fe [3.169], the trend that TMR[Fe(211)]>TMR[Fe(110)]>TMR[Fe(100)] can perhaps be justified in some way, though it is expected that the Fe(100) tunneling spin polarization should be much larger. Still, a detailed calculation of the spin polarization of the *participating* bands for each orientation is needed. A more detailed analysis of the transport properties (such as conductance and TMR bias dependence, as in Sect. 3.6.2) has not yet been reported, and should certainly help understand the relation between the transport properties and Fe crystallographic orientation.

Fully Epitaxial Fe/MgO/CoFe(100)

Bowen et al. [3.170] have prepared *fully* epitaxial Fe/MgO/FeCo(100) MTJs. In contrast to the results of Yuasa et al. [3.167], they observe TMR values of approximately 60 %. Figure 3.15 shows the TMR as a function of applied bias at 30 K. Perhaps surprisingly, the behavior is essentially the same as that observed for

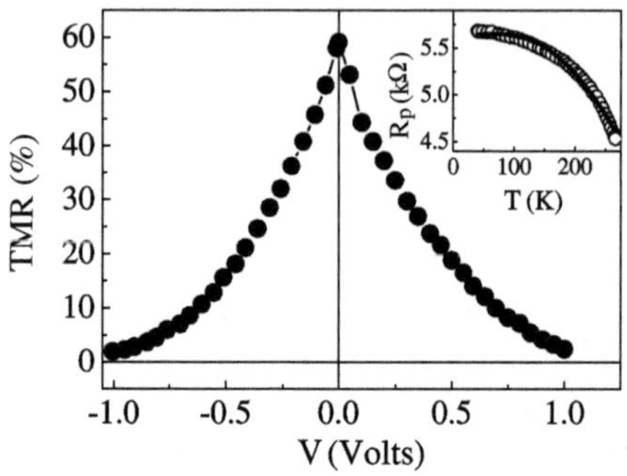

Fig. 3.15. TMR at 30 K as a function bias for a Fe/MgO 20Å/$Fe_{50}Co_{50}$(100) fully epitaxial MTJ. Inset: Parallel resistance vs. temperature for the same junction with a bias of 10 mVolts, along with a fit to the model of Shang et al. [3.149] (Sect. 3.5). From [3.170]

non-epitaxial systems (i.e., polycrystaline electrodes and amorphous barriers, see Sect. 3.5). The inset to Fig. 3.15 shows the temperature dependence of the TMR for the same junction, along with a fit to the model of Shang et al. [3.149] (see Sect. 3.5.5), which is also in line with what is observed in disordered systems. The most obvious feature is that the observed TMR is positive. Comparing the bias dependence of the TMR to that of Co/$SrTiO_3$/LSMO and Co/Al_2O_3/$SrTiO_3$/LSMO junctions [3.51, 52] (see Sect. 3.6.1), the authors conclude that the Fe/MgO and FeCo/MgO interfaces predominantly select *s*-hybridized electrons. Indeed, the sign of the tunneling spin polarization from the Fe/MgO interface was determined to be positive, using LSMO/$SrTiO_3$/MgO/Fe junctions, in the spirit of the experiments of de Teresa et al. (see Sect. 3.6.1). The observation of a positive TMR, and hence positive tunneling spin polarization, are line with previous experimental work and expected from theory, as discussed in Sect. 3.3.2. These preliminary experiments suggest that the observed TMR effects are consistent with the simplistic Julliere model, with no band structure or momentum-filtering effects yet observed.

The fact that the observed TMR is nearly a factor of 5 higher than that observed by Yuasa et al. [3.167] seems to point to the importance of the electronic structure of the *entire* electrode-barrier system. In particular, with epitaxial barriers, the role of k vector selection and exact nature of the interface bonding are of extreme importance (e.g., an Mg vs an O terminated interface should behave differently). However, the TMR values are still much lower than those predicted theoretically [3.45, 46, 56], the reason for which is yet unclear. One possibility recently pointed out by Meyerheim et al. [3.171] is perhaps that at the Fe/MgO interface an interfacial FeO_x layer is formed, considerably altering the spin dependent tunneling. According to preliminary calculations by Zhang et al. [3.171, 172], this would limit the TMR to only about 76%, rather than the expected several 1000% for perfect Fe/MgO interfaces. Further, if a similar oxidation occurs at Fe/Al_2O_3 interfaces, this could perhaps also help to explain the results of Yuasa et al. [3.167].

Thus far, only bias and temperature dependence of the TMR have been explored, and for a single crystallographic orientation, making it perhaps premature to draw detailed conclusions. Still, these experiments show great promise for not only observing novel transport behavior, but also for making a detailed comparison with state-of-the-art theoretical predictions [3.45, 46, 56]. Of particular interest is the predicted dependence of TMR and conductance on MgO thickness [3.45, 56], which should give some evidence of momentum filtering by the epitaxial barrier (see following Sect. 3.3.2). Further, novel effects arising from quantum confinement and quantum wells in thin interfacial layers may be observable (see Sect. 3.6.4). Clearly, this is an area of research which is still in its infancy.

Fully Epitaxial GaMnAs/AlAs/GaMnAs(100)

Tanaka and Higo [3.173] have taken a slightly different approach to spin polarized tunneling in epitaxial systems, fabricating $Ga_{1-x}Mn_xAs$-AlAs based structures. $Ga_{1-x}Mn_xAs$ is a ferromagnetic semiconductor whose T_C (and presumably P) may be tailored by Mn doping. The possible advantages of novel III-V based structures include not only simple integration with existing semiconductor technology, but also relatively straightforward fabrication of high-quality epitaxial structures, easily controlled physical and electronic structure, and the integration of quantum heterostructures which is easier than in any other materials system. Further, little is understood about spin polarized tunneling in these systems, creating the opportunity for interesting new physical effects.

In order to achieve antiparallel alignment of the GaMnAs layers, structures of $Ga_{1-x}Mn_xAs$/GaAs/AlAs/GaAs/$Ga_{1-y}Mn_yAs$ were fabricated, with two different Mn concentrations ($x = 3.3\%$ and $y = 4.0\%$) yielding two different coercivities (see Sect. 3.5.2). In this case the GaAs layers were to prevent Mn diffusion into the AlAs barrier as well as to create smooth interfaces. Using wedge-shaped AlAs barriers, they were able to show an exponential thickness dependence of the device resistance, consistent with tunneling (see Sect. 3.1). Up to 75% TMR was observed at low temperatures. However, the magnetizations of the GaMnAs layers were not fully antiparallel, which anticipates even larger effects for a fullly antiparallel alignment.

They were able to explain the barrier thickness dependence of the TMR by considering k conservation and the allowed (imaginary) k vectors within the AlAs barrier. As discussed (briefly) in Sect. 3.3.2, strict $k_{||}$ conservation is expected to hold for a fully epitaxial system, while in systems with amorphous barriers (e.g., the experiments of Yuasa et al. described above) it is not. Though they were able to explain the thickness dependence of the TMR based on k conservation considerations, the fact that a true antiparallel alignment was not reached perhaps calls this analysis into question. The temperature dependence of the TMR was also investigated, and was for the first time in MTJs studied through the T_C of the ferromagnetic electrode (see Sect. 3.5.5) showing good agreement with the model of Shang et al. [3.149] The role of the interfacial GaAs layers (see following Sect. 3.6.4), however, was not addressed, though it may perhaps expected to be minimal. If these results can be verified with a clear demonstration of antiparallel magnetization alignment, this may indeed represent a near model system for investigating not only the temperature and thickness dependence of the TMR, but also the effect of novel quantum heterostructures which may be relatively easily introduced. Given the recent interest in spin-polarized transport in semiconductors, these results are almost certain to be extended in future investigations.

3.6.4 Interface Dusting

From the discussions in Sect. 3.2.3, it is anticipated that MTJs are extremely interface sensitive. In fact, the crucial role of interface bonding as discussed in Sect. 3.6.1 is one way to see this. Another route to exploring interface sensitivity is the insertion of ultrathin layers (often called "dusting" layers) at the electrode-barrier interface. As seen in Sect. 3.3.1, using this method Tedrow and Meservey [3.41] showed with SPT that only a few monolayers of ferromagnetic material are needed for full tunneling spin polarization. Similarly, one may also utilize non-magnetic interlayers at a ferromagnet-insulator interface to study the decay of the tunneling spin polarization. By studing the evolution of MTJ features, particularly, $\mathrm{d}I/\mathrm{d}V(V)$ and TMR, as a function of interlayer thickness, one can gain insight on the interfacial sensitivity of MTJs.

Widely used in investigations of Giant Magnetoresistance, non-magnetic interface layers have proven extremely valuable in tunneling experiments as well. Moodera et al. [3.174] first applied used this method with the SPT technique in Al/Al_2O_3/Au/Fe junctions in order to measure the spin polarization as a function of Au interlayer thickness. They found that polarization decreased exponentially for the first two monolayers Au but decreased as $1/d$ at larger thicknesses. In the context of MTJ's, Parkin investigated TMR as a function of the thickness of a nonferromagnetic layer grown on Al_2O_3 [3.175]. In these experiments, a large tunneling spin polarization was surprisingly maintained over distances in excess of 10 nm, in striking contrast with the earlier experiments of Moodera for Au on Al_2O_3 [3.174], as well as later experiments of Sun and Freitas for Cu on Al_2O_3 [3.176]. As a first attempt to clarify these conflicting results, Zhang and Levy [3.177] have argued that the behavior of TMR in the presence of an interfacial nonmagnetic layer depends critically on the quality of the interfacial layer, with the shorter length scale resulting from thickness fluctuations. Mathon and Umerski [3.178] have recently found that quantum well states in the metallic interlayer are necessary for nonzero TMR, even in the limit of coherence loss (i.e., loss of k conservation during transport). In general, most theoretical approaches predict an oscillitory dependence of TMR on interlayer thickness [3.177–179]. Even among experimental results, at this stage there was little apparent consistency, and nearly all of the results were at odds with theory.

The recent work by LeClair et al. [3.33] showed that this apparent discrepancy is growth related. Most early dusting studies investigated junctions where the non-magnetic layer was grown on top of the Al_2O_3 barrier. In this case, however, they grew Cu interlayers both above *and* below the Al_2O_3 barrier, which resulted in two different TMR decay lengths, as shown in Fig. 3.16a. For Cu above the Al_2O_3 barrier, the length scale was roughly three times larger than for Cu below the Al_2O_3 barrier. Using in situ scanning Auger spectroscopy, they were able to show that Cu grows on Al_2O_3 in a highly three-dimensional manner. This gives rise to an artificially inflated TMR decay length when Co is grown on top of this Cu, since in some places the Co is

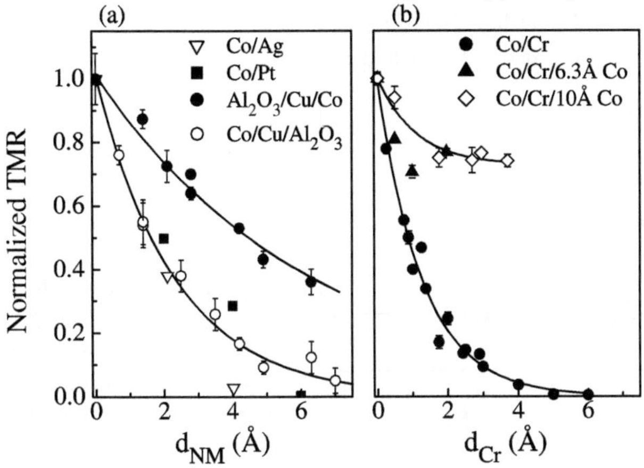

Fig. 3.16. **(a)** Normalized TMR at 10 K as a function of Cu thickness for Co/Cu d_{Cu}/Al_2O_3/Co and Al_2O_3/Cu d_{Cu}/Co. Also shown for comparison are results with Ag and Pt interlayers. **(b)** Normalized TMR at 10 K as a function of Cr thickness for Co/Cr d_{Cr}/Al_2O_3/Co and Co/Cr d_{Cr}/Co d_{Co}/Al_2O_3/Co tunnel junctions. Adding only a few monolayers of Co on Cr almost completely restores the TMR, demonstrating the interfacial sensitivity of MTJs. Lines are fits to an exponential decay. From [3.32, 33, 180]

in direct contact with Al_2O_3. It was further shown that in Co/Cu/Al_2O_3/Co junctions, where Cu was grown on Co rather than Al_2O_3, near-ideal Cu growth resulted. In this latter system, it was found that the normalized TMR (i.e., TMR(d_{Cu})/TMR($d_{Cu}=0$)) decayed approximately exponentially with increasing Cu thickness, $\exp(-d_{Cu}/\xi)$. Fitting the TMR decay gave $\xi \approx 2.6$ Å, equivalent to just more than one monolayer Cu. As discussed in Sect. 3.2.3, Appelbaum and Brinkman [3.22, 23] pointed out that tunneling in non-superconducting junctions should be sensitive to the density of states within a few Fermi wavelengths (i.e., k_F^{-1}) of the electrode-barrier interface. In this case, $\xi \approx 3.5 k_F^{-1}$, in good agreement, and suggesting that indeed k_F^{-1} may be the relevant length scale. This is supported by the results of Moodera et al. [3.180] with Ag and Au (Fig. 3.17) interlayers, which have almost the same value of k_F^{-1} and give a length scale similar to Cu interlayers, as do Pt interlayers (Fig. 3.16a). The relative success of many-body approaches, such as those of Appelbaum, and the failure of more recent independent electron models [3.177–179, 181] demonstrates again the true many-body nature of the tunneling process. However, one must also consider that the independent electron models consider strict k conservation in tunneling, which is not realistic in junctions with non-epitaxial electrodes and amorphous barriers (see Sect. 3.2.3), and may also account for the failure of these models [3.33].

A further demonstration of interface sensitivity was subsequently obtained by LeClair et al. [3.32] using ultrathin Cr layers in Co/Cr/Al_2O_3/Co MTJs. The TMR decay was again approximately exponential, and in this case was even more rapid, (see Fig. 3.16b) with $\xi \approx 1.0$ Å, or only ~0.5 monolayers Cr. In these experiments, they also added an additional Co layer on top of the Cr dusting layer, i.e., Co/Cr

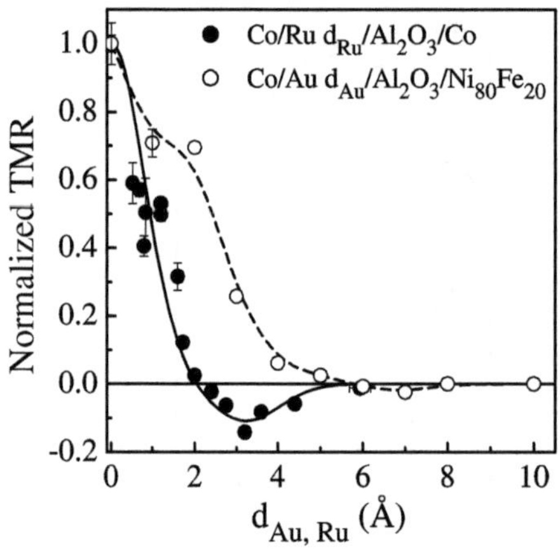

Fig. 3.17. TMR vs d_{NM} for NM = Au (*open*) at 77 K and for NM = Ru (*closed*) at 5 K. In both cases, a negative TMR is observed for a small range of NM thicknesses. Lines are only guides to the eye. From [3.101, 183]

d_{Cr}/Co 6.3, 10 Å/Al_2O_3/Co, shown in Fig. 3.16b. Strikingly, the TMR was almost completely restored with only these 3–5 monolayers of Co. This further confirms that only the few monolayers of the electrode adjacent to the ferromagnet-insulator interface dominate MTJ properties, in very good agreement with earlier SPT work on ultrathin magnetic layers (see Sect. 3.3.1). More detailed experiments analyzing the $\mathrm{d}I/\mathrm{d}V(V)$ characteristics in MTJs with multiple dusting layers further conclusively established the importance of interfacial density of states, using the observed zero-bias anomalies caused by the interfacial density of states modification as a probe. In this way, the importance of density of states modifications and band matching (or $s-d$ scattering [3.182]) at Co/NM interfaces was demonstrated, in agreement with experiments on bulk dilute transition metal alloys and magnetic multilayers. However, a discussion of these results is beyond the scope of this review, and we refer the reader to [3.32].

Two last examples of other interesting effects possible with dusted MTJs are demonstrated with Au and Ru dusting of Co/Al_2O_3 interfaces. Moodera et al. [3.183] studied Co/Au/Al_2O_3/$Ni_{80}Fe_{20}$ junctions, and found that the TMR actually became *negative* for some values of d_{Au}, Fig. 3.17. This result was explained in terms of quantum-well states in the Au layer [3.177–179, 181], which correctly predicted the negative TMR, but incorrectly predicted the observed bias dependence of the TMR. LeClair et al. [3.101] found similar results in Co/Ru/Al_2O_3/Co junctions, see Fig. 3.17. In this case, the unusual thickness and voltage dependence of the TMR could be well explained, at least qualitatively, by considering the band matching and density of states modification at the Co/Ru interface. ^{59}Co NMR experiments (see Sect. 3.4.2) confirmed the presence of an interfacial Co-Ru alloy, rather than a pure Ru layer. Furthermore, ab initio calculations of the density of states of Co-Ru alloys are consistent with a negative *sp* spin polarization (see Sect. 3.3.2). Quantum well states in this case could be ruled out, since the "interlayer" was in reality an interface alloy, and also essentially polycrystalline. Given the similarity of the Au and Ru dusting results, further work is needed in these systems to fully clarify both results. In any case, the fact that the sign of the spin polarization can be changed not only by interface bonding (Sect. 3.6.1) but by modifying the interfacial *electrode* density of states clearly points toward the unique opportunity of engineering MTJ interfaces to create novel properties.

3.6.5 Hybrid Spin Filter – MTJ Devices

As mentioned in Sect. 3.3.3, the combination of a spin dependent density of states with a spin filter tunnel barrier may lead to interesting effects. One such effect, viz., a new and large magnetoresistance effect, has recently been observed.

From Sect. 3.3.3, we have seen that using a magnetic tunnel barrier, such as EuS, results in a highly spin polarized tunnel current via the spin filter effect. If we now add a magnetic electrode to the spin filter structure, we must consider the role of the spin polarized density of states in the electrode as well. Since the tunnel current depends on the number of filled states in the first electrode as well as the number of available states in the second, with one magnetic electrode, see Fig. 3.18a,b, the tunnel current will depend on the relative orientation of the filtered spins (i.e., the EuS magnetization direction) and the electrode magnetization. For parallel alignment, see Fig. 3.18a, spin polarized electrons tunnel from the magnetic electrode through the filter, which selects only the *majority* spin component of the DOS, leading to a relatively large current. For the antiparallel case, Fig. 3.18b, only the minority component of the ferromagnetic electrode DOS is selected, leading to a relatively small current. One may consider this device as analogous to a traditional magnetic tunnel junction with one half-metallic electrode and a *nonmagnetic* barrier. The magnitude of the expected TMR-like effect may then be estimated using the Julliere model (see Sect. 3.5.1): $\Delta R/R_p = 2P_m P_f/(1 - P_m P_f)$, where P_m is the spin polarization of the ferromagnetic

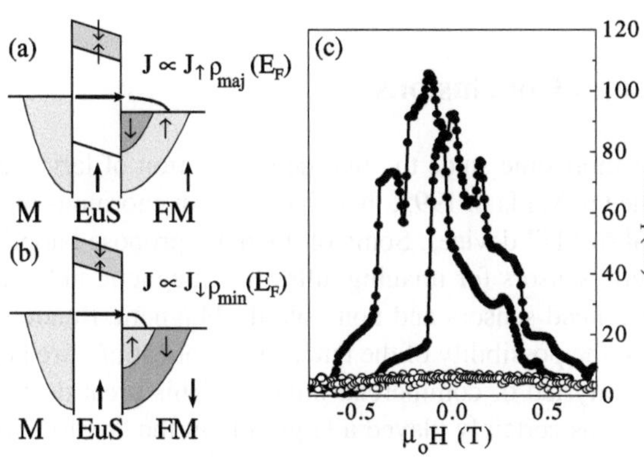

Fig. 3.18. Explanation of spin filter magnetoresistance. Below the T_C of EuS, the tunnel barrier is spin-split, resulting in a highly spin polarized tunnel current. With a ferromagnetic (FM) electrode, the tunnel current depends on the relative magnetization orientation. For parallel alignment (P), **(a)**, a large current results, while for antiparallel alignment (AP), **(b)**, a small current results. **(c)** Magnetoresistance in a Al/EuS/Gd junction as a function of magnetic field at $T = 2$ K (*closed, below the EuS* T_C), and $T = 30$ K (*open*, above the EuS T_C). From [3.184]

electrode, P_f is the efficiency (polarization) of the spin filter, and $\Delta R/R_p$ is defined as previously.

LeClair et al. [3.184] have fabricated Al/EuS/Gd junctions with a view to observe this magnetoresistance effect. As a first indication of the presence of spin filtering in this structure, a sizable decrease in junction resistance below the T_C of the EuS has been observed, which is, as expected, thermally smeared upon the application of an external magnetic field [3.62, 63]. Fig. 3.18c shows resistance versus magnetic field, measured at a dc bias of $\approx$5 mVolts, for a Ta 5 nm/Al 3 nm/EuS 5 nm/Gd 15 nm structure at 2 K (well below the T_C of EuS) and at 30 K (well above T_C). At 2 K, a magnetoresistance effect of $\sim$100% (in some cases more than 130%) is observed, clearly indicating efficient spin filtering. However, at 30 K, almost no magnetoresistance (<5%) is observed, proving the decisive role of the ferromagnetic spin filter barrier (the small persisting effect is due only to the field-induced magnetization in the EuS layer). As expected, for parallel alignment of the EuS and Gd magnetizations, a low resistance state is observed (at high fields), while for antiparallel alignment a high resistance is observed (at low fields). However, complete antiparallel alignment was not reached, where in an even larger effect can be anticipated for truly antiparallel alignment. Given the observed effects of more than 130%, it can be determined (using the above model) that for a Gd polarization of 45% [3.185], the filter efficiency is 88%, in line with previous investigations [3.62].

The concept of spin filtering may have potential utility for spin injection into semiconductors or for other novel hybrid devices. The almost 100% [3.62] filter spin polarization, as well as the lack of any impedance mismatch problem [3.64] with semiconductors, makes spin filtering a nearly ideal method for spin injection into semiconductors, enabling novel spintronic devices [3.135]. These hybrid devices, besides having application potential, illustrate that rich physics remains in this area.

3.7 Outlook and Conclusions

In the relatively short time since the first demonstration of large room temperature magnetoresistance in MTJs in 1995, there has been an enormous interest in the commercial potential of MTJ devices. Some of the main proposed applications are large arrays of magnetic sensors for imaging, ultra-low magnetic field sensors, and most prominently, read-head sensors and non-volatile Magnetic Random Access memories. In particular, the possibility of the latter two applications are being aggressively reasearched by a myriad of commercial interests. This great demand for improved magnetic devices has certainly played a large role in the "rebirth" of spin-polarized tunneling.

More striking from a physics point of view, however, is that despite three decades of research in spin polarized tunneling, it is only in the last few years that the influence of the insulating barrier and the ferromagnet-insulator interface on tunneling spin polarization has been recognized. Certainly, many of the experiments presented point the way toward "engineered" TMR structures which utilize novel electrode/barrier *combinations* to achieve large magnetoresistances and tailored bias dependencies.

The demonstration of novel spin filter-MTJ hybrid devices is also an intruiging direction for future spin electronic devices, though whether these devices ultimately have application potential remains to be seen. The development of spin-filter devices operating at room temperature would pave the way for a great number of new effects and devices, the possibilities for which are only beginning to be explored. However, many basic fundamental aspects of MTJs remain incompletely understood, such as the true origin of the bias and temperature dependence of the TMR, or even an accurate prediction of the magnitude of the TMR effect in realistic structures.

Many possible future directions for spin polarized tunneling are beyond the scope of this review and have not been covered here. We attempt now to name only a few. The development of the spin polarized STM [3.186, 187] has only in the last few years opened up the entirely new field of spin polarized tunneling with *nanometer lateral resolution* in addition to providing a new view of atomic scale magnetism. Certainly this technique holds many fascinating insights in the future. The development of three terminal spin-valve [3.188] and tunnel transistor devices also provides a novel view on spin-polarized tunneling and spin-polarized hot-electron transport, and these techniques should benefit from the recent advances in more traditional spin polarized tunneling devices. Perhaps attracting the most attention as of late is spin injection into semiconductors for novel semiconducting spin electronic devices. The current viewpoint is that tunneling from a ferromagnet through a Schottky or tunnel barrier into a semiconductor is perhaps the best approach to this problem, in which case this field should benefit greatly from the field of spin polarized tunneling.

Theoretical results have mostly not been presented here, except where required. A great body of theoretical predictions for MTJ and hybrid devices has developed in the last years, a large number of which await verification, to name only a few, novel transport properties in epitaxial systems and double junctions, half metallic electrodes, the interplay of the Coulomb blockade and spin polarized tunneling, and resonant- or co-tunneling effects. Certainly many promising results lie ahead.

Acknowledgement. This review was made possible by the contributions from a number of groups and individuals from groups around the world (all of whom we could not possibly name), for which we are grateful. The valuable discussions and contributions of R. Meservey, P.M. Tedrow, J. Nowak, R. Jansen, C. Tanaka, R.J.M. van de Veerdonk, C.H. Shang, and G.P. Berera to the MIT group have been essential. In Eindhoven, W.J.M. de Jonge, J.T. Kohlhepp, B. Koopmans, J.K. Ha, O. Kurnosikov, G.W.M. Baselmans, J.J.A.P. Noijen, K. Flipse, and H. Dalderop have provided invaluable contributions. Special gratitude from J.S.M. to the undergraduates at MIT and from P.L. and H.J.M.S. to the undergraduate and Ph.D. students in Eindhoven for their contributions and enthusiasm. We also thank S. Yuasa, T. Nagahama, and M. Bowen for providing data and unpublished results, and A.H. Davis and J.M. MacLaren for their collaboration and interest in this work. Finally, P.L. thanks E.A. Verduijn for his assistance in the manuscript preparation. The research program at MIT is supported by NSF grant DMR 9730908, ONR grant N00014-J-92-1847 as well as TDK Research Award for JSM, and the Eindhoven University is supported by the Dutch technology foundation STW and is part of the research program of the "Stichting voor Fundamenteel Onderzoek der Materie (FOM)," which is financially supported by the "Nederlandse Organisatie voor Wetenschappelijk Onderzoek."

References

3.1. A. Sommerfeld, H. Bethe: *Handbuch der Physik* (Verlag Julius Springer, Berlin, 1933), Vol. 24, pt. 2, p. 333.
3.2. J.C. Slonczewski: IBM Technical Disclosure Bulletin **19**, 2328 (1976).
3.3. M. Julliere: Phys. Lett. **54A**, 225 (1975).
3.4. J.S. Moodera, L.R. Kinder, T.M. Wong, R. Meservey: Phys. Rev. Lett. **74**, 3273 (1995).
3.5. C.B. Duke: *Tunneling in Solids* (Academic Press, New York, 1969).
3.6. E.L. Wolf: *Principles of Electron Tunneling Spectroscopy* (Oxford University Press, London, 1985).
3.7. We note here that for the entirety of this review, except where specifically noted, we restrict ourselves to *disordered* systems, where strict k selection rules for the tunneling process are broken [3.54] (i.e., breakdown of $k_{||}$ conservation).
3.8. J.G. Simmons: J. Appl. Phys. **34**, 1763 (1963).
3.9. W.F. Brinkman, R.C. Dynes, J.M. Rowell: J. Appl. Phys. **41**, 1915 (1970).
3.10. W.A. Harrison: Phys. Rev. **123**, 85 (1961).
3.11. J. Bardeen: Phys. Rev. Lett. **6**, 57 (1961).
3.12. S. Zhang, P.M. Levy: Eur. Phys. J. B **10**, 599 (1999).
3.13. I. Giaever: Phys. Rev. Lett. **5**, 147 (1960).
3.14. I. Giaever: Phys. Rev. Lett. **5**, 464 (1960).
3.15. J. Nicol, S. Shapiro, P.H. Smith: Phys. Rev. Lett. **5**, 461 (1960).
3.16. I. Giaever: *Nobel Lectures in Physics, 1971-1980* (World Scientific, Singapore, 1992), p. 144.
3.17. I. Giaever, K. Megerle: Phys. Rev. **122**, 1101 (1961).
3.18. I. Giaever, H. Hart, K. Megerle: Phys. Rev. **126**, 941 (1962).
3.19. C. Kittel: *Introduction to Solid State Physics*, 7 ed. (John Wiley and Sons, Inc., New York, 1996), Chap. 12, p. 364.
3.20. M.H. Cohen, L.M. Falicov, J.C. Phillips: Phys. Rev. Lett. **8**, 316 (1962).
3.21. J.R. Schrieffer, D.J. Scalapino, J.M. Wilkins: Phys. Rev. Lett. **10**, 336 (1963).
3.22. J.A. Appelbaum, W.F. Brinkman: Phys. Rev. **186**, 464 (1969).
3.23. J.A. Appelbaum, W.F. Brinkman: Phys. Rev. B **2**, 907 (1970).
3.24. F. Mezei, A. Zawadowski: Phys. Rev. B **3**, 167; 3127 (1971).
3.25. E.Y. Tsymbal, D.G. Pettifor: J. Phys. Cond. Matt. **9**, L411 (1997).
3.26. H. Itoh, J. Inoue, S. Maekawa, P. Bruno: J. Magn. Soc. Japan **23**, 52 (1999).
3.27. J.S. Moodera, J. Nassar, J. Mathon: J. Magn. Magn. Mater. **200**, 248 (1999).
3.28. C. Uiberacker, K. Wang, C. Heide, P.M. Levy: J. Appl. Phys. **89**, 7561 (2001).
3.29. C. Uiberacker, P.M. Levy: Phys. Rev. B **64**, 193404 (2001).
3.30. H. Itoh, J. Inoue: Surf. Sci. **493**, 748 (2001).
3.31. R. Meservey, P.M. Tedrow: Phys. Rep. **238**, 173 (1994).
3.32. P. LeClair, J.T. Kohlhepp, H.J.M. Swagten, W.J.M. de Jonge: Phys. Rev. Lett. **86**, 1066 (2001).
3.33. P. LeClair, H.J.M. Swagten, J.T. Kohlhepp, R.J.M. van de Veerdonk, W.J.M. de Jonge: Phys. Rev. Lett. **84**, 2933 (2000).
3.34. D.J. Monsma, S.S.P. Parkin: Appl. Phys. Lett. **77**, 720 (2000).
3.35. P.M. Tedrow, R. Meservey: Phys. Rev. Lett. **26**, 192 (1971).
3.36. Fitting program by E.A. Verduijn, 2001.
3.37. J.S. Moodera, R.J.M. van de Veerdonk, Proceedings of the 15th ICMFS, Sunshine Coast, Queensland, Australia, 1997.
3.38. D.C. Worledge, T.H. Geballe: Appl. Phys. Lett. **76**, 900 (2000).

3.39. D.C. Worledge, T.H. Geballe: Phys. Rev. Lett. **85**, 5182 (2000).
3.40. P.M. Tedrow, R. Meservey: Phys. Rev. Lett. **27**, 919 (1971).
3.41. P.M. Tedrow, R. Meservey: Solid State Commun. **16**, 71 (1975).
3.42. D.A. Papaconstantopoulos: *Handbook of the Band Structures of Elemental Solids* (Plenum, New York, 1986).
3.43. I.I. Mazin: Phys. Rev. Lett. **83**, 1427 (1999).
3.44. M.B. Stearns: J. Mag. Mag. Mater. **8**, 167 (1977).
3.45. W.H. Butler, X.G. Zhang, T.C. Schulthess, J.M. MacLaren: Phys. Rev. B **63**, 055416 (2001).
3.46. W.H. Butler, X.G. Zhang, T.C. Schulthess: Phys. Rev. B **63**, 092402 (2001).
3.47. P. Mavropoulos, N. Papanikolaou, P.H. Dederichs: Phys. Rev. Lett. **85**, 1088 (2000).
3.48. J.M. MacLaren, X.G. Zhang, W.H. Butler, X. Wang: Phys. Rev. B **59**, 5470 (1999).
3.49. I.I. Mazin: Europhys. Lett. **55**, 404 (2001).
3.50. M.J. Pilsch, J.L. Chang, J. Silcox, R.A. Buhrman: Appl. Phys. Lett. **79**, 391 (2001).
3.51. J.M. de Teresa, A. Barthélémy, A. Fert, J.P. Contour, F. Montaigne, P. Seneor: Science **286**, 507 (1999).
3.52. J.M. de Teresa, A. Barthélémy, A. Fert, J.P. Contour, R. Lyonnet, F. Montaigne, P. Seneor, A. Vaurès: Phys. Rev. Lett. **82**, 4288 (1999).
3.53. M. Sharma, S.X. Wang, J.H. Nickel: Phys. Rev. Lett. **82**, 616 (1999).
3.54. J.E. Dowman, M.L.A. MacVicar, J.R. Waldram: Phys. Rev. **186**, 452 (1969).
3.55. K. Wang, S. Zhang, P.M. Levy, L. Szunyogh, P. Weinberger: J. Magn. Magn. Mater. **189**, L131 (1998).
3.56. J. Mathon, A. Umerski: Phys. Rev. B **63**, 220403 (2001).
3.57. J. Mathon: Phys. Rev. B **56**, 11810 (1997).
3.58. J.M. MacLaren, W.H. Butler, X.G. Zhang: J. Appl. Phys. **83**, 6521 (1998).
3.59. X. Hao, J.S. Moodera, R. Meservey: Phys. Rev. B **42**, 8235 (1990).
3.60. P. Wachter: *Handbook on the Physics and Chemistry of Rare Earths* (North-Holland, Amsterdam, 1979), Vol. I, Chap. 19.
3.61. A. Mauger, C. Godart: Phys. Rep. **141**, 51 (1986).
3.62. J.S. Moodera, X. Hao, G.A. Gibson, R. Meservey: Phys. Rev. Lett. **61**, 637 (1988).
3.63. L. Esaki, P.J. Stiles, S. von Molnar: Phys. Rev. Lett. **19**, 852 (1967).
3.64. R. Fiederling, M. Keim, G. Reuscher, W. Ossau, G. Schmidt, A. Waag, L.W. Molenkamp: Nature **402**, 787 (1999).
3.65. J.S. Helman, B. Abeles: Phys. Rev. Lett. **37**, 1429 (1976).
3.66. S. Maekawa, U. Gäfvert: IEEE Trans. Magn. **MAG-18**, 707 (1982).
3.67. Y. Suezawa, F. Takahashi, Y. Gondo: Jpn. J. Appl. Phys **31**, L1415 (1992).
3.68. R. Kabani, J.S. Moodera, P.M. Tedrow: Mater. Res. Soc. Ext. Abstr. EA (1990).
3.69. J. Nowak, J. Rauluszkiewicz: J. Magn. Magn. Mater. **109**, 79 (1992).
3.70. P. LeClair, J.S. Moodera, R. Meservey: J. Appl. Phys. **76**, 6546 (1994).
3.71. J.A. Appelbaum: Phys. Rev. Lett. **17**, 91 (1966).
3.72. J.A. Appelbaum: Phys. Rev. B **154**, 633 (1967).
3.73. P.W. Anderson: Phys. Rev. Lett. **17**, 95 (1966).
3.74. R. Jansen, J.S. Moodera: J. Appl. Phys. **83**, 6682 (1998).
3.75. R. Jansen, J.S. Moodera: Phys. Rev. B **61**, 9047 (2000).
3.76. T. Miyazaki, N. Tezuka: J. Magn. Magn. Mater. **139**, 231 (1995).
3.77. J.S. Moodera, L.R. Kinder, J. Nowak, P. LeClair, R. Meservey: Appl. Phys. Lett. **69**, 708 (1996).
3.78. J.C. Slonczewski: Phys. Rev. B **39**, 6995 (1989).
3.79. M. Klaua, D. Ullmann, J. Barthel, W. Wulfhekel, J. Kirschner, R. Urban, T.L. Monchesky, A. Enders, J.F. Cochran, B. Heinrich: Phys. Rev. B **64**, 134411 (2001).

3.80. P. LeClair, H.J.M. Swagten, J.T. Kohlhepp, A.A. Smits, B. Koopmans, W.J.M. de Jonge: J. Appl. Phys. **87**, 6070 (2000).
3.81. H.A.M. de Gronckel, H. Kohlstedt, C. Daniels: Appl. Phys. A **70**, 435 (2000).
3.82. V. Kottler, M.F. Gillies, A.E.T. Kuiper: J. Appl. Phys. **89**, 3301 (2001).
3.83. M. Sharma, S.X. Wang, J.H. Nickel: J. Appl. Phys. **85**, 7803 (1999).
3.84. R. Schad, K. Mayer, J. McCord, D. Allen, D. Yang, M. Tondra, D. Wang: J. Appl. Phys. **89**, 6659 (2001).
3.85. W. Zhu, C.J. Hirschmugl, A.D. Laine, B. Sinkovic, S.S.P. Parkin: Appl. Phys. Lett. **78**, 3103 (2001).
3.86. L. Seve, W. Zhu, B. Sinkovic, J.W. Freeland, I. Coulthard, W.J. Antel, S.S.P. Parkin: Europhys. Lett. **55**, 439 (2001).
3.87. K. Knechten, P. LeClair, J.T. Kohlhepp, H.J.M. Swagten, B. Koopmans, W.J.M. de Jonge: J. Appl. Phys. **90**, 1675 (2001).
3.88. A.T.A. Wee, K. S.S.X. Wang: Appl. Phys. Lett. **74**, 2528 (1999).
3.89. O. Kurnosikov, F.C. de Nooij, P. LeClair, J.T. Kohlhepp, B. Koopmans, H.J.M. Swagten, W.J.M. de Jonge: Phys. Rev. B. **64**, 153407 (2001).
3.90. W. Wulfhekel, M. Klaua, D. Ullmann, F. Zavaliche, J. Kirschner, R. Urban, T. Monchesky, B. Heinrich: Appl. Phys. Lett. **78**, 509 (2001).
3.91. Y. Ando, H. Kameda, H. Kubota, T. Miyazaki: J. Appl. Phys. **87**, 5206 (2000).
3.92. T. Dimopoulos, V. da Costa, C. Tiusan, K. Ounadjela, H.A.M. van den Berg: J. Appl. Phys. **89**, 7371 (2001).
3.93. V. da Costa, C. Tiusan, T. Dimopoulos, K. Ounadjela: Phys. Rev. Lett. **85**, 876 (2000).
3.94. W.H. Rippard, A.C. Perrella, R.A. Buhrman: Appl. Phys. Lett. **78**, 1601 (2001).
3.95. O. Kurnosikov, J. de Jong, H.J.M. Swagten, W.J.M. de Jonge: Appl. Phys. Lett. (2002), accepted.
3.96. A.E.T. Kuiper, M.F. Gillies, V. Kottler, G. W. 't Hooft, J.G.M. van Berkum, C. van der Marel, Y. Tamminga, J.H.M. Snijders: J. Appl. Phys **89**, 1965 (2001).
3.97. R.C. Sousa, J.J. Sun, V. Soares, P.P. Freitas, A. Kling, M.F. da Silva, J.C. Soares: J. Appl. Phys **85**, 5258 (1999).
3.98. P. Shang, A.K. Petford-Long, J.H. Nickel, M. Sharma, T.C. Anthony: J. Appl. Phys. **89**, 6874 (2001).
3.99. A.C.C. Yu, R. Doole, A. Petford-Long, T. Miyazaki: Jpn. J. Appl. Phys. **40**, 5058 (2001).
3.100. P. LeClair, J.T. Kohlhepp, C.H. van de Vin, H. Wieldraaijer, H.J.M. Swagten, W.J.M. de Jonge, A.H. Davis, J.M. MacLaren, R. Jansen, J.S. Moodera: Phys. Rev. Lett. (2001), submitted, see cond-mat/0107569.
3.101. P. LeClair, B. Hoex, H. Wieldraaijer, J.T. Kohlhepp, H.J.M. Swagten, W.J.M. de Jonge: Phys. Rev. B **64**, 100406 (2001).
3.102. P.C. Riedi, T. Thomson, G.J. Tomka: *Handbook of Magnetic Materials Vol. 12*, edited by K.H.J. Buschow (Elsevier, Amsterdam, 1999), pp. 97–258.
3.103. W.J.M. de Jonge, H.A.M. de Gronckel, K. Kopinga: *Ultrathin Magnetic Structures II*, edited by B. Heinrich, J.A.C. Bland (Springer-Verlag, Berlin, 1994), p. 279.
3.104. R. Jansen, J.S. Moodera: Appl. Phys. Lett. **75**, 400 (1999).
3.105. A. Vedyayev, D. Bagrets, A. Bagrets, B. Dieny: Phys. Rev. B **63**, 064429 (2001).
3.106. J.S. Moodera, E.F. Gallagher, K. Robinson, J. Nowak: Appl. Phys. Lett. **70**, 3050 (1997).
3.107. R.C. Sousa, J.J. Sun, V. Soares, P.P. Freitas, A. Kling, M.F. da Silva, J.C. Soares: Appl. Phys. Lett. **73**, 3288 (1998).
3.108. Z. Zhang, S. Cardosa, P.P. Freitas, P. Wei, N. Barradas, J.C. Soares: Appl. Phys. Lett. **78**, 2911 (2001).

3.109. Z. Zhang, S. Cardosa, P.P. Freitas, X. Battle, P. Wei, N. Barradas, J.C. Soares: J. Appl. Phys. **89**, 6665 (2001).
3.110. X.-F. Han, A.C.C. Yu, M. Oogane, J. Murai, T. Daibou, T. Miyazaki: Phys. Rev. B. **63**, 224404 (2001).
3.111. Y. Lu, X.W. Li, G.Q. Gong, G. Xiao, A. Gupta, P. Lecoeur, J.Z. Sun, Y.Y. Wang, V.P. Dravid: Phys. Rev. B. **54**, R8357 (1996).
3.112. M. Viret, M. Drouet, J. Nassar, J.P. Contour, C. Fermon, A. Fert: Europhys. Lett. **39**, 545 (1997).
3.113. S.S.P. Parkin (unpublished).
3.114. X.-F. Han, M. Oogane, H. Kubota, Y. Ando, T. Miyazaki: Appl. Phys. Lett. **77**, 283 (2000).
3.115. P. LeClair: Ph.D. thesis, Eindhoven University of Technology, Eindhoven, The Netherlands, 2002.
3.116. J.S. Moodera (unpublished).
3.117. W.E. Pickett, J.S. Moodera: Phys. Today **5**, 39 (2001).
3.118. R.A. de Groot, F.M. Mueller, P.G. van Engen, K.H.J. Buschow: Phys. Rev. Lett. **50**, 2024 (1983).
3.119. S. Ishida, S. Fujii, S. Kawhiwagi, S. Asano: J. Phys. Soc. Japan **64**, 2152 (1995).
3.120. K. Schwarz: J. Phys. F: Met. Phys. **16**, L211 (1986).
3.121. A. Yanase, K. Siratori: J. Phys. Soc. Jpn. **53**, 312 (1984).
3.122. S.F. Alvarado, M. Erbudak, P. Munz: Phys. Rev. B **14**, 2740 (1976).
3.123. J.Y.T. Wei, N.C. Yeh, R.P. Vasques: Phys. Rev. Lett. **79**, 5150 (1997).
3.124. Y. Okimoto, K. Katsufuji, T. Ishikawa, A. Urushibara, T. Arima, Y. Tokura: Phys. Rev. Lett. **75**, 109 (1995).
3.125. I.I. Mazin: Appl. Phys. Lett. **77**, 3000 (2000).
3.126. J.H. Park, E. Vescovo, H.J. Kim, C. Kwon, R. Ramesh, T. Venkatesan: Nature **392**, 794 (1998).
3.127. D. Ristoiu, J. P. Nozières, C.N. Borca, T. Komesu, H. k. Jeong, P.A. Dowben: Europhys. Lett. **49**, 624 (2000).
3.128. Y. Ji, G.J. Strijkers, F.Y. Yang, C.L. Chien, J.M. Byers, A. Anguelouch, G. Xiao, A. Gupta: Phys. Rev. Lett. **86**, 5585 (2001).
3.129. M.H. Jo, N.D. Mathur, J.E. Evetts, M.G. Blamire: Appl. Phys. Lett. **77**, 3803 (2000).
3.130. M.-H. Jo, N.D. Mathur, N.K. Todd, M.G. Blamire: Phys. Rev. B **61**, R14905 (2000).
3.131. C.T. Tanaka, J. Nowak, J.S. Moodera: J. Appl. Phys. **86**, 6329 (1999).
3.132. T. Shinjo, H. Yamamoto: J. Phys. Soc. Japan **59**, 3061 (1990).
3.133. C. Dupas, P. Beauvillain, C. Chappert, J.P. Renard, F. Trigui, P. Veillet, E. Velú, D. Renard: J. Appl. Phys. **67**, 5680 (1990).
3.134. J.S. Moodera, L.R. Kinder: J. Appl. Phys. **79**, 4724 (1996).
3.135. G.A. Prinz: J. Magn. Magn. Mater. **200**, 57 (1999), for a review of magnetoelectronic applications.
3.136. A.E. Berkowitz, K. Takano: J. Magn. Magn. Mater. **200**, 552 (1999), for a review of exchange anisotropy and exchange biasing.
3.137. J.C.S. Kools: I.E.E.E. Trans. Magn. **32**, 3165 (1996), for a review of exchange bias applied to spin valves.
3.138. S. Gider, B.U. Runge, A.C. Marley, S.S.P. Parkin: Science **281**, 797 (1999).
3.139. J.S. Moodera, J. Nowak, R.J.M. van de Veerdonk: Phys. Rev. Lett. **80**, 2941 (1998).
3.140. R.J.M. van de Veerdonk, J.S. Moodera, W.J.M. de Jonge: J. Magn. Magn. Mater. **198-199**, 152 (1999).
3.141. X.-F. Han, J. Murai, Y. Ando, H. Kubota, T. Miyazaki: Appl. Phys. Lett. **78**, 2533 (2001).

3.142. S. Zhang, P.M. Levy, A.C. Marley, S.S.P. Parkin: Phys. Rev. Lett **79**, 3744 (1997).
3.143. A.M. Bratkovsky: Phys. Rev. B **72**, 2334 (1998).
3.144. To be more precise, the tunnel resistance represents the quantity V/I, while $\mathrm{d}V/\mathrm{d}I \equiv 1/(\mathrm{d}I/\mathrm{d}V)$ represents the *tangent* of the V(I) curve. For a nonlinear V(I) characteristic, these two quantities are necessarily not the same. In general, if we require that difference between the parallel and antiparallel V(I) characteristics must vanish at large biases, it is easy to see that $\Delta R/R_{\mathrm{p}}$ must tend smoothly to zero at large bias, while $\Delta G/G_{\mathrm{a}}$ *must* for some biases exhibit a sign change. In most cases, these two quantities are identical *only* at zero bias.
3.145. P. LeClair, H.J.M. Swagten, J.T. Kohlhepp, W.J.M. de Jonge: Appl. Phys. Lett. **76**, 3783 (2000).
3.146. A.H. Davis, J.M. MacLaren: J. Appl. Phys. **87**, 5224 (2000).
3.147. A.M. Bratkovsky: Phys. Rev. B **56**, 2344 (1997).
3.148. A. Vedyayev, D. Bagrets, A. Bagrets, B. Dieny: Phys. Rev. B **63**, 064429 (2001).
3.149. C.H. Shang, J. Nowak, R. Jansen, J.S. Moodera: Phys. Rev. B **58**, R2917 (1998).
3.150. D.T. Pierce, R.J. Celotta, J. Unguris, H.C. Siegman: Phys. Rev. B **26**, 2566 (1982).
3.151. H.C. Siegmann: J. Phys.: Condens. Matter **4**, 8395 (1992).
3.152. A.H. MacDonald, T. Jungwirth, M. Kasner: Phys. Rev. Lett. **81**, 705 (1998).
3.153. B. Nadgorny, R.J. Soulen, M.S. Osofsky, I.I. Mazin, G. Laprade, R.J.M. van de Veerdonk, A.A. Smits, S.F. Cheng, E.F. Skelton, S.B. Qadri: Phys. Rev. B **61**, R3788 (2000).
3.154. A.H. Davis, J.M. MacLaren, P. LeClair: J. Appl. Phys. **89**, 7567 (2001).
3.155. T. Kreutz, T. Greber, P. Aebi, J. Osterwalder: Phys. Rev. B **58**, 1300 (1998).
3.156. B. Kim, A.B. Andrews, J.L. Erskine, K.J. Kim, B.M. Harmon: Phys. Rev. Lett. **68**, 1931 (1992).
3.157. M. Shimizu, A. Katsuki, H. Yamada, K. Terao: J. Phys. Soc. Jpn. **21**, 1654 (1966).
3.158. D. Nguyen-Mahn: Mater. Res. Soc. Symp. Proc. **492**, 319 (1998).
3.159. I.I. Oleinik, E.Y. Tsymbal, D.G. Pettifor: Phys. Rev. B **65**, 020401(R) (2001).
3.160. M. Sugiyama, J. Hayakawa, K. Itou, H. Asano, M. Matsui, A. Sakuma, M. Ichimura: J. Magn. Soc. Jpn. **25 pt. 2**, 795 (2001).
3.161. J.Z. Sun, K.P. Roche, S.S.P. Parkin: Phys. Rev. B **61**, 11244 (2000).
3.162. I.I. Oleinik, E.Y. Tsymbal, D.G. Pettifor: Phys. Rev. B **62**, 3952 (2000).
3.163. J.M. Rowell: J. Appl. Phys. **40**, 1211 (1969).
3.164. W. Oepts, M.F. Gillies, R. Coehoorn, R.J.M. van de Veerdonk, W.J.M. de Jonge: J. Appl. Phys. **89**, 8038 (2001).
3.165. R. Jansen, J. Nowak, J.S. Moodera (unpublished).
3.166. A. Davis: Ph.D. thesis, Tulane University, New Orleans, La, 2000.
3.167. S. Yuasa, T. Sato, E. Tamura, Y. Suzuki, H. Yamamori, K. Ando, T. Katayama: Europhys. Lett. **52**, 344 (2000).
3.168. Note that in this work, the authors reactively deposited Al in a partial O_2 environment, allowing them to easily alter the Al_2O_3 thickness.
3.169. J. Callaway, C.S. Wang: Phys. Rev. B **16**, 2095 (1977).
3.170. M. Bowen, V. Cros, F. Petroff, A. Fert, C.M. Boubeta, J.L. Costa-Kr amer, J.V. Anguita, A. Cebollada, F. Briones, J.M. de Teresa, L. Morellòn, M.R. Ibarra, F. G uell, F. Peirò, A. Cornet: Appl. Phys. Lett. **79**, 1655 (2001).
3.171. H.L. Meyerheim, R. Popescu, J. Kirschner, N. Jedrecy, M. Sauvage-Simkin, B. Heinrich, R. Pinchaux: Phys. Rev. Lett. **87**, 076102 (2001).
3.172. X. Zhang, W.H. Butler (unpublished).
3.173. M. Tanaka, Y. Higo: Phys. Rev. Lett. **87**, 026602 (2001).

3.174. J.S. Moodera, M.E. Taylor, R. Meservey: Phys. Rev. B **40**, R11980 (1989).
3.175. S.S.P. Parkin, see U.S. Patent No. 5,764,567 (1998) (unpublished).
3.176. J.J. Sun, P.P. Freitas: J. Appl. Phys. **85**, 5264 (1999).
3.177. S. Zhang, P.M. Levy: Phys. Rev. Lett. **81**, 5660 (1998).
3.178. J. Mathon, A. Umerski: Phys. Rev. B **60**, 1117 (1999).
3.179. V. Vedyaev, N. Ryzhanova, C. Lacroix, L. Giacomoni, B. Dieny: Europhys. Lett. **39**, 219 (1997).
3.180. J.S. Moodera, T.H. Kim, C. Tanaka, C.H. de Groot: Phil. Mag. B **80**, 195 (2000).
3.181. W. Zhang, B. Li, Y. Li: Phys. Rev. B **58**, 14959 (1998).
3.182. A. Vedyayev, N. Ryzhanova, R. Vlutters, B. Dieny: Europhys. Lett. **46**, 808 (1999).
3.183. J.S. Moodera, J. Nowak, L.R. Kinder, P.M. Tedrow, R.J.M. van de Veerdonk, A.A. Smits, M. van Kampen, H.J.M. Swagten, W.J.M. de Jonge: Phys. Rev. Lett. **83**, 3029 (1999).
3.184. P. LeClair, J.K. Ha, H.J.M. Swagten, J.T. Kohlhepp, C.H. van de Vin, W.J.M. de Jonge: Appl. Phys. Lett. (2002), accepted.
3.185. C.H. Kant, unpublished Andreev reflection results. The Gd tunneling spin polarization is yet unknown with EuS barriers.
3.186. S. Heinze, M. Bode, O. Pietzsch, A. Kubetzka, X. Nie, S. Blugel, , R. Wiesendanger: Science **288**, 1805 (2000), and references therein.
3.187. W. Wulfhekel, J. Kirschner: Appl. Phys. Lett. **75**, 1944 (1999).
3.188. D.J. Monsma, J.C. Lodder, T.J.A. Popma, B. Dieny: Phys. Rev. Lett. **74**, 5260 (1995).

4

Interlayer Exchange Coupling

M. D. Stiles

4.1 Introduction

Experiments in 1986 [4.1–3] demonstrated coupling between the magnetizations of two ferromagnetic layers separated by a non-magnetic spacer layer. Subsequent discoveries in these systems, including giant magnetoresistance [4.4, 5], led to an explosion in measurements and theories. The greatest interest has been in the simplest form of the coupling,

$$\frac{E}{A} = -J\hat{\boldsymbol{m}}_1 \cdot \hat{\boldsymbol{m}}_2 \,, \tag{4.1}$$

called bilinear because the energy per area is linear in the directions of both magnetizations $\hat{\boldsymbol{m}}_i$. With this form of the interaction, positive values of the coupling constant J favor parallel alignment of the magnetizations and negative values favor antiparallel alignment. The discovery [4.6] that the sign of J and hence the preferred alignment of the magnetizations oscillates as the thickness of the spacer layer is varied accelerated the interest started by the discovery of giant magnetoresistance. The sign has been observed to change as many as sixty times [4.7] as the spacer layer thickness is varied between zero and eighty monolayers (see Fig. 4.3 below).

Chapter 2 of *Ultrathin Magnetic Structures II*, published in 1994, contains articles by Hathaway; Fert and Bruno; Pierce, Unguris, and Celotta; and Parkin reviewing aspects of interlayer exchange coupling [4.8]. These articles were written at a time of great evolution in our understanding. Shortly thereafter, a consensus developed on a theoretical model that unified many of the existing models. Simultaneously, high precision experiments were carried out enabling stringent tests of the theoretical models. In this chapter, I briefly summarize the situation that existed when these previous articles were written and then focus on the model that developed and measurements that were made shortly thereafter. For more details, see the earlier chapters [4.8] or other review articles [4.9–15].

By 1993, oscillatory interlayer exchange coupling had been measured in a large variety of systems. There had been measurements of a number of related effects including biquadratic coupling and quantum well states. Progress in growing high quality samples and measuring them accurately was evolving in parallel with the development of theoretical models. Systematic studies of multilayers grown by sputtering, typically with Co as the ferromagnet, had revealed oscillation periods in the range of 0.9 nm to 1.2 nm for V, Cu, Mo, Ru, Rh, Re, and Ir spacer layers [4.16–20] and longer periods of 1.5 nm for Os [4.21] and 1.8 nm for Cr [4.16]. Studies on lattice-matched systems grown by molecular-beam epitaxy (MBE) showed more complicated behaviors including much shorter periods. Examples include Co/Cu [4.22–24], Cr/Fe [4.25–27], Ag/Fe [4.28], Au/Fe [4.29].

There had been progress toward addressing two issues complicating the comparison between theory and experiment. One complication is the significant disorder that is present in real systems but absent in theoretical models. For one form of disorder, namely thickness fluctuations, averaging the interlayer exchange coupling over the growth front had been proposed [4.30] as a solution. Unfortunately, the growth front had not been measured and this correction had not been made quantitative. Progress on treating other types of disorder, like interdiffusion, was still to come. Another complication is that the thickness of the spacer layer can be both difficult to control and to measure with the desired accuracy. The use of wedge-shaped spacer layers [4.23, 25, 31], in which the thickness varies continuously, had been developed as a practical solution to this problem. Starting around the time the earlier reviews were written, there were a number of high precision measurements on wedge samples in which the sample quality was measured. These measurements have allowed stringent comparisons between theory and experiment.

Some samples had been shown to have a perpendicular alignment of the magnetizations [4.31, 32]. Phenomenologically, this alignment can be explained by the presence of a coupling between the magnetizations of the form

$$\frac{E}{A} = -J_2(\hat{\boldsymbol{m}}_1 \cdot \hat{\boldsymbol{m}}_2)^2 , \tag{4.2}$$

called biquadratic because it is quadratic in both of the magnetization directions. All measured values of J_2 are negative, favoring perpendicular orientation of the two magnetizations. Biquadratic coupling has a separate origin from the bilinear coupling, in general coming from the presence of disorder [4.33]. In Sect. 4.4, I discuss this coupling and several of the models that have been proposed to explain it.

In addition to measurements of the coupling, there had been measurements of magnetic multilayers using photoemission and inverse photoemission [4.34–37]. For these measurements, the sample consisted of a nonmagnetic layer on top of a magnetic layer. Strong peaks in the photoemission intensity were identified as arising from quantum well states. As the thickness of the top layer was varied, the quantum well states were observed to shift in energy. The periodicity in thickness at which these states crossed the Fermi level established a connection between the quantum well states and the interlayer exchange coupling.

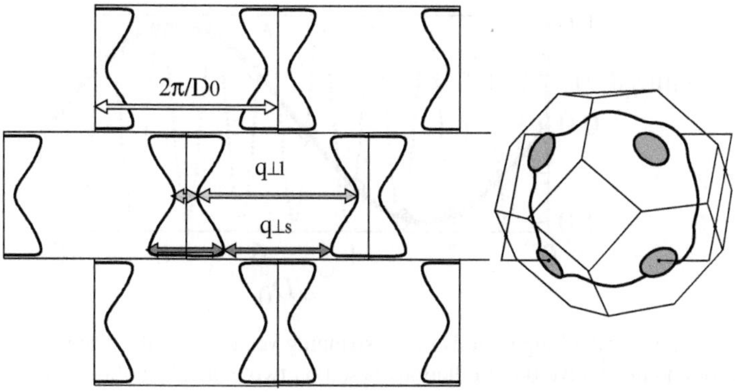

Fig. 4.1. Critical spanning vectors for Cu in the 100 direction. The *right panel* shows a representation of the Cu Fermi surface and a rectangular slice (*solid line*) through it. The necks in the (111) directions are indicated in gray. The dashed lines indicate the bulk Brillouin zone. In the *left panel*, the rectangular slice from the right panel is repeated periodically using an extended zone scheme. There, the heavy curves show the Fermi surface of Cu in that slice. The white arrow gives a 1-dimensional reciprocal lattice vector in the interface direction. The *gray arrows* give the critical spanning vectors. An equivalent pair of vectors is shown for both the long (l) and short (s) period spanning vectors

By 1993, there were a number of theoretical models for interlayer exchange coupling. All gave the result that the Fermi surface of the spacer layer material [4.38] determined the coupling periods. In metals, the Fermi surface is a sharp cut-off in momentum space between filled states and unfilled states. In many contexts, the existence of this sharp cut-off gives rise to spatial oscillations. Well-known examples include Friedel oscillations of the charge density due to a localized perturbation and the Ruderman-Kittel-Kasuya-Yosida (RKKY) interaction between magnetic impurities [4.39]. The early models for oscillatory interlayer exchange coupling showed it to be an additional example.

The models showed that the critical spanning vectors of the Fermi surface of the spacer-layer material determine the oscillation periods. A spanning vector of the Fermi surface is a vector parallel to the interface normal that connects two points on the Fermi surface, one point having a positive component of velocity in the interface direction and the other a negative component. A *critical* spanning vector is a spanning vector that connects two sheets of the Fermi surface at a point where they are parallel to each other, see Fig. 4.1. For noble metal spacer layers, Bruno and Chappert [4.40] found the critical spanning vectors of Fermi surfaces that had been previously determined in de Haas-van Alphen measurements. The periods agree with the oscillations seen experimentally.

Since the spacer layer material is periodic, the Fermi surface is defined in a periodic Brillouin zone. After the discovery of oscillatory coupling, there was some doubt about whether the coupling was a Fermi surface effect because some early models ignored the lattice periodicity. The periods expected from free electron Fermi wave vectors are much smaller than the typically observed periods of 0.9 nm to 1.2 nm.

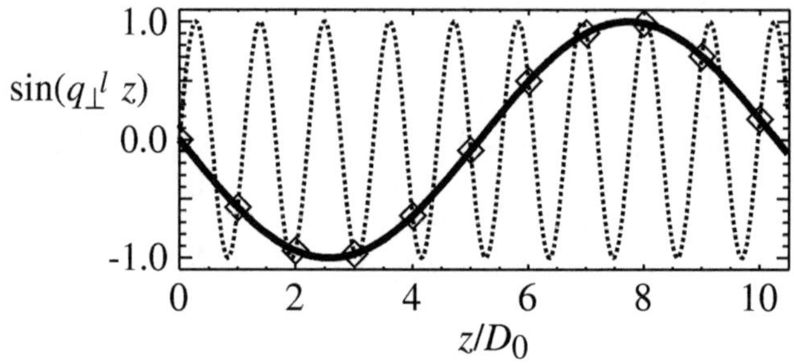

Fig. 4.2. Aliasing. For the long period critical spanning vectors, $q_\perp{}^l$, illustrated in Fig. 4.1, the solid and dotted curves give the oscillations based on two equivalent choices for this vector. The squares gives the values of both curves evaluated at the discrete layer thicknesses. At these points, the two curves are completely equivalent

Several groups [4.41–43] pointed out that *aliasing* would bring the periods into better agreement. Aliasing describes the result of periodically sampling an oscillating function. Here, the oscillating function is the oscillatory coupling and the periodic sampling occurs at the discrete layers thicknesses of nD_0 where D_0 is the individual layer thickness, and n is an integer. Figure 4.2 shows that the oscillation determined by the long period spanning vector $q_\perp{}^l$ in Fig. 4.1 is equivalent to the oscillation determined by $|q_\perp{}^l - g_0|$ where $g_0 = 2\pi/D_0$ when both oscillations are evaluated at discrete layers $z = nD_0$. The wave vector g_0 gives the appropriate length of the Brillouin zone in the interface direction. Some confusion can be avoided by always reducing the spatial frequency to its smallest equivalent magnitude.

If models are based on different descriptions of the electronic structure, the critical spanning vectors of the Fermi surface are different and the models predict different periods and different coupling strengths. When two models use the same electronic structure for the spacer layer and hence predict the same periods, they predict different coupling strengths if they use different descriptions for the electronic structure of the ferromagnetic layers.

In 1993, there was no consensus on the appropriate model for the interlayer exchange coupling because there was no consensus on the appropriate description of the electronic structure of transition metal multilayers. In the delocalized limit, all of the electrons are in band states. In the ferromagnet, the energy bands for majority and minority electrons are split by the exchange interaction. In the localized limit, the electrons are grouped into two sets, localized electrons that are spin polarized and delocalized electrons that couple to the localized magnetic electrons. These two limits emphasize different parts of the physics that are necessary for a correct description of ferromagnetism. One part is the hybridization between the electron states on different sites, and the other is the electron-electron interaction which is predominantly local.

The limit of completely localized magnetism was represented by models, related to the RKKY interaction between impurities, based on perturbative calculations of the $s-d$ (or $s-f$) Hamiltonian. In these models [4.30, 40, 44, 45], there are hybridized band states that interact with localized d states. The electron-electron interactions among the d states give rise to the magnetism. A more sophisticated approach to this limit is the Anderson model [4.46, 47], where the d levels become partially delocalized due to the hybridization with the s levels.

In the opposite limit of completely delocalized states, the simplest model is the free electron model [4.48, 49], which has the virtue of being analytically solvable. In this approximation, the bands in the ferromagnet are shifted in energy with respect to each other. The simplest case is to have the majority bands aligned with the bands of the spacer layer and the minority bands shifted infinitely high in energy. This limit is frequently called the infinite-U limit. Edwards et al. [4.50] carried out closely related calculations for simple tight-binding models. Because the Fermi surfaces in tight-binding models are not spherical, their calculations make it clear that it is the critical spanning vectors, and not just the Fermi wave vectors that determine the oscillation periods. After the earlier reviews were written, a consensus developed that an accurate description of interlayer exchange coupling requires an accurate Fermi surface for the ferromagnet. This requirement necessitates a delocalized description (see Sect. 4.3).

The rest of this chapter focuses on developments since the previous volume of this series was written. Section 4.2 describes growth of the samples and some of the experimental techniques used to measure the coupling. Section 4.3 presents the model used to describe the interlayer coupling. Predictions of this model are compared to experiment. Disorder is discussed from an experimental point of view in Sect. 4.2 and from a theoretical point of view in Sect. 4.3. Disorder can introduce other types of coupling into these systems, in particular biquadratic coupling. Section 4.4 describes these other forms of coupling. Finally, Sect. 4.5 discusses calculations and measurements on specific systems including Co/Cu(100), Au/Fe(100), and Cr/Fe in terms of this model.

4.2 Experiment

4.2.1 Sample Growth

To make measurements that quantitatively test our understanding of interlayer exchange coupling, it is important to grow samples that are as close as possible to those assumed in theoretical treatments. These assumptions are usually quite restrictive. A typical calculation of interlayer exchange coupling assumes that the ferromagnet and the spacer layer are coherent, that is, they share a common lattice in the plane of the interface between them. Further, most (but not all) calculations assume that there are no defects either at the interface or in the bulk. Growing samples that approach this ideal is quite challenging. An alternative to eliminating defects is to quantify them. If the type and distribution of defects is well measured, then the burden is on the theorists to treat the imperfect system rather than the perfect one. However, it is also quite difficult to measure the defects in sufficient detail.

The first consideration in choosing an experimental system is to find a spacer layer that is close to lattice matched with a transition metal ferromagnet, Fe, Co, or

Ni. When thin enough, a metal layer frequently assumes the in-plane lattice constant of the substrate on which it is deposited. This is coherent growth. As the thickness of the deposited layers increases, the strain energy associated with its modified lattice structure becomes too large and the film relaxes by introducing dislocations at the interface. To avoid dislocations, it is necessary to start with a pair of materials that have very similar lattice constants. Unfortunately, there are very few.

The most commonly studied systems are Fe/Cr and Co/Cu. Cr and Fe, which both take the body-centered cubic crystal structure, have less than one percent lattice mismatch, so can grow coherently to fairly large thicknesses. Since they have the same crystal structure they can be studied in several interface orientations, particularly (100), (110), and (211). Co naturally takes the hexagonal-close-packed (hcp) structure. However, when it grows on Cu, it frequently grows in the face-centered-cubic (fcc) structure of Cu (pseudomorphic growth) with a lattice mismatch of less than one percent, and so can also grow coherently to large thicknesses. Here also, a number of interface orientations have been investigated, particularly (100), (110), and (111). However, in the (111)-direction the energy difference between fcc growth and hcp growth comes only from different stacking of hexagonal planes and hence is very small. Both types of growth tend to occur, leading to extended defects between different regions and very poor growth. For a review of growth in this system see [4.51].

Two other systems for which coherent growth can be achieved up to large thicknesses are Ag/Fe and Au/Fe. At first glance, these pairs are not only poorly lattice matched, but they do not have the same crystal structure. However, the (100) interface lattices of the noble metal and of the Fe differ by less than one percent, if the noble metal lattices are rotated by 45° with respect to that of Fe [4.29, 52]. If the starting substrates are sufficiently flat, very good growth can be achieved. However, the presence of steps leads to extended defects through the layer because the growth is not pseudomorphic and the layer thicknesses are quite different.

Other material pairs have been studied. An interesting pair is Fe/Cu. For thin enough layers, Fe grows in a face-centered-tetragonal structure (almost pseudomorphic) on Cu. Alternatively, Cu grows in a distorted body-centered-cubic (bcc) structure on Fe for thin enough layers. Thus, it is possible to study interlayer exchange coupling for bcc Cu [4.23, 28]. Transition metal spacer layers such as Fe/Pd [4.53], Fe/Nb [4.54], and Fe/Mo [4.55] have been grown by MBE. The same quality of growth has not been achieved in these systems as in the systems with much smaller lattice mismatch.

Even when the lattice mismatch is close to zero, the multilayers are still not perfect. The starting substrate is never perfectly flat and the growth is never perfectly layer by layer, so there are always variations in the thickness of the layers, typically called thickness fluctuations. If the growth front is measured, it is possible to account for the variations of the spacer layer thickness when comparing theory and experiment by averaging over the contributions from different thicknesses. However, when the interlayer exchange coupling has rapid oscillations, the measured coupling is significantly modified by the thickness fluctuations. Even for reasonably good growth, thickness fluctuations can completely obscure the experimental signature of short period oscillations in the coupling. For this reason, it is desirable to grow the sample in a layer-by-layer mode to keep the growth front as narrow as possible. Since layer-by-layer growth is determined by the competition between nucleation of islands and diffusion of deposited adatoms, it tends to require higher substrate temperatures during growth. Unfortunately, higher growth temperatures tend to promote interdiffusion at the interfaces. Such interdiffusion has been observed both for Cr growth on Fe [4.56–59] and Fe growth on Cu [4.60]. Interdiffusion, which gives rise to scattering centers, is more difficult to treat theoretically than thickness fluctuations. It also can be more difficult to measure.

In addition to the lattice matching between materials in the multilayer, the choice of substrate plays a large role in the quality of the growth. If substrates of one of the materials in the multilayer are available, they are frequently the best choice. Iron whiskers, which can be extremely flat [4.61], and copper single crystals are examples. However, these choices are not appropriate for transport measurements, like giant magnetoresistance, because the substrate provides a short circuit that prohibits any measurement of the transport properties of the multilayer. In this case, insulating substrates like MgO or semiconducting substrates like Si or GaAs can be used. For these substrates, great care is required to get really high quality growth. See [4.62] and [4.63] for descriptions of the complexity of growing a Fe/Au multilayer on a GaAs substrate.

As mentioned in the introduction, the difficulties in growing samples with particular spacer-layer thicknesses and in measuring those thicknesses are most easily addressed by growing wedge-shaped samples. To grow such samples, a shutter between the sample and the evaporator is moved to expose different parts of the sample to different total fluxes. After the wedge has been grown, Reflection High Energy Electron Diffraction (RHEED) (see Chapter 5 of Volume I of this series for details) can be used to determine the thickness of the sample at various positions along the wedge. RHEED is a commonly used technique for monitoring the quality and amount of growth. A high energy electron beam is reflected from the surface at glancing angles. The resulting diffraction pattern is sensitive to the details of the surface, in particular the presence of steps. If the growth is layer by layer, there are fewer steps when layers are close to complete and more when the layer is half filled. In this case, the intensity of different spots in the RHEED pattern oscillate with a period of one layer. In typical use, RHEED is used to monitor the thickness of a film during growth. However, for wedge samples, it is typically used after growth, when the RHEED beam is scanned along the wedge and the RHEED oscillations are monitored as a function of position to give the thickness at that position, see Fig. 4.3.

4.2.2 Measurement Techniques

Interlayer exchange coupling is not measured directly. Rather, some magnetic property of a sample, like its hysteresis loop, is measured and the exchange coupling is inferred by comparing the measured property with a model. Some or all of the parameters of the model, including the interlayer layer coupling constant, are varied until the predictions of the model agree with the measurement. The reliability of the

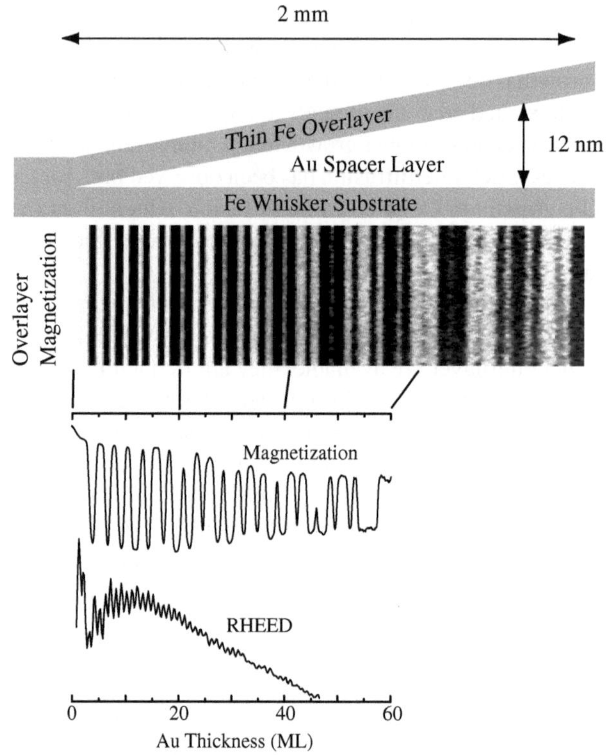

Fig. 4.3. SEMPA image of coupling in Fe/Au/Fe with geometry and RHEED. A schematic view of the wedge-shaped sample is shown at the top of the figure. The approximate dimensions give an indication of the very small slope of the wedge. Immediately below is a SEMPA image of the magnetization of the Fe overlayer. White and black indicate parallel and antiparallel alignment to the substrate, and hence ferromagnetic and antiferromagnetic coupling. Below that is a line scan through the image and then a measurement of the RHEED intensity along the wedge. The oscillations are used to determine the thickness of the spacer layer along the wedge. Examination of the oscillations indicate that the slope of the wedge is not constant, but the wedge is slightly curved. The RHEED and the magnetization curves have been corrected for this curvature, but the image has not, hence the variation of the lines connecting the image with the line scan

results depend on the accuracy of the model, in particular whether it includes all of the physics necessary to describe the experiment. In addition to the interlayer coupling, the models usually include the magnetic anisotropy of each of the layers. Typically, models assume coherent rotation, at least within each layer, so that the intralayer exchange interaction does not play a role. Magnetostatic effects are typically not included in the models. Hysteretic effects are usually ignored. That is, the "hysteresis loop" is calculated from a global energy minimum so there is no hysteresis in the model. At the same time, hysteresis that shows up in the measurement is averaged over to give a similar "loop" to compare to.

Measurements of interlayer exchange coupling fall into two broad classes. In the first class, the hysteresis loop (or some part) is measured. For example, Parkin et al. first observed oscillatory interlayer exchange coupling [4.6] using the giant magnetoresistance to determine the relative orientation of the magnetic layers. Here the resistance of the film in zero field was compared with the resistance in large field. If the coupling is ferromagnetic, there is no change, and if the coupling is antiferromagnetic the change can be substantial. The field required to saturate the resistance is used to estimate the strength of the antiferromagnetic coupling.

One technique commonly used to measure the hysteresis loop is the Magneto-Optic Kerr Effect (MOKE) (see Chapter 4 of Volume II of this series or [4.64] for details). MOKE can be used in an imaging mode by scanning the focused spot of a laser across the surface or by imaging a wide area of illumination. It is not particularly surface sensitive and has the advantage that it is sensitive to the magnetic state of both layers. Using the sensitivity to both layers, MOKE images [4.31] have directly identified perpendicular alignment of two layers.

Another imaging technique that has contributed significantly to understanding interlayer exchange coupling is Scanning Electron Microscopy with Polarization Analysis (SEMPA) (see Chapter 2 of Volume II of this series for details). Since this technique is based on measuring secondary electrons, it is generally not used with an applied field, limiting it to studies of the remnant state. On the other hand, it has greater spatial resolution than optical techniques like MOKE, and can be used on smaller wedges, requiring smaller areas of sample perfection. Since it can only measure the remnant state, SEMPA has not been used to measure coupling strengths, but it has been used to determine the sign of the coupling for enough oscillations of the coupling to allow high precision measurements of the periods.

The other class of measurements is based on determining the curvature of the energy with respect to small variations in the magnetization direction. Two such techniques are Ferromagnetic Resonance (FMR) and Brillouin Light Scattering (BLS) (see Chapter 3 of Volume II of this series for details of both). In a multilayer structure, certain resonances, whether as a function of field at fixed frequency (FMR) or frequency at fixed field (BLS), depend on on the strength of the interlayer coupling. The resonance positions can be used to determine the coupling.

4.3 Physical Mechanism for Bilinear Coupling

In 1993, a model for interlayer exchange coupling [4.65–67] based on spin-dependent reflection at interfaces provided a framework to unify previous models for exchange coupling. This model is closely related to the model of Mathon et al. [4.68], formulated in terms of normalized spectral densities. Since the electron states in 3*d* transition metals are strongly hybridized, a delocalized description is necessary to accurately compute the interface reflection amplitudes. This fact led to a consensus that a delocalized description of the electronic structure of the 3*d* transition metals is more appropriate for models of interlayer exchange coupling. In this section, I present the model that led to this consensus.

First, the necessity of a delocalized description of transition metal ferromagnetism raises the question that frequently starts many discussions of interlayer exchange coupling: Is the coupling RKKY? The answer depends on what is meant by RKKY. If the most general sense is meant, that the oscillations of the coupling are due to the sharp cut-offs in momentum space due to the Fermi surface of the spacer layer, then the answer is yes. If the more restrictive sense is meant, that a perturbative treatment of an $s-d$ Hamiltonian gives an adequate description of transition metal ferromagnetism, then the answer is no. The model I describe below is an RKKY model in the first, general sense, but not in the second, more restrictive sense.

From (4.1), the interlayer coupling constant is given by the difference in energy between the antiparallel alignment of the magnetizations and the parallel alignment

$$J = \frac{1}{2A}(E_{\text{anti}} - E_{\text{par}}) \ . \tag{4.3}$$

Computing the interlayer exchange coupling is reduced to computing the energy difference between two configurations. Some calculations of interlayer exchange coupling evaluate these total energies directly using the local-spin-density approximation (LSDA) [4.69, 70] or various tight-binding models [4.71, 72]. Because the exchange coupling is much smaller than the total electronic energy for either alignment, it can be difficult to converge these calculations numerically.

Calculation of the energy differences in (4.3) can be significantly simplified by using the "force theorem". First, let me stress the difficulty of carrying out self-consistent calculations of the total energies in (4.3) to sufficient accuracy to make the energy difference meaningful. The difficulty arises from treating the electron-electron interaction, even though it is only in mean field theory. The force theorem provides a way to avoid these self-consistent calculations. It states that the energy difference between two configurations is approximately given by the difference in the single particle energies in the two configurations. One simply develops approximations for the potential and then computes the sum of the single particle energies for each configuration. This approach ignores any explicit calculation of the electron-electron interactions. The force theorem states that if the approximations for the potential are good enough, the difference in energy of the single-particle-energy sums is very close to the difference in the fully self-consistent energies. Fortunately, it is fairly simple to develop good enough approximations for the potential because of the short range of the screening in metals. For thick enough layers, the potential in the middle of each layer can be treated as independent of the rest of the system and the potential near each interface as independent of the other interfaces. Thus, a good approximate potential for a multilayer can be constructed piece-wise from much smaller systems. Alternately, the potential can be found for the case of parallel alignment and then simply modified to give an approximate potential for antiparallel alignment.

Another feature of the force theorem is its pedagogical value. In the rest of the section, I develop a model for interlayer exchange coupling based on differences in the single particle energies. The force theorem allows me to ignore (to a good approximation) the complications of the electron-electron interactions.

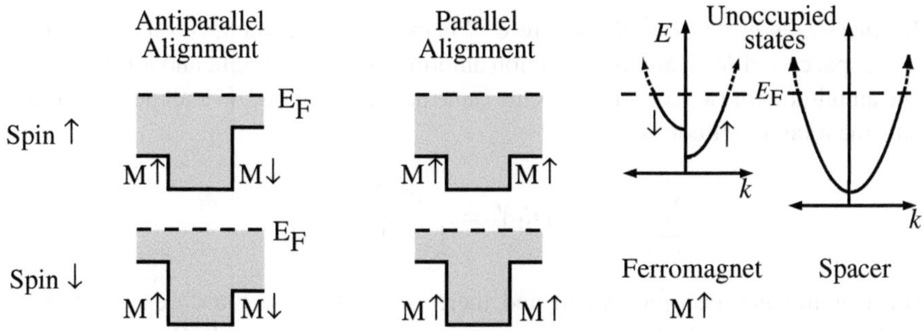

Fig. 4.4. Quantum wells used to compute interlayer exchange coupling. On the *right*, the two panels give typical band structures for free electron models of interlayer exchange coupling. On the *left*, the four panels give the quantum wells for spin up and spin down electrons for parallel and antiparallel alignment of the magnetization. The *gray regions* designate the occupied states

The calculation is further simplified when spin-orbit coupling is neglected. Then, since the magnetizations are collinear for both the antiparallel and the parallel alignments, the majority and minority spins do not interact, and each can be computed separately. At this point, the calculation has been reduced to computing the single particle energies for the four quantum wells shown schematically in Fig. 4.4. The appropriate sums and differences of the energies give the interlayer exchange coupling.

For a simple trilayer, in which both ferromagnetic layers are semi-infinite, all the sums of single particle energies are infinite. However, the infinite contributions from the two ends cancel out. It is important to treat this cancellation correctly, which is most easily done by considering the "cohesive energy" of each quantum well. The cohesive energy is the energy required to make the quantum well out of its bulk constituent materials. For a quantum well with total ferromagnetic thickness L and spacer layer thickness D the cohesive energy is

$$\frac{\Delta E_{\mathrm{qw}}}{A} = \frac{E_{\mathrm{tot}}}{A} - L\varepsilon_{\mathrm{FM}} - D\varepsilon_{\mathrm{NM}} , \tag{4.4}$$

where ε is the energy density of each bulk material that makes up the layers. When using the force theorem, the energy densities are just the single particle energy densities.

4.3.1 Quantum Well States Due to Spin-polarized Reflection

Electrons reflect from interfaces between two materials. For free electron models, the interface is a simple (spin-dependent) step in the potential and the reflection amplitudes are straightforward to compute. For real material interfaces, such a calculation is not so straightforward, but it is still feasible [4.73–76].

In trilayers, electrons reflect from both interfaces and the multiply reflected waves interfere with each other. The amplitude for one round trip in a spacer layer of thickness D is $e^{ikD} R_R e^{ikD} R_L$, where e^{ikD} is the phase accumulated on one traversal of the spacer and $R_{R/L}$ are the reflection amplitudes from the right and left interfaces. The amplitude for two round trips is the same quantity squared. For all possible round trips the total amplitude is

$$\sum_{n=1}^{\infty}[e^{i2kD} R_R R_L]^n = \frac{e^{i2kD} R_R R_L}{1 - e^{i2kD} R_R R_L} . \tag{4.5}$$

The denominator becomes small and there is constructive interference whenever $2kD + \phi_R + \phi_L = 2n\pi$, where n is an integer and $\phi_{R/L} = \mathrm{Im}[\ln(R_{R/L})]$ is the change in the phase of a reflected electron. The constructive interference inside the spacer layer gives rise to resonances, frequently referred to as quantum well states. Whenever the reflection probability is one, these quantum well states are true bound states. Otherwise they are like bound states that are broadened by transmission into the ferromagnetic layers. When the reflection amplitude is close to one, $|R| \approx 1$, the resonances are sharp and when the reflection amplitude is close to zero, they are broad. For a free electron model, these are shown schematically in Fig. 4.5.

As the thickness D of the spacer layer is varied, the resonances and bound states shift in energy. If there is a resonance at the Fermi energy at a thickness D, then resonances cross the Fermi energy whenever the thickness is $D + 2n\pi/2k_F$, where k_F is the Fermi wave vector of the spacer layer and n is an integer. This periodic crossing of the Fermi energy by quantum well resonances is the origin of the oscillations in the interlayer exchange coupling. In this 1-d model, the Fermi surface consists of two points, $k = \pm k_F$. The period is determined by the spanning vector of the Fermi surface, $2k_F$.

For closely related samples, these quantum well states have been seen in photoemission and inverse photoemission. For reviews, see [4.77, 78]. In these experiments, one of the ferromagnetic layers is left off so that the density of states in the spacer layer could be measured. Since the reflection amplitude from the surface is different from the interface with the ferromagnet, the resonances are sharper and slightly displaced in energy compared to the quantum well. Nevertheless, the peaks seen in the photoemission spectra of these samples show the expected behavior as a function of energy and thickness. In addition, the quantum well resonances seen in photoemission

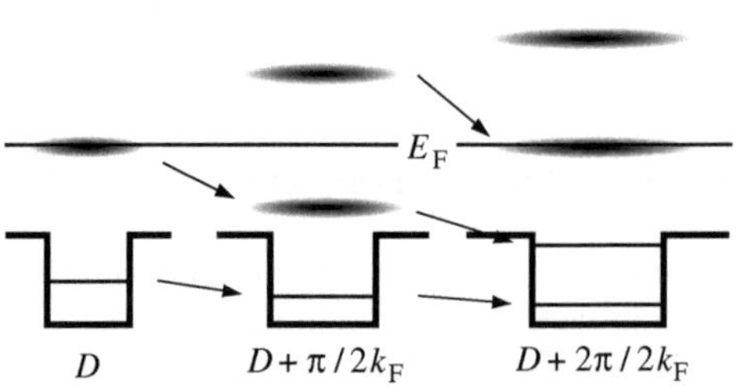

Fig. 4.5. Evolution of quantum well resonances with spacer layer thickness. The three panels illustrate the bound states (*lines*) and resonances (*fuzzy ellipses*) for quantum wells of increasing thickness. The arrows indicate how each resonance evolves as the thickness is increased

cross the Fermi level with the same periodicity as is found for the interlayer coupling in related trilayers. Since the reflection at the interface with the ferromagnetic layer is spin dependent, the quantum well states are spin dependent [4.79].

The quantum well states affect the density of states in trilayer, and hence the cohesive energy of the quantum well. The change in the density of states of the quantum well *for each spin*, defined similarly to the cohesive energy of the quantum well [4.14], is

$$\Delta n(E,D) = -\frac{1}{\pi}\mathrm{Im}\left[\frac{\mathrm{d}}{\mathrm{d}E}\ln\left(1-\mathrm{e}^{\mathrm{i}2kD}R_\mathrm{R}R_\mathrm{L}\right)\right] . \tag{4.6}$$

The derivative of the ln gives a factor that is the multiple scattering amplitude in the quantum well, (4.5). The cohesive energy of the quantum well is the sum of the single particle energies or equivalently, the integral over the change in the density of states

$$\Delta E_\mathrm{qw} = -\int_{-\infty}^{E_\mathrm{F}} \mathrm{d}E(E-E_\mathrm{F})\Delta n(E,D) . \tag{4.7}$$

For the time being, this is a one-dimensional model, so there is no factor of the area. This expression can be integrated by parts to give

$$\Delta E_\mathrm{qw} = \frac{1}{\pi}\mathrm{Im}\int_{-\infty}^{E_\mathrm{F}} \mathrm{d}E\ln\left(1-\mathrm{e}^{\mathrm{i}2kD}R_\mathrm{R}R_\mathrm{L}\right) . \tag{4.8}$$

This result is valid for the case when there is only one state going in each direction in the spacer layer. For the more general case, see [4.66, 80].

For fixed thickness, the integrand in (4.8) oscillates rapidly through the energy dependence of k. All of these oscillations cancel out in the integration, except those close to the Fermi energy, where there is a sharp cut-off. The only contribution is from a range of states near E_F of width proportional to $\hbar v_\mathrm{F}/D$, where v_F is the Fermi velocity. Over this energy range, the energy dependence of $R_{\mathrm{R/L}}$ can be ignored and the wave vector can be assumed to vary linearly with energy, $k \approx k_\mathrm{F} + E/(\hbar v_\mathrm{F})$, so that in the limit of a thick spacer layer

$$\lim_{D\to\infty}\Delta E_\mathrm{qw} = \frac{\hbar v_\mathrm{F}}{2\pi D}\sum_n \frac{1}{n}\mathrm{Re}\left[(R_\mathrm{R}R_\mathrm{L})^n\,\mathrm{e}^{\mathrm{i}2k_\mathrm{F}nD}\right] . \tag{4.9}$$

Including the full energy dependence gives terms with higher order in D^{-1}, called preasymptotic corrections [4.81, 82]. When the reflection amplitudes are small only the first term in the sum over n contributes, giving

$$\lim_{D\to\infty}\Delta E_\mathrm{qw} \approx \frac{\hbar v_\mathrm{F}}{2\pi D}|R_\mathrm{R}R_\mathrm{L}|\cos\left[2k_\mathrm{F}D+\phi_\mathrm{R}+\phi_\mathrm{L}\right] . \tag{4.10}$$

The higher powers of the reflection amplitudes give higher harmonics of the oscillation. They change the shape and weakly change the amplitude without changing the period. In the asymptotic limit, the cohesive energy for this one-dimensional model decays like one over the spacer-layer thickness, and oscillates with a period set by the (one-dimensional) Fermi spanning vector $2k_{\mathrm{F}}$ with a strength that goes like the product of the two reflection amplitudes. The cohesive energy decays because the energy range of states that contribute without cancellation decreases as $\hbar v_{\mathrm{F}}/D$.

The interlayer exchange coupling is then the sum and difference of the cohesive energies for the four different quantum wells seen in Fig. 4.4. All four quantum wells have cohesive energies that oscillate with the same period because their oscillation period is determined by the Fermi surface spanning vector of the spacer layer material. The oscillations do not cancel because they have different amplitudes and possibly phases. For this one-dimensional case, in the large D, small R limit, the result is

$$\begin{aligned}\lim_{D\to\infty} J &\approx \frac{\hbar v_{\mathrm{F}}}{4\pi D}\mathrm{Re}\left[(R_\uparrow R_\downarrow + R_\downarrow R_\uparrow - R_\uparrow^2 - R_\downarrow^2)\,\mathrm{e}^{\mathrm{i}2k_{\mathrm{F}}D}\right] \\ &\approx -\frac{\hbar v_{\mathrm{F}}}{4\pi D}\mathrm{Re}\left[(R_\uparrow - R_\downarrow)^2\,\mathrm{e}^{\mathrm{i}2k_{\mathrm{F}}D}\right] . \end{aligned} \tag{4.11}$$

Here, I have assumed that the reflection amplitudes are the same for the left and right interfaces but they depend on whether the spin is majority ($\uparrow$) or minority ($\downarrow$). The first form above shows the contribution for each of the quantum wells, the two asymmetric quantum wells for antiparallel alignment of the magnetizations minus the majority and minority quantum wells for parallel alignment. The second form shows that the interlayer exchange coupling depends on the spin dependence of the reflection amplitudes.

4.3.2 Critical Spanning Vectors

Real multilayers are three dimensional, not one dimensional. However, if the interface is coherent and there are no defects, the multilayer is periodic in the two directions parallel to the interface. Then, the crystal momentum parallel to the interface, $\boldsymbol{K}$, is conserved. In this case, the problem simplifies to a two-dimensional set of *independent* one-dimensional quantum wells. This simplification is the reason that all (to my knowledge) theoretical treatments of interlayer exchange coupling assume coherent interfaces. For an ideal interface, the cohesive energy (now per area) is just the integral over the interface Brillouin zone (IBZ) of a series of one-dimensional quantum well energies.

$$\frac{\Delta E_{\mathrm{qw}}}{A} = \frac{1}{\pi}\int_{\mathrm{IBZ}} \frac{\mathrm{d}^2K}{(2\pi)^2}\mathrm{Im}\int_{-\infty}^{E_{\mathrm{F}}} \mathrm{d}E\,\ln\left(1 - \mathrm{e}^{\mathrm{i}2k_z(\boldsymbol{K})D}R_{\mathrm{R}}(\boldsymbol{K})R_{\mathrm{L}}(\boldsymbol{K})\right) . \tag{4.12}$$

In the small R, large D limits, integrating over energy as above gives

$$\lim_{D\to\infty}\frac{\Delta E_{\mathrm{qw}}}{A} \approx \frac{\hbar v_{\mathrm{F}}}{2\pi D}\int_{\mathrm{IBZ}} \frac{\mathrm{d}^2K}{(2\pi)^2}\mathrm{Re}\left(\mathrm{e}^{\mathrm{i}2k_{\mathrm{F}z}(\boldsymbol{K})D}R_{\mathrm{R}}(\boldsymbol{K})R_{\mathrm{L}}(\boldsymbol{K})\right) . \tag{4.13}$$

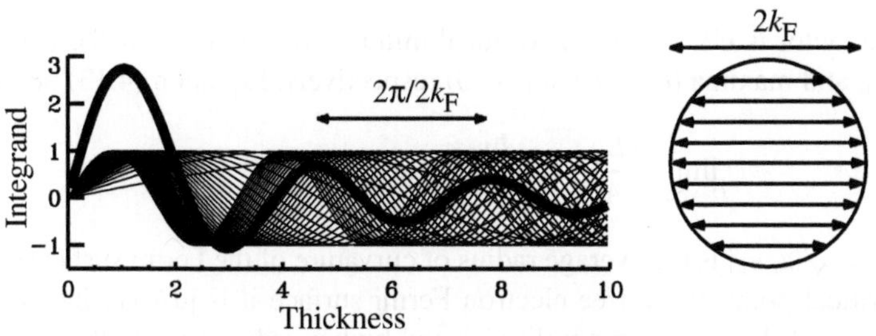

Fig. 4.6. Selection of critical spanning vectors. The *right panel* shows a slice through a free electron Fermi surface and some spanning vectors. The *left panel* shows the oscillating functions from each of these spanning vectors (*light curves*) and the integral over all of them (*heavy curve*). The peaks in the heavy curve come from constructive interference among some of the light curves. Fewer and fewer of the light curves interfere constructively as the thickness increases

Each parallel wave vector has a different one-dimensional Fermi surface given by $k_{\mathrm{F}z}(\boldsymbol{K})$. For each parallel wave vector, the energy integration gives an oscillation as a function of thickness with a period set by its Fermi surface spanning vector, $2k_{\mathrm{F}z}(\boldsymbol{K})$.

For large D, the integrand oscillates rapidly as a function of $\boldsymbol{K}$, through the $\boldsymbol{K}$ dependence of $2k_{\mathrm{F}z}(\boldsymbol{K})$. As is the case above for the energy integration, the oscillations all tend to cancel out, except where there is a sharp cut-off. For the parallel wave vector integration, these cut-offs occur where two sheets of the Fermi surface become parallel in the direction of the interface normal, see Fig. 4.1. The vector connecting the two parallel points is called a critical spanning vector. The projection into the interface Brillouin zone of the points where the Fermi surface sheets become parallel is called the critical point. The regions of the Fermi surface close to the critical point have oscillations that contribute in phase with each other, see Fig. 4.6. As the spacer layer thickness increases, the area of the Fermi surface that contributes in phase decreases as κ/D, where κ is the radius of curvature of the Fermi surface at the critical point. This behavior gives an additional factor of D^{-1} to the decay of the cohesive energy.

For thick spacer layers, the $\boldsymbol{K}$ dependence of the reflection amplitudes is ignored and the $\boldsymbol{K}$ dependence of k_z is expanded to quadratic order. The Fermi surface spanning vectors, q close to the critical point satisfy

$$q = 2k_{\mathrm{F}} + \frac{1}{\kappa_u}k_u^2 + \frac{1}{\kappa_v}k_v^2 \,, \tag{4.14}$$

where u and v are the principle axes of this paraboloid. For simplicity, I have taken the critical point to be at the center of the Interface Brillouin zone. In the large D limit, the parallel wave vector integration in (4.13), reduces to the form

$$\int_{-\infty}^{\infty} \mathrm{d}x \int_{-\infty}^{\infty} \mathrm{d}y\, \mathrm{e}^{\mathrm{i}\left(\frac{x^2}{a}+\frac{y^2}{b}+c\right)} = \pi\, \mathrm{e}^{\mathrm{i}c}\sqrt{|ab|}\,\mathrm{e}^{\mathrm{i}\chi} \,. \tag{4.15}$$

The final factor is $e^{i\chi} = i, 1, -i$ for local minima ($a > 0$ and $b > 0$), saddle points ($ab < 0$), and maxima ($a < 0$ and $b < 0$) respectively. Equation (4.13) becomes

$$\lim_{D\to\infty} \frac{\Delta E_{\mathrm{qw}}}{A} \approx \frac{\hbar v_{\mathrm{F}}\kappa}{8\pi^2 D^2} \mathrm{Re}\left[R_{\mathrm{R}} R_{\mathrm{L}} \mathrm{e}^{\mathrm{i}2k_{\mathrm{F}}D} \mathrm{e}^{\mathrm{i}\chi}\right] , \tag{4.16}$$

where $\kappa = \sqrt{|\kappa_u \kappa_v|}$ is the average radius of curvature of the Fermi surface difference at the critical point. For a free electron Fermi surface it is just k_{F}. Including the $\boldsymbol{k}$ dependence of the reflection amplitudes and higher order terms in the expansion of q gives terms in the cohesive energy that decay as higher powers of D^{-1}. These are additional preasymptotic corrections [4.82].

4.3.3 Asymptotic Form

Each critical spanning vector makes a contribution to the cohesive energy as in (4.16). The contributions from each quantum well in Fig. 4.4 are added and subtracted to give the interlayer exchange coupling. With the critical spanning vectors indexed by α, the interlayer exchange coupling takes its asymptotic form

$$\begin{aligned}\lim_{D\to\infty} J(D) &\approx \sum_{\alpha} \frac{J^\alpha}{D^2} \sin(q_\perp^\alpha D + \phi^\alpha) \\ &\approx \sum_{\alpha} \frac{\hbar v_\perp^\alpha \kappa^\alpha}{16\pi^2 D^2} \mathrm{Re}\left[(R_\uparrow^\alpha - R_\uparrow^\alpha)^2 \mathrm{e}^{\mathrm{i}q_\perp^\alpha D} \mathrm{e}^{\mathrm{i}\chi^\alpha}\right] ,\end{aligned} \tag{4.17}$$

where $q_\perp^\alpha$ is the length of the critical spanning vector, $\mathrm{e}^{\mathrm{i}\chi^\alpha}$ is the phase from the type of critical point (maximum, minimum, saddle point), and ϕ^α is a phase determined by χ^α and the phases of the reflection amplitudes. The effective Fermi velocity is defined as

$$\frac{2}{v_\perp^\alpha} = \frac{1}{|v_{\mathrm{F}\perp}^{\mathrm{r}}|} + \frac{1}{|v_{\mathrm{F}\perp}^{\mathrm{l}}|} , \tag{4.18}$$

where $v_{\mathrm{F}\perp}^{\mathrm{r/l}}$ are the components of the Fermi velocities in the interface direction for the right and left going states at the critical point. Equation (4.17) differs by factors of two from other versions of the same result. These factors depend on the definition of J, (4.1), and the definitions of $v_\perp^\alpha$ and κ^α. Here, the latter two have been chosen so that they reduce to v_{F} and k_{F} for free electron models. Also, in some versions the difference in the reflection amplitudes is written $\Delta R = (R_\uparrow - R_\downarrow)/2$, which introduces additional factors of two. The asymptotic form, (4.17) is the main result of this model. In common with previous models, it shows that the oscillation periods of the interlayer exchange coupling are determined by the critical spanning vectors of the Fermi surface. This model shows that the strength of the oscillation for each critical spanning vector is determined by properties of the spacer layer Fermi surface, $v_\perp^\alpha$, and κ^α, and by the spin difference in the reflection amplitudes. In the context of this result, other models can be interpreted as different approximations for the reflection amplitudes. In RKKY models, the reflection amplitudes depend on the spin-dependent scattering by the localized impurities. In infinite-U models, the reflection amplitudes are either zero or one. In free electron models, they are determined by scattering from potential steps.

Generally speaking, the reflection amplitudes in (4.17) are the reflection amplitudes for all of the material on either side of the the spacer layer. Thus, if the ferromagnetic layers have a finite thickness, they should be the reflection amplitudes for a finite thickness of the ferromagnet. If there is a capping layer, it affects the reflection amplitudes as well. If the outer surfaces of the ferromagnetic layers are flat enough, there are resonances in the ferromagnet (and capping layer) that cause the coupling to oscillate. Both have been seen experimentally [4.83–85]. Different ferromagnetic layers, e.g. different thicknesses, can lead to differences between theoretical results and between theoretical and experimental results.

When spacer layers adopt their bulk structure, their critical spanning vectors can be determined from the Fermi surfaces that have been measured by de Haas-van Alphen experiments, as shown in Table 4.1. When compared with periods determined from high precision measurements of wedge-shaped spacer layers with SEMPA, the agreement is quite remarkable. The comparison of results for the long period of Cu(001) spacer layers is not as good as other results. Possible reasons for this are discussed below. For the Cr spacer layer, the entries in the Table are only for the long period. The periods extracted from the de Haas-van Alphen measurements for Cr are for the Fermi surface pocket centered at the N-point in the Brillouin zone. Justification for this choice, in addition to the *a posteriori* agreement in the Table, is discussed below.

While the agreement between the measured periods and the experimental Fermi surface properties is quite good, the agreement between measured and calculated coupling strengths is not as good. I discuss several examples in Sect. 4.5. There are several reasons for the lack of agreement. On the experimental side, the oscillation

Table 4.1. Comparison of oscillation periods measured in magnetic multilayers with those expected from the critical spanning extracted from Fermi surfaces measured in de Haas-van Alphen measurements (dHvA)

Interface	Period (ML)	Period (ML)	Technique	Reference
Ag/Fe(100)	2.38	5.58	dHvA	[4.40]
	2.37 ± 0.07	5.73 ± 0.05	SEMPA	[4.86]
Au/Fe(100)	2.51	8.60	dHvA	[4.40]
	2.48 ± 0.05	8.6 ± 0.3	SEMPA	[4.7]
Cu/Co(100)	2.56	5.88	dHvA	[4.40]
	2.60 ± 0.05	8.0 ± 0.5	MOKE	[4.23]
	2.58 to 2.77	6.0 to 6.17	SEMPA	[4.87]
Cr/Fe(100)	11.1		dHvA	[4.88, 89]
	12 ± 1		SEMPA	[4.25]
	12.5		MOKE	[4.90]
Cr/Fe(112)	14.4		dHvA	[4.88]
	15.4		MOKE	[4.90]

periods are not affected by disorder to lowest order, but the oscillation amplitudes are. Also, the highest precision measurements of the periodicity were made with SEMPA on wedge-shaped samples. Since SEMPA is generally not compatible with applied fields, SEMPA has not been used to measure coupling strengths. On the theoretical side, the experimentally measured Fermi surfaces are not sufficient to compute the coupling strengths. There remain two theoretical approaches, both of which have disadvantages. One approach is to go back to (4.3) and compute the total energies using the LSDA or a tight-binding approach. The other approach is to compute the coupling strengths using (4.17) by calculating the spin-dependent reflection amplitudes for realistic band structures. Such calculations also require the LSDA or a tight-binding approach.

Formally, the LSDA is justified for computing total energies of different configurations, but not for computing band structures or related quantities like reflection amplitudes. However, to study interlayer exchange coupling it is useful to view LSDA as a single-particle approximation that treats the electron-electron interaction in mean field theory. In this approach, LSDA does a good, but imperfect, job of predicting band structures for transition metals. However, this approach makes it possible to connect the total energy calculations with the model calculations. Hopefully, the discussion of the model has made it clear that the properties of the LSDA band structure (even though it may not be formally justified) determine the results of the total energy calculations.

The differences between real Fermi surfaces and those calculated using the LSDA complicate the comparison between theory and experiment. The results from total energy calculations cannot be compared directly to experiment because the underlying Fermi surface gives oscillation periods that are at least slightly off. After several oscillations, theory and experiment become out of phase and impossible to compare. On the other hand, very few measurements are made for films thick enough that the asymptotic limit is reached. Comparison with the asymptotic results is difficult when preasymptotic corrections are important.

4.3.4 Disorder

All measurements are made at finite temperature, T. A number of calculations [4.50, 67] have shown that the effect of a thermal electron distribution in the spacer layer is to multiply each term in the asymptotic form, (4.17), by the factor

$$\frac{2\pi k_{\mathrm{B}} T D/(\hbar v_{\perp}^{\alpha})}{\sinh\left[2\pi k_{\mathrm{B}} T D/(\hbar v_{\perp}^{\alpha})\right]} \,. \tag{4.19}$$

For a typical free electron Fermi velocity (1.5×10^{15} nm/s) and a spacer thickness of $D = 4\,\mathrm{nm}$, (4.19) is approximately 0.998 at room temperature (compared to 1 at zero temperature) and can frequently be ignored. This correction accounts for the temperature dependence of the electrons in the spacer layer, but not the temperature dependence of the ferromagnetic layers. The low-lying excitations of a ferromagnet are spin waves, which correspond to time-dependent local rotations in the direction of the magnetization rather than its value. These rotations change the relative orientation of the magnetizations on either side of the interface and hence change the net coupling between them [4.91]. However, in most measurements, the measurement temperature is significantly below the Curie temperature so there is not much spin wave excitation. Thus, it is often a good approximation to ignore the temperature dependence of the bilinear coupling.

The simplest type of disorder to treat is the effect of thickness fluctuations. In many samples the spacing between steps at the interfaces is in an intermediate regime. The spacing is large enough that the description of the coupling for ideal interfaces holds for the regions between the steps. At the same time, it is small enough compared to the intralayer domain wall widths in the ferromagnetic layers that the magnetization directions do not rotate significantly between steps. The consequence of this slight rotation – biquadratic coupling – is discussed in Sect. 4.4. In this regime, the effective bilinear coupling is given by the average of the couplings over the different thicknesses present in the spacer layer. If the coupling for an ideal thickness of n layers of thickness D_0 is $J(n)$ and the probability of having a thickness nD_0 for a nominal deposited thickness of D is $P(n, D)$, the effective coupling strength is

$$J(D) = \sum_n P(n, D) J(n) \,. \tag{4.20}$$

If the width of the growth front is measured by scanning tunneling microscopy, x-ray diffraction or some other technique, theoretical coupling strengths can be averaged to compare with measured strengths.

Steps are not only associated with thickness fluctuations, they are also scattering centers that cause diffuse scattering, see Fig. 4.7. Steps are one type of interfacial defect, other types are dislocations that form to relieve heteroepitaxial stress and interdiffused atoms. The effect of localized interdiffusion has been studied [4.92–97], but, to my knowledge, the effect of extended defects, like steps and dislocations

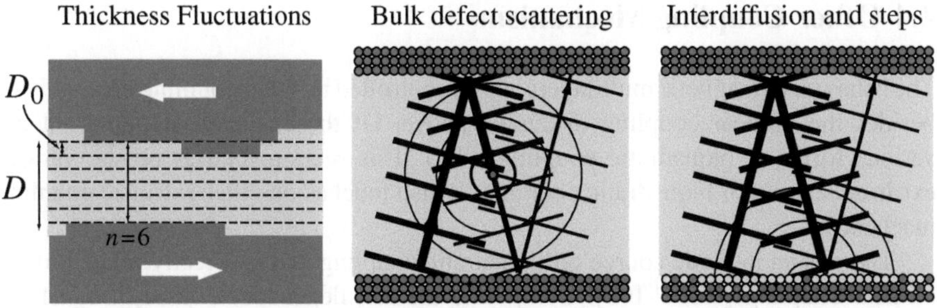

Fig. 4.7. Schematic types of disorder. The *left panel* shows a multilayer with thickness fluctuations in the spacer layer. Several thicknesses are indicated, D_0 is the thickness of an individual layer, $D = 6.5\ D_0$ is the average thickness of the spacer layer, and $n = 6$ is the number of layers at one point. The *middle panel* shows an electron wave reflecting from both interfaces. Scattering from a bulk defect reduces the amplitude, reducing the interference and hence the amplitude of the quantum well state. The *right panel* shows a similar process for a defect located at one of the interfaces

has not. Calculations that treat interfacial disorder typically use either a periodic array of defects or the coherent potential approximation.

Inspection of the asymptotic formula, (4.17) can give a qualitative understanding of the effects of interdiffusion and defect scattering. When the interdiffusion is localized to the interface, the properties of the spacer layer do not change, so only the reflection amplitudes are changed. If the interdiffused atoms are far apart, they mainly contribute to the incoherent scattering at the expense of the coherent scattering. This process reduces the reflection amplitudes. The reduction can depend strongly on spin [4.98] and even on which part of the Fermi surface is involved [4.96]. As the intermixing becomes greater, the interface starts to behave more like an averaged material and the incoherent scattering from each defect can decrease. Bulk defects, on the other hand, affect the properties of the spacer layer, and not the reflection amplitudes (at least to a first approximation). In the asymptotic formula, (4.17), the effect of bulk scattering could be modeled by adding an imaginary part to the critical spanning vector. The imaginary part comes from the reduction of the coherent part of the wave function due to incoherent scattering. One consequence of incoherent scattering of either type is to reduce the importance of multiple trips through the spacer and hence make the higher harmonics neglected in (4.10) less important.

The only type of bulk defect that has been treated theoretically is alloying in the spacer layer. The effect of the defect scattering due to the alloying is usually of secondary interest. The main focus of the these studies is to systematically vary the electronic structure, and hence critical spanning vectors, of the spacer layer to develop a better understanding of the coupling. Comparison between measurements [4.52, 99–103] and calculations [4.101–105] of the variation of the coupling periods as a function of composition gives evidence as to the origin of the coupling. I discuss this point more below for the case of Fe/Cr.

4.4 Other Coupling Mechanisms

The behavior of magnetic multilayers can be controlled by other coupling mechanisms besides the bilinear coupling discussed above. Of these, the most important are various forms of biquadratic coupling, (4.2). This section focuses on the various extrinsic sources of biquadratic coupling. It also touches on other extrinsic coupling mechanisms.

There is an intrinsic source of biquadratic coupling that is directly related to the bilinear coupling [4.106–108]. Computing the full dependence on $\hat{\boldsymbol{m}}_1 \cdot \hat{\boldsymbol{m}}_2$ of the model discussed in Sect. 4.3 rather than just the endpoints, $\hat{\boldsymbol{m}}_1 \cdot \hat{\boldsymbol{m}}_2 = \pm 1$, reveals that the coupling depends on higher orders of the angle between the magnetizations. It turns out however, that the higher order terms are much smaller than the bilinear term, $n = 1$, and are generally negligible. Further, the biquadratic contribution that comes from this expansion oscillates between favoring collinear and perpendicular alignment of the magnetizations. In experiment, biquadratic coupling appears to always favor perpendicular alignment. This discrepancy, and the small size of the intrinsic biquadratic coupling argue in favor of an extrinsic origin for observed biquadratic coupling.

4.4.1 Thickness-fluctuation Biquadratic Coupling

Typically, intralayer exchange coupling prevents the magnetization direction in the ferromagnetic layers from rotating too rapidly in space. Thus, in the presence of variations in the spacer layer thickness, the bilinear coupling gets averaged over the growth front as described in (4.20). However, the intralayer exchange is not infinite. The finite intralayer exchange coupling gives a second consequence of thickness variations – biquadratic coupling.

In this section, I present a version of the result due originally to Slonczewski [4.11, 33] demonstrating the generation of biquadratic coupling by thickness variations. The basic idea is quite simple. The biquadratic coupling is an effective coupling due to fluctuations in the layer magnetizations around their average directions. There are regions of the multilayer with different coupling strengths (due to different thicknesses). Since the intralayer exchange coupling is strong enough that the layers each have well-defined average magnetization directions, each average direction is determined by the coupling, the external field, and whatever anisotropies are present. With the average directions so determined, regions where the coupling would tend towards greater alignment between the magnetizations balance the regions where the coupling would tend towards less alignment. However, the intralayer exchange coupling is not so strong that the system cannot lower its energy by allowing the magnetization in each region to rotate slightly away from the average towards its preferred direction. It turns out that these fluctuations give the greatest energy gain when the magnetizations are perpendicular and no gain when they are collinear. Thus, these fluctuations give an effective interaction between the average magnetization directions that favors perpendicular alignment of the average directions.

In the presence of thickness variations, the interlayer exchange coupling $J(\boldsymbol{R})$ depends on the position $\boldsymbol{R}$ in the plane of the interface. The average bilinear coupling is the average of the local bilinear coupling

$$J_1 = \frac{1}{A} \int \mathrm{d}^2 R J(\boldsymbol{R}) \,, \tag{4.21}$$

where A is the area. The variations in the coupling give an effective biquadratic coupling between the average magnetizations with the strength [4.109]

$$J_2 = -\frac{1}{4A} \int \frac{\mathrm{d}^2 K}{(2\pi)^2} \frac{J(\boldsymbol{K}) J(-\boldsymbol{K})}{A_{\mathrm{ex}} K} \,, \tag{4.22}$$

where A_{ex} is the intralayer exchange coupling constant, $K = |\boldsymbol{K}|$, and $J(\boldsymbol{K})$ is the Fourier transform of the interlayer exchange coupling constant

$$J(\boldsymbol{K}) = \int \mathrm{d}^2 R \, \mathrm{e}^{\mathrm{i}\boldsymbol{K}\cdot\boldsymbol{R}} (J(\boldsymbol{R}) - J_1) \,. \tag{4.23}$$

Equation (4.22) can be more easily understood in simple limits.

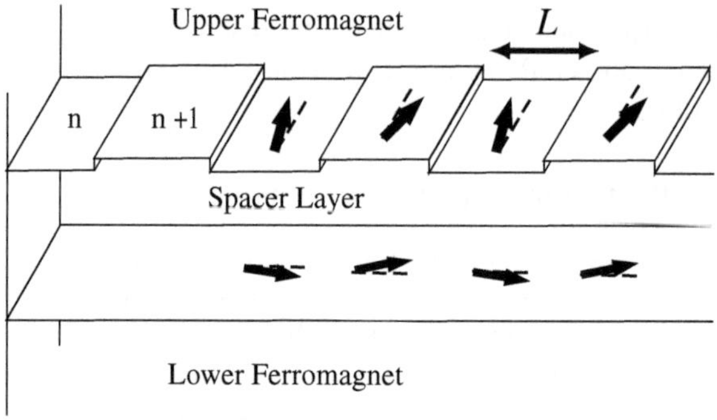

Fig. 4.8. Thickness variations and biquadratic coupling. For a trilayer in which the thickness of the spacer layer varies periodically between n and $n+1$ layers, the heavy arrows show the local variation in the magnetization direction. In the regions with an n-layer thick spacer, in which the coupling is taken to be antiferromagnetic, the magnetizations rotate a little away from each other. For the $n+1$-layer thick regions, in which the coupling is taken to be ferromagnetic, they rotate toward each other

Consider the simple model in which the spacer layer consists of parallel strips of width L with alternating thicknesses and hence coupling strengths J^n and J^{n+1}, see Fig. 4.8. For this model, the relative angle of the magnetizations is $\theta = \theta_0 + \delta\theta \sin(\pi x/L)$, where $\delta\theta$ is the size of the fluctuations. Over the region from 0 to L, where the coupling is J^n, the energy change due to the fluctuations is proportional to $J^n \delta\theta$. Over the region from L to $2L$, the sine function changes sign and the energy change due to the fluctuations is proportional to $-J^{n+1}\delta\theta$. The net coupling energy per area due to the fluctuations is proportional to $-\Delta J\, \delta\theta$, where $\Delta J = J^n - J^{n+1}$. Fluctuations in the right direction lower the energy of the system. The energy gain is balanced by the cost in intralayer exchange energy because the magnetization now varies spatially. Since the intralayer exchange coupling depends on the square of the gradient of the magnetization, for this simple model, it is proportional to $(A_{\rm ex}/L)\,\delta\theta^2$. Combining the changes due to the fluctuations for the interlayer exchange coupling and the intralayer exchange and finding the minimum with respect to the amplitude of the fluctuations gives $\delta\theta \propto -\Delta J/(A_{\rm ex}/L)$. For this fluctuation amplitude, the change in the energy per area due to the fluctuations gives the strength of the biquadratic coupling

$$J_2 \sim -\frac{(\Delta J)^2 L}{A_{\rm ex}} \,. \tag{4.24}$$

This form is also found from (4.22) by inserting the specific form of $J(\boldsymbol{R})$.

While (4.24) is quite simple, its qualitative features generalize to more realistic situations. In real systems, the stripes are replaced by arbitrary-shaped terraces of different thicknesses, still, there is usually a characteristic length scale L that determines the biquadratic coupling. The biquadratic coupling strength increases when this length scale increases because the fluctuations can get larger. However, if the terraces get too large, the fluctuations get large enough that the average direction loses its meaning. Also, more than two thicknesses are generally present in the growth front, but that just introduces an effective ΔJ. The coupling increases as the difference in the coupling for the different terraces get larger. The differences tend to be largest when the coupling is oscillating rapidly, that is when there is short period coupling. The coupling gets weaker as the intralayer exchange interaction increases because exchange suppresses the fluctuations in magnetization direction that lower the energy.

The form for the biquadratic coupling in (4.22) has been derived for semi-infinite ferromagnetic layers. In the opposite limit, in which the thicknesses of the upper and lower ferromagnetic layers, $t_{\mathrm{U/L}}$, are thinner than the length scale of the important fluctuations, L, an additional factor of $K^{-1}(t_{\mathrm{U}}^{-1} + t_{\mathrm{L}}^{-1})/2$ should be included in the integrand of the coefficient of the biquadratic coupling. The biquadratic coupling gets stronger as the films become thinner because the exchange-energy penalty gets smaller, allowing the fluctuations to get larger. In this limit, the result for the simple model, (4.24) gets an additional factor of $(L/t_{\mathrm{U}}) + (L/t_{\mathrm{L}})$.

While short period oscillations in the bilinear coupling favor the generation of biquadratic coupling, short period coupling is not necessary. In fact, the bilinear coupling does not even need to change sign to give a biquadratic contribution to the coupling. This last point may seem surprising, because on first glance it does not appear that there is any frustration in the system. However, when the applied field is such that the magnetizations are not collinear, there is still a balance between regions that would rotate closer to perpendicular and those that would rotate further away.

Finally, to be in the limit that the average magnetization direction is well defined, the biquadratic coupling must be weaker than the absolute value of the unaveraged bilinear coupling. However, the biquadratic coupling can be stronger than the *averaged* bilinear coupling when couplings of different sign are present in the growth front, because the effective bilinear coupling is averaged over the growth front.

Strong evidence for this mechanism of biquadratic coupling comes from wedged Cr spacer layers on Fe whiskers [4.110]. Cr has a short period bilinear coupling that is almost commensurate with the lattice. There is a node in the short period coupling when the number of layers is $n_0 \approx 24$ where the coupling strength behaves roughly like $J_n \sim (n - n_0)(-1)^n$ [4.30, 110]. In samples that show these short period oscillations, regions of parallel alignment are separated from regions of antiparallel alignment by regions of perpendicular alignment. If we assume perfect layer by layer growth, the average bilinear coupling varies linearly between complete layers. Somewhere between the two thicknesses of complete layers the coupling goes through zero. Near this thickness, the biquadratic coupling dominates the averaged bilinear coupling. The width of this region is roughly determined by the strength of the biquadratic coupling at the halfway point divided by the slope of the linearly varying bilinear coupling. For bilinear coupling that goes to zero near the node, the width of the biquadratically coupled regions goes to zero linearly around the node in the coupling, $W \sim |n - n_0|$.

If the sample is grown at a lower temperature, the thickness fluctuations are greater and the short period bilinear coupling is obscured. A long period coupling reveals itself. In the regions where the long period coupling goes through zero, the average bilinear coupling also varies linearly, but with a slope that is not related to the short period coupling. In this case, the width of the transition region goes to zero quadratically near the node in the otherwise obscured bilinear coupling, $W \sim (n - n_0)^2$. Within experimental uncertainty, both the linear and quadratic variation of the transition widths are observed in the appropriate samples [4.110].

4.4.2 Pin-hole Coupling

The simplest coupling that competes with the interlayer exchange coupling is the coupling due to the presence of pinholes. A pinhole, in this context, is a break in the spacer layer giving direct exchange coupling between the two ferromagnetic layers. Since there is direct contact between the ferromagnetic layers, pin-hole coupling is ferromagnetic. Historically, early attempts to separate ferromagnetic layers by a non-magnetic layer were frustrated by the presence of pinholes. It was the observation of antiferromagnetic coupling in 1986 [4.1] that gave convincing evidence that some coupling besides pin-holes was dominant. Even when pin-hole coupling does not dominate the bilinear coupling, it can cause biquadratic coupling when the bilinear coupling is antiferromagnetic [4.111]. This mechanism is closely related to the thickness-fluctuation-induced biquadratic coupling, but requires an appropriate distribution of pinholes rather than a distribution of thicknesses with appropriate couplings.

4.4.3 Magnetostatic Coupling

Magnetostatic interactions make two types of contributions to the properties of magnetic multilayers, macroscopic and microscopic. The main macroscopic contribution is a demagnetizing factor that gives a strong anisotropy to keep the magnetizations in plane. Since other macroscopic magnetostatic effects are negligible on thin film samples, all macroscopic magnetostatic effects are typically replaced by an effective uniaxial anisotropy that pushes the moments in plane. On the other hand, in patterned thin film samples, with reduced lateral dimensions, the macroscopic magnetostatic interactions can become quite important.

There are two coupling mechanisms that arise from magnetostatic interactions on a microscopic scale. Both are related to the roughness of the thin films. A ferromagnetic coupling, originally described by Néel [4.112], is frequently called "orange peel" coupling [4.113–117]. More recently, Demokritov et al. [4.118] demonstrated that roughness and magnetostatic interactions can give a biquadratic coupling. Both of these couplings come from the fringing fields that exist outside the surface due to roughness.

When a surface is rough, there are magnetic poles at the surface because the intralayer exchange is strong enough to prevent the magnetization from rotating and following the surface profile. A useful limit, at least from a pedagogical point of view, is where the roughness of the interfaces is slowly varying and much smaller in amplitude than the separation of the interfaces. Slowly varying implies that the local surface normal is always close to the average surface normal. Then, it is possible to replace the magnetostatics of the rough interfaces by a distribution of magnetic "charges" on a flat surface

$$\sigma(\boldsymbol{R}) = \boldsymbol{M} \cdot \boldsymbol{n}(\boldsymbol{R}) \ . \tag{4.25}$$

Here, the magnetization $\boldsymbol{M}$ is assumed to be uniform, and the local normal to the interface $\boldsymbol{n}$ varies with local (two-dimensional) position $\boldsymbol{R}$ on the surface. This surface charge density describes a flat approximation for a rough interface. By analogy with electrostatics, the interaction between two such interfaces separated by a non-magnetic spacer layer of thickness D is

$$E = \mu_0 \int \mathrm{d}^2 R \int \mathrm{d}^2 R' \frac{\sigma_{\mathrm{U}}(\boldsymbol{R})\sigma_{\mathrm{L}}(\boldsymbol{R}')}{\sqrt{D^2 + (\boldsymbol{R} - \boldsymbol{R}')^2}} \ , \tag{4.26}$$

where U(L) denotes the upper (lower) interface. Orange peel coupling arises in situations where the spacer layer has a uniform thickness. In this case, frequently described as having conformal or correlated roughness, the interface normals are locally opposite, $\boldsymbol{n}_{\mathrm{U}} = -\boldsymbol{n}_{\mathrm{L}} = \boldsymbol{n}$. When there is no preferred direction to the roughness, the interaction becomes [4.109]

$$J_1 = \frac{\mu_0 M_{\mathrm{s}}^2}{4\pi} \int \mathrm{d}^2 K \, |\boldsymbol{n}_{\perp}(\boldsymbol{K})|^2 \, \frac{\mathrm{e}^{-KD}}{K} \ , \tag{4.27}$$

where $\boldsymbol{n}_{\perp}(\boldsymbol{K})$ is the Fourier transform of the part of the local normal that is transverse to the average normal. Since the magnetizations are assumed to be in-plane, only the transverse component contributes. This bilinear coupling is the orange-peel coupling described by Néel.

Equation (4.27) can be made more intuitive by considering the case of sinusoidally varying interfaces, $z_{\mathrm{L}} = \delta \cos(2\pi x/L)$ and $z_{\mathrm{U}} = D + \delta \cos(2\pi x/L)$, see Fig. 4.9. When the magnetization in one layer is perpendicular to the corrugation, i.e. along $\hat{\boldsymbol{x}}$ in this case, there is ferromagnetic coupling to the other layer with strength

$$J_1 \sim \mu_0 M_{\mathrm{s}}^2 \frac{\delta^2}{L} \mathrm{e}^{-2\pi D/L} \ . \tag{4.28}$$

When the magnetization in one layer is along the corrugation, i.e. along $\hat{\boldsymbol{y}}$, rather than perpendicular, there is no coupling to the other layer. The dependence on the absolute directions of the magnetizations, rather than just the relative orientation is a consequence of directionality of the roughness in this simple model. Fully isotropic roughness gives a true ferromagnetic bilinear coupling. Since this coupling depends exponentially on the thickness of the spacer layer, the dominant roughness will tend to have periods somewhat greater than the spacer layer thickness. As the roughness increases, the coupling energy increases quadratically, just as the electrostatic interaction between two equal charges increases as the square of the charges. Both the general expression and the model result break down when the thickness of the spacer layer goes to zero because the roughness is no longer small compared to the spacer layer thickness.

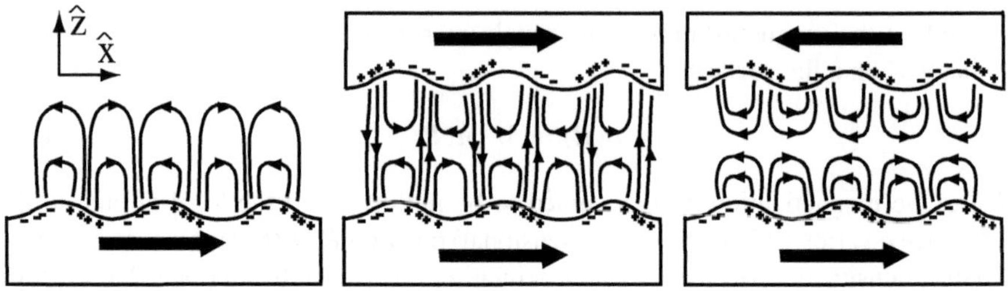

Fig. 4.9. Orange peel coupling from correlated roughness. The *left panel* shows the fringing field outside a rough surface of a material with a uniform magnetization. The pluses and minuses are the effective magnetic "charges." The *middle panel* shows how the fringing fields change in the presence of another interface with correlated roughness for the case of parallel magnetizations. The *right panel* shows that for antiparallel magnetizations, none of the field lines cross the center line, raising the field energy

When the roughness is uncorrelated, the orange-peel coupling is zero, but there is still a microscopic magnetostatic contribution to biquadratic coupling. In the limit that the magnetizations are uniform, the energy is independent of the relative magnetization directions when the roughness is uncorrelated and there is no coupling. However, when the intralayer exchange coupling is finite so that the magnetization directions can fluctuate, there will be biquadratic coupling. Above, I argued that when thickness fluctuations give spatial variations in the bilinear coupling, fluctuations in the magnetization direction can lower the energy. Here also, fluctuations in the magnetization direction can lower the energy. In both cases, the fluctuations lower the energy the most when the two magnetizations are nominally perpendicular.

Above a rough surface, magnetic charges give rise to a magnetic field which couples to the magnetization of the upper film. The magnetic field due to roughness in the lower interface is

$$\boldsymbol{B}(\boldsymbol{r}) = \mu_0 \int \frac{\mathrm{d}^2 K}{(2\pi)^2} \frac{\boldsymbol{k}}{iK} \mathrm{e}^{\mathrm{i}\boldsymbol{K}\cdot\boldsymbol{R}} \mathrm{e}^{-Kz} \int \mathrm{d}^2 R' \, \mathrm{e}^{-\mathrm{i}\boldsymbol{K}\cdot\boldsymbol{R}'} \sigma_{\mathrm{L}}(\boldsymbol{R}') \,, \tag{4.29}$$

where $\boldsymbol{K}$ is a two-dimensional wave vector in the plane of the interface and $\boldsymbol{k} = (\boldsymbol{K}, i|\boldsymbol{K}|)$ is a complex three dimensional wave vector. The interaction between the rough lower layer and a semi-infinite upper layer is

$$E = -\int \mathrm{d}^2 R \int_D^{\infty} \mathrm{d}z \boldsymbol{M}_{\mathrm{U}}(\boldsymbol{r}) \cdot \boldsymbol{B}_{\mathrm{L}}(\boldsymbol{r}) \,. \tag{4.30}$$

Note that this interaction is not between the charges on the two interfaces, but between the charges on one interface and the whole volume of the other layer. Responding to variations in $\boldsymbol{B}_{\mathrm{L}}(\boldsymbol{r})$, fluctuations in $\boldsymbol{M}_{\mathrm{U}}(\boldsymbol{r})$ lower the magnetostatic energy at the expense of raising the exchange energy. The net lowering of the energy due to the fluctuations is greatest when the magnetizations are perpendicular. Assuming that the roughness has no preferred direction, the resulting biquadratic coupling has the form [4.109]

$$J_2 = -\frac{\mu_0^2 M_s^4}{64 A_{ex} A} \int \frac{d^2 K}{(2\pi)^2} \frac{e^{-2KD}}{K^3} |\boldsymbol{n}_\perp(\boldsymbol{K})|^2 \ . \tag{4.31}$$

This energy gives a biquadratic coupling that favors perpendicular alignment of upper and lower magnetizations. If the upper surface is rough, there will be an equivalent contribution from the magnetic poles in the upper interface coupling to the lower layer.

For a simple model with one corrugated interface, $z_L = \delta \cos(2\pi x/L)$, and one smooth interface $z_U = D$, see Fig. 4.10, (4.31) gives

$$J_2 \sim -\frac{\mu_0^2 M_s^4 L \delta^2}{A_{ex}} e^{-4\pi D/L} \ , \tag{4.32}$$

provided the lower magnetization is perpendicular to the corrugation. This interaction favors very long period roughness. When the ferromagnetic films have finite thicknesses, this contribution to the biquadratic coupling becomes weaker, the opposite behavior from that found for the biquadratic coupling in (4.24). In the latter case, the thinner ferromagnetic films allow the fluctuations to increase because the driving force for the fluctuations acts only at the interface. On the other hand, in (4.31) the driving force for the fluctuations acts throughout film, so that making the film thinner reduces the net effect of the fluctuations.

Evidence for this mechanism of biquadratic coupling comes from measurements of its thickness dependence [4.119] in Fe/Ag/Fe samples grown on GaAs. The authors compared the experimental results with a model equivalent to (4.31) and found good agreement using roughness parameters consistent with the measured roughness. Additional evidence comes from the observation that for most wedges grown on Fe whiskers, biquadratic coupling dominates in the thickest parts of the wedge [4.120]. Here, other mechanisms are expected to become weaker as spacer layers become

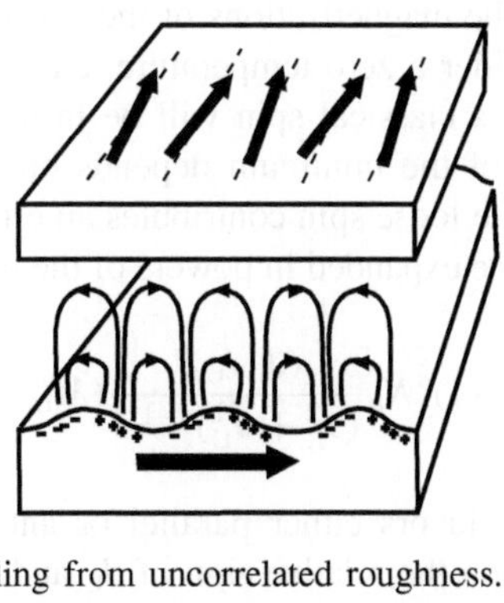

Fig. 4.10. Biquadratic coupling from uncorrelated roughness. The fringing fields from the bottom, corrugated surface couple to the perpendicular magnetization in the smooth upper layer giving small amplitude oscillations in the magnetization direction

thicker. The magnetostatic coupling is dominated by long wavelength roughness making the exponential damping less effective at limiting the coupling. For thick spacer layers, where other interactions have become small, this form of biquadratic coupling is likely to be the dominant interaction.

A study of an amorphous spacer layer showed evidence for both orange-peel coupling and magnetostaic biquadratic coupling [4.121]. For thin layers, the magnetizations of the two layers are parallel in the ground state, but as the thickness of the spacer increases, the alignment crosses over to perpendicular. The authors were able to fit the results to a combination of these two coupling measurements using a roughness consistent with that measured by scanning tunneling microscopy.

4.4.4 Loose Spins

Many measurements of biquadratic coupling show a very strong temperature dependence. On the other hand, the bilinear coupling and the magnetization generally vary slowly with temperature, so that inserting their temperature dependence into the expressions for biquadratic coupling, (4.22) and (4.31), gives only weak dependence. Slonczewski [4.122] has developed a "loose spin" mechanism that does show a strong temperature dependence. While the argument that the other types of coupling do not show a strong temperature dependence may be overly simplistic, the loose spin model has been established in samples constructed specifically to test for it.

To produce a strong temperature dependence, Slonczewski postulates the existence of magnetic moments (spins) that are only weakly coupled to the magnetic layers. There is no direct exchange coupling, only the indirect exchange coupling mediated by the spacer layer electrons. These weakly coupled spins are paramagnetic. However, they see a potential that is anisotropic due to their coupling to the magnetic layers

$$V(\hat{\boldsymbol{m}}) = J_{\mathrm{L}}\widehat{\boldsymbol{M}}_{\mathrm{L}} \cdot \hat{\boldsymbol{m}} + J_{\mathrm{U}}\widehat{\boldsymbol{M}}_{\mathrm{U}} \cdot \hat{\boldsymbol{m}} \,, \tag{4.33}$$

where L (U) refers to the lower (upper) layer. Note that the potential depends on the relative orientation of the magnetizations of the two layers. To see the origin of biquadratic coupling, consider a zero-temperature, classical treatment of the loose spin. At zero temperature, a classical spin will be in its minimum energy configuration. Since the energy of the minimum depends on the relative orientation of the layer magnetizations, the loose spin contributes an effective interaction between magnetizations which can be expanded in powers of the relative orientation

$$J = \sum_{n=1}^{\infty}(-1)^n N_n \frac{(J_{\mathrm{L}}J_{\mathrm{U}})^n}{(J_{\mathrm{L}}^2 + J_{\mathrm{U}}^2)^{n-1/2}}(\widehat{\boldsymbol{M}}_{\mathrm{L}} \cdot \widehat{\boldsymbol{M}}_{\mathrm{U}})^n \,. \tag{4.34}$$

The bilinear term, $n = 1$ favors either parallel or antiparallel alignment of the magnetizations depending on the relative sign of J_{L} and J_{U}. The biquadratic piece always favors perpendicular alignment. The numerical factors N_n are such that the bilinear term always wins. A single loose spin always gives a dominant bilinear coupling. However, for a collection of loose spins, the signs of J_L and J_U can vary. In this case, the bilinear contributions may cancel out, leaving a dominant biquadratic coupling.

At finite temperature, the fraction of the time each spin spends close to its minimum energy orientation depends strongly on temperature. Its paramagnetic behavior gives a strong temperature dependence to the coupling, (4.34). At temperatures high compared to the potential minimum, the spin will have an isotropic orientation distribution, and the effective coupling mediated by the loose spin will go to zero. Since the loose spin is not directly coupled to the magnetic layer, the temperature scale of the coupling will be much lower than the Curie temperature of the magnetic layers.

The existence of this coupling mechanism has been tested in experiments in which magnetic atoms have been intentionally introduced at low densities into the spacer layer [4.123–125]. Agreement with the loose spin model was found in [4.125]; this experiment used the lowest density of loose spins. The authors speculate that the higher densities used in the other publications obscured the behavior expected from the model.

4.4.5 Torsion Model

If the spacer layer is antiferromagnetic, none of the models discussed above are sufficient to describe the coupling. For ideal multilayers, the coupling is simply the result of the direct coupling from layer to layer in the antiferromagnet. However, it is impossible to avoid thickness fluctuations so the coupling will be frustrated. Slonczewski introduced the "torsion model" [4.11] to describe frustrated antiferromagnetic spacer layers.

The torsion model is based on the following model of the behavior for an ideal system of two ferromagnetic layers separated by an antiferromagnetic spacer. All moments are assumed to remain in plane, and can be described by an azimuthal angle ϕ_j. When the relative directions of the magnetizations in the ferromagnets are rotated way from their minimum energy configuration, a spiral winds up in the antiferromagnet. For an $n+1$ layer spacer, taking n to be even for the present, the minimum energy angles for the ferromagnetic layer are $\phi_L = \phi_U$. When the magnetization of the top layer is rotated through an angle $\Delta\phi$, the magnetizations in the jth layer in the antiferromagnet are

$$\phi_j = j\frac{\Delta\phi}{n+2} + \phi_j^{AF} \ . \tag{4.35}$$

The first term gives the spiraling of the magnetization, and the second term, $\phi_j^{AF} = 0,\ \pi$ for j even or odd, gives the antiferromagnetic reversal from layer to layer. I have assumed that the exchange interaction between the ferromagnetic layer and the antiferromagnetic layer is the same as the exchange between layers in the antiferromagnet. The exchange energy due to the spiral structure is $-(n+2)J_{AF}\cos(\Delta\phi/(n+2))$. When n is large enough, a small angle expansion of this energy gives a constant term plus

$$E_{n+1} = \frac{J_{\rm AF}}{2(n+2)} (\Delta\phi)^2 \,. \tag{4.36}$$

For an n layer spacer, the minimum energy configuration has $\phi_{\rm L} = \phi_{\rm U} + \pi$. If the magnetization of the upper layer is rotated so that it makes an angle $\Delta\phi$ with respect to magnetization of the lower layer, the energy is

$$E_n = \frac{J_{\rm AF}}{2(n+1)} (\pi - \Delta\phi)^2 \,. \tag{4.37}$$

When n is large enough that this model does not break down, both of these energies vary quadratically in the rotation angle rather than as the cosine of the rotation angle.

If the growth front has areas of thickness n and $n+1$, the coupling will be frustrated. In terms of the fractional area of thickness n, θ_n, the net coupling energy is

$$J = \theta_{n+1} \frac{J_{\rm AF}}{2(n+2)} (\Delta\phi)^2 + \theta_n \frac{J_{\rm AF}}{2(n+1)} (\pi - \Delta\phi)^2 \,. \tag{4.38}$$

The quadratic dependence leads to different behavior in the presence of thickness fluctuations than was found for the cosine dependence of the usual bilinear coupling. For the case of bilinear coupling, the net bilinear coupling can dominate the biquadratic coupling giving parallel or antiparallel alignment. Such collinear alignment will persist up to some finite amount of interfacial roughness. In the torsion model, any roughness leads to non-collinear coupling, although the degree of non-collinearity can be small.

The torsion model is based on the assumption that nearest-neighbor direct exchange adequately describes the coupling in the antiferromagnet and the coupling of the antiferromagnet to the ferromagnet. It has been shown [4.126] that it is necessary to include many distant neighbors to describe the spin-wave spectrum of transition metal ferromagnets. It is also clear that such a description is not adequate for an incommensurate antiferromagnet like Cr [4.110, 127]. On the other hand, this model appears to describe Mn spacer layers well [4.128–134] even though a simple nearest-neighbor-exchange model may not be valid.

4.5 Specific Systems

In this section I discuss selected systems that illustrate many of the important issues in interlayer exchange coupling. For a more complete comparison between theory and experiment, see [4.13]. For a comprehensive compilation of experimental and theoretical results, see [4.15]. Both those reviews emphasize transition metal systems. For reviews of work on rare earths, see [4.135, 136].

4.5.1 Co/Cu

One of the most extensively studied system is Co/Cu, particularly the (100) orientation. It exhibits many of the difficulties encountered when comparing theory and experiment. In spite of these difficulties, a general consensus of the theoretical behavior has emerged based on total energy calculations as well as asymptotic (and preasymptotic) analyses [4.67, 69, 73, 74, 81, 82, 137–140]. As is known from the analysis of the experimental Fermi surface [4.40], there are two different critical spanning vectors, one associated with the interface zone center that gives a long period oscillation, and one associated with the necks that gives a short period. In the asymptotic regime, the long period oscillation is weak and the short period oscillation is relatively strong. However, for smaller thicknesses, where most experimental measurements are made, preasymptotic corrections become very important and change the asymptotic results quite dramatically.

At the critical point for the long period coupling, the reflection amplitudes for both spins are quite small, hence the weak asymptotic coupling. However for parallel wave vectors *close* to the critical point, the minority-electron reflection becomes close to unity, see Fig. 4.11. As mentioned above in the discussion following (4.13), the parallel wave vectors around the critical point within a region $\approx \kappa/D$ contribute to the coupling. Thus as D becomes smaller, the regions with strong reflection contribute to the long period oscillation with preasymptotic corrections, which decay as higher powers of D^{-1}. For thin spacer layers, the long period coupling strength is found to be substantial. In addition, since the parts of the Fermi surface making the dominant contribution to the oscillatory coupling have shorter spanning vectors, the apparent period is changed for thin spacer layers as well. Such variation in the periods are generally present, but become more important when the asymptotic coupling is weak and the preasymptotic corrections are significant.

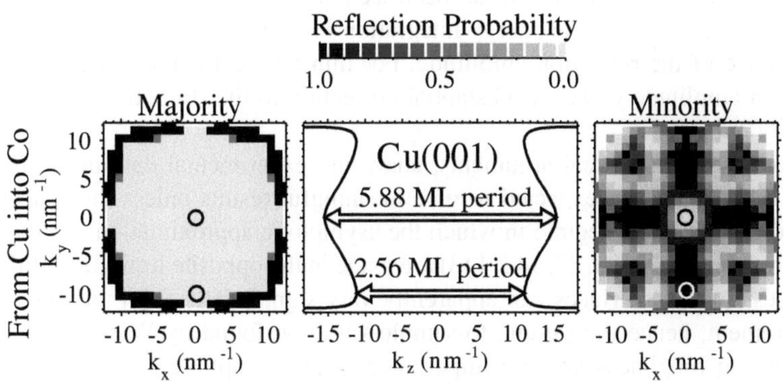

Fig. 4.11. Spin-dependent reflection from Co/Cu(001). The *left* and *right panels* show the Fermi surface of Cu shaded based on the reflection probability from Co for majority and minority electrons respectively. The shading scale is given at the top. The Fermi surfaces are projected into the Interface Brillouin zone. The *middle panel* shows a slice through the Fermi surface and indicates the critical spanning vectors. The projection of the critical spanning vectors are indicated by *circles* in the left and right panels

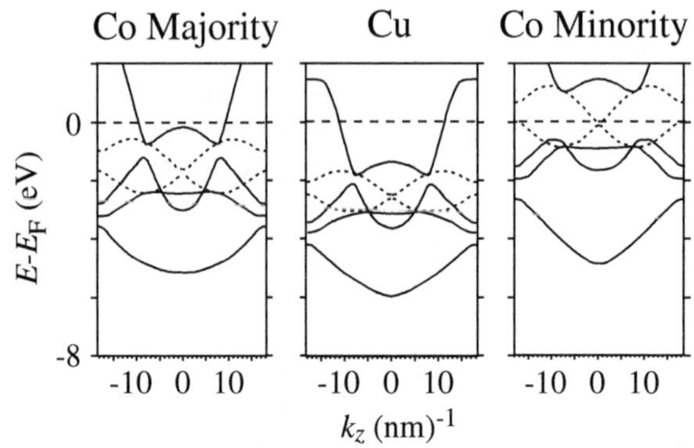

Fig. 4.12. Band structures of Co and Cu. The band structures are shown along the line in the (100) direction through the critical point for the short period coupling. States along this line are either even (*solid lines*) or odd (*dotted lines*) with respect to a mirror plane. The two do not couple to each other during reflection and transmission. Thus, electrons at the Fermi energy in Cu see a symmetry gap if they have minority spin, and completely reflect. The phase of the reflection amplitude varies by π from the bottom to the top of the gap. Since the gap is less than 2 eV, the phase variation is substantial

Different preasymptotic corrections are important for the short period oscillation in the coupling. At the necks, the reflection probability for the minority electrons is quite strong because there is a symmetry gap for the states in the Co with the symmetry of those on the Cu Fermi surface. However, the gap is fairly narrow in energy. Since the phase of the reflection amplitude changes by π as a function of energy when going from the bottom of a gap to the top, the reflection amplitude (but not the reflection probability) has a strong energy dependence. Evidence for this rapid variation of the phase of the reflection amplitude has been seen in photoemission studies of Cu on Co [4.141]. In carrying out the energy integration in (4.9), I have neglected the energy dependence of the reflection amplitude. For thin layers, the phase variation of the reflection amplitude makes a substantial correction to the strength of the coupling, typically reducing it.

From these theoretical arguments, analyzing experimental data in terms of the form (4.17) cannot be expected to give meaningful results unless the spacer layer thicknesses are in the regime in which the asymptotic approximation is valid. Most measurements [4.22–24, 87, 142, 143] are made in the opposite limit. In addition, the roughness of the interfaces can apparently vary significantly from measurement to measurement, hence the sample-to-sample variation found by Weber et al. [4.87]. The varying roughness and the importance of preasymptotic corrections explains the wide variation in the measured periods for the long period oscillation given in Table 4.1. Due to the unmeasured effect of interface roughness, it is not surprising that the measured values of the coupling strength are at least a factor of three smaller than the equivalent theoretical results.

4.5.2 Au/Fe and Ag/Fe

As shown in Table 4.1, the periods measured for Au/Fe and Ag/Fe trilayers [4.7, 86] grown on Fe whiskers are in remarkable agreement with those predicted from the experimental Fermi surface [4.40]. To measure the coupling strengths, Unguris et al. [4.144] used MOKE to measure the coupling strength for Au/Fe on similar samples. The samples were first characterized in zero field by SEMPA, including a measurement of the RHEED intensity oscillations to determine the width of the growth front. Using the measured width of the growth front in their analysis, they were able to correct for the averaging of the coupling due to roughness, (4.20). Fitting the measured coupling strength to the asymptotic form (4.17), they determined the coupling strengths [4.145], $J^{\mathrm{S}}/(1\ \mathrm{nm})^2 = 1.29\,\mathrm{mJ/m^2} \pm 0.16\,\mathrm{mJ/m^2}$ and $J^{\mathrm{L}}/(1\ \mathrm{nm})^2 = 0.18\,\mathrm{mJ/m^2} \pm 0.02\,\mathrm{mJ/m^2}$. The coupling strength has been measured for Ag/Fe [4.146], but has not been analyzed in terms of the asymptotic form.

There have been far fewer calculations of the coupling for Au/Fe than for Co/Cu [4.74, 96, 147, 148]. Calculations, which predated the experimental results, of the asymptotic coupling strengths [4.74] gave $2.0\,\mathrm{mJ/m^2}$ and $1.1\,\mathrm{mJ/m^2}$ for the short and long periods respectively. The short period coupling strength is within a factor of two the experimental result, but the long period coupling is about a factor of six larger. Recent total energy calculations [4.96] have cast doubt even on this agreement for the short period coupling. These calculations were analyzed, like the experiment, to extract the effective asymptotic coupling strengths. The extracted values [4.149], $3.4\,\mathrm{mJ/m^2}$ and $1.1\,\mathrm{mJ/m^2}$ were considerably larger than the experiment for both periods.

There are several possible reasons for the discrepancy in the calculations of the short period coupling strength. At the critical point for the short period, the reflection amplitudes are changing rapidly, which introduces two possible reasons for disagreement. As there has not been an analysis of the preasymptotic corrections for this system, it may be the total energy calculations are not in the asymptotic limit. More likely, the small errors in the Fermi surface as computed in the layer matching calculation [4.74] may give much larger errors in the coupling due to the rapid variation in the reflection amplitudes. If this explanation is correct, some uncertainty remains is both calculations due to the small errors introduced by the local spin density approximation. Alternatively, the difference may be due to the different treatment of the ferromagnetic layers, semi-infinite for the asymptotic calculation and finite for the total energy calculation. However, total energy calculations [4.96] in which the ferromagnetic thickness is varied at fixed spacer thickness do not show significant variation.

The common explanation for disagreement between theory and experiment, thickness fluctuations, does not apply to the measurements on this system for two reasons. First, they have been accounted for in the extraction of the coupling strengths from experiment. In addition, thickness fluctuations reduce the short period coupling more than the long period coupling. But to bring the theoretical results into agreement with the measured results, the long period strength needs to be reduced more than that of the short period. Optiz et al. [4.96] looked at the effect of interfacial defects on the coupling. They found that Au atoms that are interdiffused into the top Fe layer suppress the long period coupling much more strongly than the short period coupling. Such behavior could bring the theoretical results into better agreement with experiment. Au "floating" on top of Fe during growth has been observed [4.150], making it plausible that there is interdiffusion at the interface.

Other total energy calculations [4.147, 148] for this system have not been analyzed in terms of the asymptotic form making it difficult to compare them with experiment or other calculations. Since the effective periods extracted from [4.148] disagree by up to 10% from the results of [4.96], it is difficult to compare total energies directly, because the calculations get out of phase with each other. The authors find that the calculated peak positions and relative heights agree well (overall heights disagree by an order of magnitude) with those measured for Au/Fe multilayers grown on GaAs [4.29]. However, there is substantial disagreement between the results for samples grown on GaAs and those grown on an Fe whisker.

There has been one calculation of the asymptotic coupling for Ag/Fe [4.74]. Since the experimental results [4.146] have not been analyzed in terms of the asymptotic form, a detailed comparison is difficult. Roughly, the measured and calculated couplings are close in magnitude.

4.5.3 Cr/Fe

The other intensively studied system is Cr/Fe, particularly the (100) orientation. Cr/Fe was the first transition metal multilayer system to show interlayer exchange coupling [4.1], one of the first the show oscillatory interlayer exchange coupling [4.6], and the first to show short period coupling [4.25–27]. Much of the interest In Cr/Fe is due to the close lattice match between Cr and Fe, better than any other transition metal pair (excluding the noble metals discussed above). In addition, Cr/Fe exhibits extremely complicated behavior, much of which arises from the presence of spin-density-wave antiferromagnetism in the Cr. There have been three review articles [4.110, 151, 152] devoted at least in part to this issue. It is interesting to read all three to compare the interpretation of the same symphony by three different conductors.

In Cr, there are a number of competing phases: paramagnetic, commensurate spin-density wave, incommensurate spin-density wave, and helical spin-density wave. The incommensurate spin-density wave can be longitudinal or transverse. The order propagation vector can be perpendicular to the interface or in the plane of the interface. In bulk Cr, the equilibrium phases are (in order of decreasing temperature): paramagnetic, transverse incommensurate spin-density wave, and longitudinal incommensurate spin-density wave. The equilibrium phase is sensitive to alloying and strain as well as coupling to one or more layers of ferromagnetic Fe. It can be very difficult to distinguish between the different phases experimentally.

Taken at face value, many measurements of Cr/Fe are contradictory. However, it has become clear that the differences arise from the sensitivity of the antiferromagnetic order in Cr to the presence of disorder, particularly at the interfaces. It is possible to construct a consistent picture to explain much of the various behaviors, but it is very difficult to check the picture with detailed theoretical calculations.

Approaches, like model free energies [4.153], that are capable of describing the temperature dependence of the antiferromagnetism in bulk Cr are not flexible enough to realistically describe the behavior near defects. Calculations that can microscopically treat the effect of disorder, like tight-binding approximations [4.71, 154, 155] or LSDA calculations [4.156–159], are either not capable of describing the bulk behavior (tight-binding) or far too computationally expensive to treat relevant system sizes including defects (LSDA).

There are several measurements on this system that make it clear that the model discussed in Sect. 4.3 for the interlayer exchange coupling is not adequate to describe the short period coupling in Cr/Fe multilayers. Antiferromagnetism in the Cr has been seen directly by neutron scattering [4.160, 161]. The temperature and thickness dependence of the coupling [4.25, 153] has established that the spin-density wave state plays a role in the coupling for at least some systems. While many of the other measurements can be qualitatively understood, the lack of quantitative test of the explanations leaves much of the behavior open to interpretation. I would expect that experiments that test our understanding of Cr/Fe will give many more surprises. However, such experiments can be quite difficult.

Thickness fluctuations obscure the short period oscillations when samples are grown in less than optimal conditions, frequently revealing a long period coupling [4.25, 26, 162]. The long period coupling is of interest because it is the only well characterized oscillatory coupling in a lattice-matched system with something besides a noble metal as a spacer layer. Cr has a complicated Fermi surface with a large number of critical spanning vectors [4.66, 88, 89, 163]. There have been a number of proposals for the origin of the long period coupling [4.101, 157, 164]. Based on asymptotic coupling calculations, analysis of the Fermi surface measured in de Haas-van Alphen measurements, and studies of alloyed spacer layers, a consensus has formed that the ellipsoids centered at the N-point of the Cr Brillouin zone (see Fig. 4.13) are responsible for the coupling. The periods extracted from the de Haas-van Alphen measurements in Table 4.1 are from the N-centered ellipsoids.

Studies [4.102, 103] of $Cr_{1-x}V_x$ alloy spacer layers are based on the fact that this alloy modifies the electronic structure (and hence the Fermi surface) without introducing so much bulk scattering that the interlayer exchange coupling is eliminated. Then, the evolution of the period of the oscillatory coupling can be compared with the calculated evolution of the Fermi surface. Both of these studies showed that the period decreased with V concentration and that the only part of the Fermi surface that grew in an appropriate manner was the N-centered ellipsoid.

Further evidence favoring the N-centered ellipsoids was found in experiments [4.165] in which a Au layer was inserted at one of the interfaces in an Fe/Cr multilayer. The long period coupling was dramatically reduced relative to the short period coupling. The Au Fermi surface covers most of the Brillouin zone, overlapping most of the critical spanning vectors of Cr Fermi surface, but it does not overlap the critical points for the N-centered ellipsoids. The authors attribute the strong suppression of the long period coupling to the exponential decay of the quantum well states in the gap.

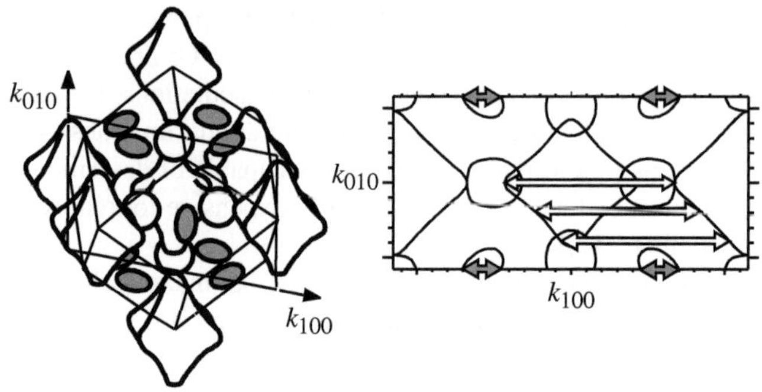

Fig. 4.13. Slice through the Cr Fermi surface. The *left panel* shows a representation of the Cr Fermi surface in its paramagnetic state. The *gray shaded parts* are the ellipsoids centered at the N points. Also indicated is a slice through the Fermi surface, which in turn is shown in the *right panel*. The *gray shaded arrows* are the critical spanning vectors at the N-centered ellipsoids and the *white arrows* indicate the nested parts of the Fermi surface that give rise to the spin density wave antiferromagnetism

The same oscillation period was observed in (112) oriented systems [4.90, 166] as in (100) oriented systems and (110) textured systems grown by sputtering [4.6]. The similarity led to speculation that the oscillation did not originate from Fermi surface properties. However, it turns out that asymptotic coupling calculations, analysis of the Fermi surface measured in de Haas-van Alphen measurements, and studies of alloyed spacer layers all suggest that the N-centered ellipsoids are the origin of the coupling in all three cases.

4.5.4 Fe/Si

While the model presented above has been derived specifically for metallic spacers layers, it has also been extended to insulating or semiconducting spacer layers [4.67]. The oscillatory coupling is replaced by an exponentially decaying coupling. However, because the preasymptotic corrections may be important, it is conceivable that the sign of the coupling may change once or even twice as a function of spacer layer thickness. The idea of studying coupling through an insulating layer started [4.167] a large effort in studying coupling through Si and related spacers layers. Here, related refers to the fact the Fe and Si have a strong propensity [4.168–171] to form silicides. Much of the effort derives from the fact that many different spacers may develop when these materials combine. Because different silicides can form, it can be very difficult to know what spacer layer is being measured.

There have been a large variety of effects measured. Coupling can be induced by light [4.172] or heat [4.173]. In some structures, biquadratic coupling dominates [4.174, 175], while in some structures with FeSi spacer layers, an oscillatory bilinear coupling has been observed [4.176]. In a series of nominally pure Si spacer layers a very strong, exponentially decaying coupling decaying coupling has been measured [4.177], consistent with predictions for insulating spacer layers, which was the original motivation for studying this system. There is apparently still significant disagreement between results in different laboratories and more work needs to be done before these systems are well understood.

4.6 Summary

This review, written from a theorist's point of view, has focused on the physical mechanisms for interlayer coupling in magnetic multilayers. A simple physical picture for interlayer exchange coupling has evolved over the last decade. Spin dependent reflection from the interfaces in multilayers sets up spin-dependent quantum well states. These quantum well states evolve in energy as the thickness of the spacer layer is varied. As these states pass through the Fermi energy, they fill or empty, changing the energy of the multilayer. These changes are periodic because the quantum well states cross the Fermi energy with a period determined by the Fermi surface of the spacer layer material. At critical points of this Fermi surface, many quantum well states have the same period giving a net oscillatory contribution to the energy. Since the reflection is spin dependent, the energy depends on the relative orientation of the layer magnetizations, i.e., there is an energy difference between parallel and antiparallel alignment of the magnetizations. This energy difference is just the interlayer exchange coupling. It has oscillatory contributions with periods determined by the critical spanning vectors of the spacer layer Fermi surface and strengths determined by the spin-dependent reflection at the interfaces.

This picture for the interlayer exchange coupling has been tested to an extent much greater than the related RKKY coupling between magnetic impurities. For some systems, the periods have been measured to 3% accuracy. The measured periods agree with those expected from the critical spanning vectors of the Fermi surfaces measured in de Haas-van Alphen experiments. The coupling strengths have been investigated by a number of theoretical techniques and have been measured in carefully prepared samples. There is no completely satisfactory comparison between theory and experiment because there is no system in which a complete characterization of the structure, including all of the defects, has been made. However, enough is understood about the likely defects and their effect on the coupling to give confidence that the physics is correctly described.

While defects confound our ability to compare theory and experiment for the bilinear coupling, they are the origin of biquadratic coupling. Biquadratic coupling, which derives from several mechanisms, appears to be present to some degree in all magnetic multilayers. Two mechanisms arise from interfacial roughness. In one case interfacial roughness gives rise to fluctuations in the strength of the bilinear coupling. In the other it gives rise to an oscillating magnetic field outside the interface. In both cases, the system can lower its energy by allowing the magnetization of the layers to fluctuate in response to the roughness induced variations. The system can lower its energy the most when the two magnetizations are perpendicular to each other, giving an effective coupling that favors perpendicular alignment of the magnetizations.

The study of interlayer exchange coupling has flourished for the past decade and a half. We have developed a clear physical description of the coupling and have successfully compared its predictions with high quality measurements. While very successful, the comparison between theory and experiment is not complete and interesting issues still remain. Successfully addressing these issues requires measuring the coupling in systems that are as close to perfect as possible and then quantitatively measuring all the defects that remain. On the theoretical side, it requires explicitly treating the measured defects. Both aspects of understanding defects are quite difficult and require substantial effort.

References

4.1. P. Grünberg, R. Schreiber, Y. Pang, M. B. Brodsky, and H. Sowers, Phys. Rev. Lett. **57**, 2442 (1986).
4.2. C.F. Majkrzak, J.W. Cable, J. Kwo, M. Hong, D.B. McWhan, Y. Yafet, J.V. Waszcak, and C. Vettier, Phys. Rev. Lett. **56**, 2700 (1986).
4.3. M.B. Salamon, S. Sinha, J.J. Rhyne, J.E. Cunningham, R.W. Erwin, J. Borchers, and C.P. Flynn, Phys. Rev. Lett. **56**, 259 (1986).
4.4. M.N. Baibich, J.M. Broto, A. Fert, F. Nguyen Van Dau, F. Petroff, P. Etienne, G. Creuzet, A. Friederich, and J. Chazelas, Phys. Rev. Lett. **61**, 2472 (1988).
4.5. G. Binasch, P. Grünberg, F. Saurenbach, and W. Zinn, Phys. Rev. B **39**, 4828 (1989).
4.6. S.S.P. Parkin, N. More, and K.P. Roche, Phys. Rev. Lett. **64**, 2304 (1990).
4.7. J. Unguris, R.J. Celotta, and D.T. Pierce, J. Appl. Phys. **75**, 6437 (1994).
4.8. Chapter 2, in *Ultrathin Magnetic Structures II,* edited by B. Heinrich and J.A.C. Bland (Springer-Verlag, Berlin, 1994) p. 45.
4.9. Y. Yafet, in *Magnetic Multilayers*, ed. L.H. Bennett and R.E. Watson (World Scientific, Singapore 1994), p. 19.
4.10. A. Fert, P. Grünberg, A. Barthelemy, F. Petroff, and W. Zinn, J. Magn. Magn. Mater. **140-144**, 1 (1995).
4.11. J.C. Slonczewski, J. Magn. Magn. Mater. **150**, 13 (1995).
4.12. B.A. Jones, IBM J. Res. Dev. **42**, 25 (1998).
4.13. M.D. Stiles, J. Magn. Magn. Mater. **200**, 322 (1999).
4.14. P. Bruno, J. Phys.-Condes. Matter **11**, 9403 (1999).
4.15. D.E. Bürgler, P. Grünberg, S.O. Demokritov, and M.T. Johnson, in *Handbook of Magnetic Materials, Vol. 13*, ed. K.H.J. Buschow (2001).
4.16. S.S.P. Parkin, Phys. Rev. Lett. **67**, 3598 (1991).
4.17. F. Petroff, A. Barthelemy, D.H. Mosca, D.K. Lottis, A. Fert, P.A. Schroeder, W.P. Pratt, R. Loloee, and S. Lequien, Phys. Rev. B **44**, 5355 (1991).
4.18. J. Fassbender, F. Nörteman, R.L. Stamps, R.E. Camley, B. Hillebrands, and G. Güntherodt, Phys. Rev. B **46**, 5810 (1992).
4.19. K. Ounadjela, D. Miller, A. Dinia, A. Arbaoui, P. Panissod, and G. Suran, Phys. Rev. B **45**, 7768 (1992).
4.20. Y. Huai and R.W. Cochrane, J. Appl. Phys. **72**, 2523 (1992).
4.21. P.J.H. Bloeman, W.J.M. de Jonge, and R. Coehoorn, J. Magn. Magn. Mater. **121**, 306 (1993).
4.22. Z.Q. Qiu, J. Pearson, and S.D. Bader, Phys. Rev. B **46**, 8659 (1992).

4.23. M.T. Johnson, S.T. Purcell, N.W.E. McGee, R. Coehoorn, J. aan de Stegge, and W. Hoving, Phys. Rev. Lett. **68**, 2688 (1992).
4.24. P.J.H. Bloemen, R. van Dalen, W.J.M. de Jonge, M.T. Johnson, and J. aan de Stegge, J. Appl. Phys. **73**, 5972 (1993).
4.25. J. Unguris, R.J. Celotta, and D.T. Pierce, Phys. Rev. Lett. **67**, 140 (1991).
4.26. S. Demokritov, J.A. Wolf, and P. Grünberg, Europhysics Letters **15**, 881 (1991).
4.27. S.T. Purcell, W. Folkerts, M.T. Johnson, N.W.E. McGee, K. Jager, J. aan de Stegge, W.B. Zeper, and W. Hoving, Phys. Rev. Lett. **67**, 903 (1991).
4.28. Z. Celinski and B. Heinrich, J. Magn. Magn. Mater. **99**, L25 (1991).
4.29. A. Fuß, S. Demokritov, P. Grünberg, and W. Zinn, J. Magn. Magn. Mater. **103**, L221 (1992).
4.30. Y. Wang, P.M. Levy, and J.L. Fry, Phys. Rev. Lett. **65**, 2732 (1990).
4.31. M. Rührig, R. Schäfer, A. Hubert, R. Mosler, J.A. Wolf, S. Demokritov, and P. Grünberg, Phys. Status Solidi A **125**, 635 (1991).
4.32. B. Heinrich, J.F. Cochran, M. Kowalewski, J. Kirschner, Z. Celinski, A.S. Arrott, and K. Myrtle, Phys. Rev. B **44**, 9348 (1991).
4.33. J.C. Slonczewski, Phys. Rev. Lett. **67**, 3172 (1991).
4.34. F.J. Himpsel, Phys. Rev. B **44**, 5966 (1991). J.E. Ortega and F.J. Himpsel, Phys. Rev. Lett. **69**, 844 (1992). J.E. Ortega, F.J. Himpsel, G.J. Mankey, and R.F. Willis, Phys. Rev. B **47**, 1540 (1993).
4.35. N.B. Brookes, Y. Chang, and P.D. Johnson, Phys. Rev. Lett. **67**, 354 (1991).
4.36. R.K. Kawakami, E. Rotenberg, E.J. Escorcia-Aparicio, H.J. Choi, J.H. Wolfe, N.V. Smith, and Z.Q. Qiu, Phys. Rev. Lett. **82**, 4098 (1999).
4.37. Z.Q. Qiu and N.V. Smith J. Phys. Condens. Matter **14**, R169 (2002).
4.38. In these models, the Fermi surface that determines the properties of the interlayer coupling is the (possibly) fictitious Fermi surface of the material that makes up the non-magnetic spacer layer. A Fermi surface is only truly defined for an infinite periodic solid. However, due to the rapid screening by metals, the potential seen by electrons in the middle of all but the thinnest spacer layers is practically indistinguishable from the potential seen in the interior of an infinite sample of the same material. Since the Fermi surface of an infinite material describes the response to perturbations of the electron system in that potential, that same Fermi surface describes the response to perturbations of the electron system inside the spacer layer. Thus, the oscillatory coupling is determined by the Fermi surface of the material that makes up the spacer layer, which is well defined, and not the Fermi surface of the spacer layer itself.
4.39. M.A. Ruderman and C. Kittel, Phys. Rev. **96**, 99 (1954); T. Kasuya, Prog. Theor. Phys. **16**, 45, 58 (1956); K. Yosida, Phys. Rev. **106**, 893 (1957).
4.40. P. Bruno and C. Chappert, Phys. Rev. Lett. **67**, 1602 (1991).
4.41. R. Coehoorn, Phys. Rev. B **44**, 9331 (1991).
4.42. D.M. Deaven, D.S. Rokhsar, and M. Johnson, Phys. Rev. B **44**, 5977 (1991).
4.43. C. Chappert and J.P. Renard, Europhys. Lett. **15**, 553 (1991).
4.44. Y. Yafet, J. Appl. Phys. **61**, 4058 (1987); Y. Yafet, J. Kwo, M. Hong, C.F. Majkrzak, and T. O'Brien, J. Appl. Phys. **63**, 3453 (1988).
4.45. Z.-P. Shi, P.M. Levy, and J.L. Fry, Phys. Rev. Lett. **69**, 3678 (1992).
4.46. C. Lacroix and J.P. Gavigan, J. Mag. Mag. Mater. **93**, 413 (1991).
4.47. P. Bruno, J. Mag. Mag. Mater. **116**, L13 (1992).
4.48. J.R. Cullen and K.B. Hathaway, J. Appl. Phys. **70**, 5879 (1991). K.B. Hathaway and J.R. Cullen, J. Magn. Magn. Mater. **104-107**, 1840 (1992).
4.49. J. Barnaś, J. Magn. Magn. Mater. **111**, L215 (1992).

4.50. D.M. Edwards, J. Mathon, R.B. Muniz, and M.S. Phan, Phys. Rev. Lett. **67**, 493 (1991).
4.51. R.F.C. Farrow, IBM J. Res. Dev. **42**, 43 (1998).
4.52. Q. Leng, V. Cros, R. Schäfer, A. Fuß, P. Grünberg, and W. Zinn, J. Magn. Magn. Mater. **126**, 367 (1993).
4.53. Z. Celinski, B. Heinrich, J.F. Cochran, W.B. Muir, A.S. Arrott, and J. Kischner, Phys. Rev. Lett. **65**, 1156 (1990); E. Fullerton, D. Stoeffler, K. Ounadjela, B. Heinrich, and Z. Celinski, Phys. Rev. B **51**, 6364 (1995).
4.54. J.E. Mattson, C.H. Sowers, A. Berger, and S.D. Bader, Phys. Rev. Lett. **68**, 3252 (1992).
4.55. M.E. Brubaker, J.E. Mattson, C.H. Sowers, and S.D. Bader, App. Phys. Lett. **58**, 2306 (1991).
4.56. A. Davies, J.A. Stroscio, D.T. Pierce, and R.J. Celotta, Phys. Rev. Lett. **76**, 4175 (1996).
4.57. D. Venus and B. Heinrich, Phys. Rev. B **53**, R1733 (1996).
4.58. R. Pfandzelter, T. Igel, and H. Winter, Phys. Rev. B 54, 4496 (1996).
4.59. V.M. Uzdin, W. Keune, H. Schror, M. Walterfang, Phys. Rev. B **6310**, 4407 (2001).
4.60. D.D. Chambliss, R.J. Wilson, and S. Chiang, IBM J. Res. Dev. **39**, 639 (1995).
4.61. J.A. Stroscio and D.T. Pierce, J. Vac. Sci. Technol. B **12**, 1783 (1994).
4.62. D.E. Bürgler, C.M. Schmidt, D.M. Schaller, F. Meisinger, R. Hofer, and H.-J. Güntherodt, Phys. Rev. B **56**, 4149 (1997).
4.63. T.L. Monchesky, B. Heinrich, R. Urban, K. Myrtle, M. Klaua, and J. Kirschner, Phys. Rev. B **60**, 10242 (1999).
4.64. B. Heinrich and J.F. Cochran, Adv. Phys. **42**, 523 (1993).
4.65. P. Bruno, J. Magn. Magn. Mater. **121**, 238 (1993).
4.66. M.D. Stiles, Phys. Rev. B **48**, 7238 (1993).
4.67. P. Bruno, Phys. Rev. B **52**, 411 (1995).
4.68. J. Mathon, M. Villeret, and D.M. Edwards, J. Phys. Condens. Matter **4** 9873, (1992).
4.69. F. Herman, J. Sticht, and M. van Schilfgaarde, J. Appl. Phys. **69**, 4783 (1991).
4.70. K. Ounadjela, C.B. Sommers, A. Fert, D. Stoeffler, F. Gautier, and V.L. Moruzzi, Europhysics Letters **15**, 875 (1991).
4.71. D. Stoeffler and F. Gautier, Progress of Theoretical Physics Supplement **101**, 139 (1990); D. Stoeffler and F. Gautier, Phys. Rev. B **44**, 10389 (1991).
4.72. H. Hasegawa, Phys. Rev. B **42**, 2368 (1990); H. Hasegawa, Phys. Rev. B **43**, 10803 (1991).
4.73. B. Lee and Y.-C. Chang, Phys. Rev. B **51**, 316 (1995); B.C. Lee, Y.C. Chang, Phys. Rev. B **62**, 3888 (2000).
4.74. M.D. Stiles, J. Appl. Phys. **79**, 5805 (1996).
4.75. K.M. Schep, J. B.A.N. van Hoof, P.J. Kelly, G.E.W. Bauer, and J.E. Inglesfield, Phys. Rev. B **56**, 10805 (1997).
4.76. I. Riedel, P. Zahn, and I. Mertig, Phys. Rev. B 63, 195403 (2001).
4.77. P.D. Johnson, Rep. Prog. Phys. **60**, 1217 (1997).
4.78. F.J. Himpsel, T.A. Jung, and P.F. Seidler, IBM J. Res. Dev. **42**, 33 (1998).
4.79. K. Garrison, Y. Chang, and P.D. Johnson, Phys. Rev. Lett. **71**, 2801 (1993).
4.80. M.S. Ferreira, J. d'Albuquerque e Castro, D.M. Edwards and J. Mathon, J. Phys. Condens. Matter **8** 11259, (1996).
4.81. J. Mathon, M. Villeret, R.B. Muniz, J. d'Albuquerque e Castro, and D.M. Edwards, Phys. Rev. Lett. **74**, 3696 (1995); J. Mathon, M. Villeret, A. Umerski, R.B. Muniz, J. d'Albuquerque e Castro, and D.M. Edwards, Phys. Rev. B **56**, 11797 (1997).

4.82. P. Bruno, Eur. Phys. J. B **11**, 83 (1999).
4.83. P.J.H. Bloemen, M.T. Johnson, M.T.H. van de Vorst, R. Coehoorn, J.J. de Vries, R. Jungblut, J. aan de Stegge, A. Reinders, and W.J.M. de Jonge, Phys. Rev. Lett. **72**, 764 (1994).
4.84. S.N. Okuno and K. Inomata, Phys. Rev. B **51**, 6139 (1995).
4.85. J.J. de Vries, A.A.P. Schudelaro, R. Jungblut, P.J.H. Bloemen, A. Reinders, J. Kohlhepp, R. Coehoorn, and W.J.M. de Jonge, Phys. Rev. Lett. **75**, 4306 (1995).
4.86. J. Unguris, R.J. Celotta, and D.T. Pierce, J. Magn. Magn. Mater. **127**, 205 (1993).
4.87. W. Weber, R. Allenpach, and A. Bischof, Europhys. Lett. **31**, 491 (1995).
4.88. M.D. Stiles, Phys. Rev. B **54**, 14679 (1996).
4.89. L. Tsetseris, B. Lee, and Y.-C. Chang, Phys. Rev. B **55**, 11586 (1997); L. Tsetseris, B. Lee, and Y.-C. Chang, Phys. Rev. B **56**, R11392 (1997).
4.90. E.E. Fullerton, M.J. Conover, J.E. Mattson, C.H. Sowers, and S.D. Bader, Phys. Rev. B **48**, 15755 (1993).
4.91. A.H. MacDonald, T. Jungwirth, and M. Kasner, Phys. Rev. Lett. **81**, 705 (1998).
4.92. J. Inoue, Phys. Rev. B **50**, 13541 (1994).
4.93. J. Kudrnovský, V. Drchal, I. Turek, M. Šob, and P. Weinberger, Phys. Rev. B **53**, 5125 (1996).
4.94. M. Freyss, D. Stoeffler, and H. Dreyssé, Phys. Rev. B **54**, 12667 (1996).
4.95. A.T. Costa, J.D.E. Castro, R.B. Muniz, Phys. Rev. B **59**, 11424 (1999).
4.96. J. Opitz, P. Zahn, J. Binder, I. Mertig, Phys. Rev. B **6309**, 4418 (2001).
4.97. P. Zahn, I. Mertig, Phys. Rev. B **6310**, 4412 (2001).
4.98. R.K. Nesbet, J. Phys. Cond. Mat. **6**, L449 (1994). R.K. Nesbet, Int. J. Quantum Chem. Symp. **28**, 77 (1995).
4.99. S.S.P. Parkin, C. Chappert, and F. Herman, Europhys. Lett. **24**, 71 (1993).
4.100. S.N. Okuno and K. Inomata, Phys. Rev. Lett. **70**, 1711 (1993).
4.101. M. van Schilfgaarde, F. Herman, S.S.P. Parkin, and J. Kudrnovský, Phys. Rev. Lett. **74**, 4063 (1995).
4.102. C.-Y. You, C.H. Sowers, A. Inomata, J.S. Jiang, S.D. Bader, and D.D. Koelling, J. Appl. Phys. (to be published).
4.103. N.N. Lathiotakis, B.L. Gyorffy, E. Bruno, B. Ginatempo, and S.S.P. Parkin, Phys. Rev. Lett. **83**, 215 (1999).
4.104. J. Kudrnovský, V. Drchal, P. Bruno, I. Turek, and P. Weinberger, Phys. Rev. B **54**, R3738 (1996).
4.105. N.N. Lathiotakis, G.L. Györffy, J.B. Staunton, and B. Újfalussy, J. Magn. Magn. Mater. **185**, 293 (1998).
4.106. R.P. Erickson, K.B. Hathaway, and J.R. Cullen, Phys. Rev. B **47**, 2626 (1993).
4.107. J. Barnas and P. Grünberg, J. Magn. Magn. Mater., **121**, 326 (1993).
4.108. D.M. Edwards, J.M. Ward, and J. Mathon, J. Magn. Magn. Mater., **126**, 380 (1993).
4.109. M.D. Stiles, Unpublished.
4.110. D.T. Pierce, J. Unguris, R.J. Celotta, M.D. Stiles, J. Magn. Magn. Mater. **200**, 290 (1999).
4.111. J.F. Bobo, H. Kikuchi, O. Redon, E. Snoeck, M. Piecuch, R.L. White, Phys. Rev. B **60**, 4131 (1999).
4.112. L. Néel, C. R. Hebd. Seances Acad. Sci. 255, 1545 (1962); L. Néel, C. R. Hebd. Seances Acad. Sci. 255, 1676 (1962).
4.113. J.C.S. Kools, W. Kula, D. Mauri, T. Lin, J. Appl. Phys. **85**, 4466 (1999).
4.114. H.D. Chopra, D.X. Yang, P.J. Chen, D.C. Parks, W.F. Egelhoff, Phys. Rev. B **61**, 9642 (2000).

4.115. B.D. Schrag, A. Anguelouch, S. Ingvarsson, G. Xiao, Y. Lu, P.L. Trouilloud, A. Gupta, R.A. Wanner, W.J. Gallagher, P.M. Rice, and S.S.P. Parkin, Appl. Phys. Lett. **77**, 2373 (2000).
4.116. S. Tegen, I. Monch, J. Schumann, H. Vinzelberg, C.M. Schneider, J. Appl. Phys. **89**, 8169 (2001).
4.117. J. Langer, R. Mattheis, B. Ocker, W. Maass, S. Senz, D. Hesse, and J. Krausslich, J. Appl. Phys. **90**, 5126 (2001).
4.118. S. Demokritov, E. Tsymbal, P. Grünberg, W. Zinn, and I.K. Schuller, Phys. Rev B **49**, 720 (1994).
4.119. U. Rücker, S.O. Demokritov, E. Tsymbal, P. Grünberg, and W. Zinn, J. Appl. Phys. **78**, 387 (1995).
4.120. J. Unguris, Private Communication.
4.121. P. Fuchs, U. Ramsperger, A. Vaterlaus, M. Landolt, Phys. Rev. B **55** 12546 (1997).
4.122. J.C. Slonczewski J. Appl. Phys. **73**, 5957 (1993).
4.123. B. Heinrich, Z. Celinski, L.X. Liao, M. From and J.F. Cochran, J. Appl. Phys. **75**, 6187 (1994).
4.124. J.J. de Vries, G.J. Strikers, M.T. Johnson, A. Reinders, and W.J.M. de Jonge, J. Magn. Magn. Mater. **148**, 187 (1995).
4.125. M. Schäfer, S.O. Demokritov, S.O. Müller-Pfeiffer, R. Schäfer, M. Schneider, P. Grünberg, and W. Zinn, J. Appl. Phys. **77**, 6432 (1995). Sc95
4.126. S.V. Halilov, H. Eschrig, A.Y. Perlov, and P.M. Oppeneer Phys. Rev. B **58**, 293 (1998).
4.127. B. Heinrich, J.F. Cochran, T. Monchesky, R. Urban, Phys. Rev. B **59**, 14520 (1999).
4.128. S.T. Purcell, M.T. Johnson, N.W.E. McGee, R. Coehoorn, and W. Hoving, Phys. Rev. B **45**, 13064 (1992).
4.129. M.E. Filipkowski, J.J. Krebs, G.A. Prinz, C.J. Gutierrez, Phys. Rev. Lett. **75**, 1847 (1995).
4.130. T. Monchesky, B. Heinrich, J.F. Cochran, and M. Klaua, J. Magn. Magn. Mat. **198–199**, 421 (1999).
4.131. S. Yan, R. Schreiber, F. Voges, C. Osthöver, and P. Grünberg, Phys. Rev. B **59**, R11641 (1999).
4.132. D.A. Tulchinsky, J. Unguris, and R.J. Celotta, J. Magn. Magn. Mater. **212**, 91–100 (2000).
4.133. D.T. Pierce, A.D. Davies, J.A. Stroscio, D.A. Tulchinsky, J. Unguris, and R.J. Celotta, J. Magn. Magn. Mater. **222**, 13 (2000).
4.134. S.S. Yan, P. Grünberg, R. Schäfer, Phys. Rev. B **62**, 5765 (2000).
4.135. C.F. Majkrzak, J. Kwo, M. Hong, Y. Yafet, D. Gibbs, C.L. Chein, and J. Bohr, Adv. Phys. **40**, 99 (1991).
4.136. J.J. Rhyne and R.W. Erwin, in Magnetic Materials, Vol. 8, ed. K.H.J. Buschow (North-Holland, Elsevier, Amsterdam, 1995) p. 1.
4.137. P. Lang, L. Nordström, R. Zeller, and P.H. Dederichs, Phys. Rev. Lett. **71**, 1927 (1993); L. Nordström, P. Lang, R. Zeller, and P.H. Dederichs, Phys. Rev. B **50**, 13058 (1994); P. Lang, L. Nordström, K. Wildberger, R. Zeller, P.H. Dederichs, and T. Hoshino, Phys. Rev. B **53**, 9092 (1996).
4.138. J. Kudrnovský, V. Drchal, I. Turek, and P. Weinberger, Phys. Rev. B **50**, 16105 (1994); V. Drchal, J. Kudrnovský, I. Turek, and P. Weinberger, Phys. Rev. B **53**, 15036 (1996). V. Drchal, J. Kudrnovský, P. Bruno, I. Turek, P.H. Dederichs, and P. Weinberger, Phys. Rev. B **60**, 9588 (1999).
4.139. K. Wildberger, R. Zeller, P.H. Dederisch, J. Kudrnovský, and P. Weinberger, Phys. Rev. B **58**, 13721 (1998).

4.140. N.N. Lathiotakis, B.L. Gyorffy, B. Ujfalussy, Phys. Rev. B **61**, 6854 (2000).
4.141. A. Danese, D.A. Arena, R.A. Bartynski, Prog. Surf. Sci. **67**, 249 (2001).
4.142. A. Cebollada, R. Miranda, C.M. Schneider, P. Schuster, and J. Kirschner, J. Magn. Magn. Mater. **102**, 25 (1991).
4.143. C. Stamm, Ch. Würsch, S. Egger, and D. Pescia, J. Magn. Magn. Mater. **177–181**, 1279 (1998).
4.144. J. Unguris, R.J. Celotta, and D.T. Pierce, Phys. Rev. Lett. **79**, 2734 (1997).
4.145. The asymptotic coupling strengths are reported in mJ/m^2 by evaluating J^α/D^2 at an effective thickness of $D = 1$ nm.
4.146. Z. Celinski, B. Heinrich, and J.F. Cochran, J. Appl. Phys. **73**, 5966 (1993).
4.147. L. Szunyogh, B. Ujfalussy, P. Weinberger, C. Sommers, Phys. Rev. B **54**, 6430 (1996).
4.148. A.T. Costa, J. d'Albuquerque e Castro, and R.B. Muniz, Phys. Rev. B **56**, 13697 (1997).
4.149. The values quoted here differ from the values given in [4.96] because of a different convention used for the definition of J.
4.150. E.R. Moog, C. Liu, S.D. Bader, and J. Zak, Phys. Rev. B **39**, 6949 (1989).
4.151. H. Zabel, J. Phys.-Condens. Matter. **11**, 9303 (1999).
4.152. R.S. Fishman, J. Phys.-Condens. Matter **13**, R235 (2001).
4.153. Z.-P. Shi and R.S. Fishman, Phys. Rev. Lett. **78**, 1351 (1997); R.S. Fishman and Z.-P. Shi, Phys. Rev. B **59**, 1384 (1999).
4.154. M. Freyss, D. Stoeffler, H. Dreysse, Phys. Rev. B **54**, 12677 (1996); M. Freyss, D. Stoeffler, H. Dreysse, Phys. Rev. B **56**, 6047 (1997); C. Cornea, D. Stoeffler, Europhys. Lett. **49**, 217 (2000); D. Stoeffler, C. Cornea, Europhys. Lett. **56**, 282 (2001).
4.155. A.T. Costa, A.C.D. Barbosa, J.D.E. Castro, R.B. Muniz, J. Phys.-Condes. Matter **13**, 1827 (2001).
4.156. M. van Schilfgaarde and F. Herman, Phys. Rev. Lett. **71**, 1923 (1993).
4.157. S. Mirbt, A.M.N. Niklasson, H.L. Skriver, and B. Johansson, Phys. Rev. B **54**, 6382 (1996).
4.158. A.M.N. Niklasson, B. Johansson, L. Nordstrom, Phys. Rev. Lett. **82**, 4544 (1999).
4.159. K. Hirai, J. Phys. Soc. Jpn. **70**, 841 (2001).
4.160. A. Schreyer, J.F. Ankner, Th. Zeidler, H. Zabel, C.F. Majkrzak, M. Schäfer, and P. Grünberg Europhys. Lett. **32**, 595 (1995); A. Schreyer, J.F. Ankner, Th. Zeidler, H. Zabel, M. Schäfer, J.A. Wolf, P. Grünberg, and C.F. Majkrzak, Phys. Rev. B **52**, 16066 (1995); A. Schreyer, C.F. Majkrzak, Th. Zeidler, T. Schmitte, P. Bödeker, K. Theis-Bröhl, A. Abromeit, J.A. Dura, and T. Watanabe Phys. Rev. Lett. **79**, 4914 (1997).
4.161. S. Adenwalla, G.P. Felcher, E.E. Fullerton, and S.D. Bader Phys. Rev. B **53**, 2474 (1996); E.E. Fullerton, S. Adenwalla, G.P. Felcher, K.T. Riggs, C.H. Sowers, S.D. Bader, and J.L. Robertson, Physica B **221**, 370 (1996); E.E. Fullerton, S.D. Bader, and J.L. Robertson, Phys. Rev. Lett. **77**, 1382 (1996).
4.162. D.T. Pierce, J.A. Stroscio, J. Unguris, and R.J. Celotta, Phys. Rev. B **49**, 14564 (1994).
4.163. D.D. Koelling, Phys. Rev. B **50**, 273 (1994); D.D. Koelling, Phys. Rev. B **59**, 6351 (1999).
4.164. M. van Schilfgaarde and W.A. Harrison, Phys. Rev. Lett. **71**, 3870 (1993).
4.165. D.E. Bürgler, F. Meisinger, C.M. Schmidt, D.M. Schaller, H.J. Guntherodt, and P. Grünberg, Phys. Rev. B **60**, R3732 (1999).
4.166. X. Bian, H.T. Hardner, and S.S.P. Parkin, J. Appl. Phys. **79**, 4980 (1996).
4.167. S. Toscano , B. Briner, H. Hopster, and M. Landolt, J. Magn. Magn. Mater. **114**, L6 (1992).

4.168. E.E. Fullerton, J.E. Mattson, S.R. Lee, C.H. Sowers, Y.Y. Huang, G. Felcher, S.D. Bader, and F.T. Parker, J. Magn. Magn. Mater. 117, L301 (1992).
4.169. A. Chaiken , R.P. Michel, and M.A. Wall, Phys. Rev. B 53, 5518 (1996).
4.170. R. Kläsges, C. Carbone, and W. Eberhardt, C. Pampuch, O. Rader, T. Kachel, and W. Gudat, Phys. Rev. B **56**, 10801 (1997).
4.171. Y. Endo , O. Kitakami, and Y. Shimada , Phys. Rcv. B bf 59, 4279 (1999).
4.172. B. Briner and M. Landolt, Z. Phys. B **92**, 137 (1993).
4.173. B. Briner and M. Landolt, Europhys. Lett. **28**, 65 (1994); M. Hunziker and M. Landolt, Phys. Rev. B **64** 134421 (2001).
4.174. Y. Saito, K. Inomata, J. Phys. Soc. Jpn. **67**, 1138 (1998).
4.175. G.J. Strijkers, J.T. Kohlhepp, H.J.M. Swagten, W.J.M. de Jonge, Phys. Rev. Lett. **84**, 1812 (2000).
4.176. R.R. Gareev, D.E. Bürgler, M. Buchmeier, D. Olligs, R. Schreiber, and P. Grünberg, Phys. Rev. Lett. **8715**, 7202 (2001).
4.177. R.R. Gareev, D.E. Bürgler, M. Buchmeier, R. Schreiber, and P. Grünberg, J. Magn. Magn. Mater. (to be published).

5

Spin Relaxation in Magnetic Metallic Layers and Multilayers

B. Heinrich

5.1 Introduction

Spintronics and high density magnetic recording employ fast magnetization reversal processes. It is currently of considerable interest to acquire a thorough understanding of the spin dynamics and magnetic relaxation processes in the nano-second time regime. The purpose of this chapter is to review the basic concepts of magnetic relaxation with emphasis on metallic ferromagnets. It is fair to say that even among people working in magnetic dynamics the concepts and understanding of spin dynamics in ferromagnetic metals are rather fragmented and lead to interpretations which are often discussed by the parties involved with a flare of passion. There is a good reason for that. The magnetic damping is sample dependent. It is not an intrinsic property of materials; it depends on noise in the system. However, we do describe the magnetic damping in terms of intrinsic and extrinsic properties. The reason is that some relaxation processes are unavoidable. At finite temperatures the scattering of spin waves (magnetic excitations) with electrons and phonons is an integral part of the system. These processes are called intrinsic. The presence of structural and compositional defects leads also to relaxation processes, and they are labelled extrinsic contributions. In the finite sized samples one investigates an assembly of modes and the dynamic response is strongly affected by mode-mode coupling. The origin of the intrinsic damping in metallic ferromagnets is often misunderstood. The reason for that is mostly historical. The scientists working in magnetic dynamics often come with a background in studying ferrites, and they have the tendency to view dynamics in metals through the eyeglasses of insulators. Magnetic relaxations in metals were studied extensively from the late fifties to mid seventies. Well, time goes fast and memory fades away even faster. A lot of things have been forgotten. The structure of this chapter will follow my personal experience which dates from the mid sixties until the present time. Hopefully, this chapter will bring some degree of general understanding of this important and fascinating field. I apologize to those who feel that their work and contributions were either insufficiently covered or perhaps even omitted. This is certainly not intentional. Unfortunately, the presentation in a relatively short chapter has to be confined to a certain flow of ideas which does not allow one to cover all diversified aspects and topics.

From the beginning of the eighties magnetic multilayers have become a very active research field with an unrestrained rate of progress. They provide a special case where dynamic interactions between the itinerant electrons and the magnetic moments in ultrathin films keep offering new exciting possibilities. The small lateral dimensions of spintronics devices and high density memory bits require the use of magnetic metallic ultrathin film structures where the magnetic moments across the film are locked together by exchange coupling. This simplifies the description of magnetic dynamics. One does not have to worry about spatial variations of the magnetic moment across the film thickness.

The chapter is organized in the following way: Sect. 5.2 describes the phenomenology of magnetic damping including some thermodynamic ideas which support a certain type of phenomenology. Section 5.3 deals with Ferromagnetic resonance (FMR) linewidth which is perhaps the simplest way to probe magnetic damping. The intrinsic damping mechanism in metals is covered in Sect. 5.4. It is shown that the spin-orbit interaction with the itinerant nature of electrons is the leading mechanism of intrinsic damping in metals. This section also includes the study of magnetic relaxations using neutron scattering, allowing one to measure the relaxation processes at high q wave vectors and around the critical point T_c where the magnetic fluctuations play an essential role. The discussion of relaxations for large angle of precession is also included in this section. In Sect. 5.5 the torque which is generated by the spin momentum transport using a dc current is reviewed and discussed. This section includes some computer simulations showing complexities which can occur in finite sized samples due to mode-mode coupling. Section 5.6 describes the non-local damping in magnetic multilayers which arises due to spin pump and spin sink effects. It is shown that spin pumping allows one to transfer information without using a net electrical current. This represents potentially a truly different approach to electronics than that employed in semiconductors. Finally, Sect. 5.7 covers the extrinsic magnetic damping which includes two magnon scattering processes and dry friction.

5.2 Magnetic Equations of Motion

The spin dynamics in the classical limit can be described by the Landau Lifshitz (L.L.) equation of motion

$$\frac{1}{\gamma}\frac{\partial \boldsymbol{M}}{\partial t} = -[\boldsymbol{M} \times \boldsymbol{H}_{\mathrm{eff}}] - \frac{\lambda}{\gamma M_s^2}(\boldsymbol{M} \times [\boldsymbol{M} \times \boldsymbol{H}_{\mathrm{eff}}]) \,, \tag{5.1}$$

where γ is the absolute value of the electron gyromagnetic ratio, M_s is the saturation magnetization (magnetization per unit volume) and λ is the L.L. damping parameter. The effective field $\boldsymbol{H}_{\mathrm{eff}}$ is given by the derivatives of the Gibbs energy density, U, with respect to the components (M_x, M_y, M_z) of the magnetization vector $\boldsymbol{M}$, see [5.1–4].

$$\boldsymbol{H}_{\text{eff}} = -\frac{\partial U}{\partial \boldsymbol{M}}, \tag{5.2}$$

U includes the Zeeman energy of the dc applied and demagnetizing fields $\boldsymbol{H}$ and the rf magnetic field $\boldsymbol{h}$, all magnetic anisotropies, and the inter-layer and intra-layer exchange interaction energies. Since U includes the Zeeman energy of the external fields it is more appropriate to call it Gibbs energy instead of Free energy that is commonly used in (5.2). However, the word of caution is in place. The internal rf magnetic field has to be evaluated by using Maxwell's equations in the presence of externally applied rf magnetic field and taking into account the spatial distribution of the rf magnetization. This way one can properly account for the long range dipolar interaction.

The first term on the righthand side of (5.1) represents the precessional torque and the second term represents the well known L.L. damping torque. For small damping, $\alpha = \lambda/\gamma M_{\text{s}} \ll 1$, the L.L. damping term can be replaced by the Gilbert damping term, resulting in the L.L. Gilbert (L.L.G.) equation of motion

$$\frac{1}{\gamma}\frac{\partial \boldsymbol{M}}{\partial t} = -[\boldsymbol{M} \times \boldsymbol{H}_{\text{eff}}] + \frac{G}{\gamma^2 M_{\text{s}}^2}\left[\boldsymbol{M} \times \frac{\partial \boldsymbol{M}}{\partial t}\right], \tag{5.3}$$

where G is the Gilbert damping parameter.

It is useful to rewrite (5.3) for an arbitrary magnetic moment M

$$\frac{1}{\gamma}\frac{\partial \mathsf{M}}{\partial t} = -[\mathsf{M} \times \boldsymbol{H}_{\text{eff}}] + \frac{\alpha}{\gamma}\left[\mathsf{M} \times \frac{\partial \boldsymbol{s}}{\partial t}\right], \tag{5.4}$$

where $\boldsymbol{s}$ is the unit vector along M. Notice that the dimensionless relaxation parameter α is a natural choice to represent the strength of Gilbert damping.

The Gilbert and L.L. equations of motion preserve the absolute magnetization, $\boldsymbol{M} \cdot \boldsymbol{M} = M_{\text{s}}^2$.

For small precession angles ($|\,\boldsymbol{m}\,| \ll M_{\text{s}}$) the magnetization vector can be linearized by setting $\boldsymbol{M} = \boldsymbol{m} + \boldsymbol{M}_{\text{s}}$, where $\boldsymbol{M}_{\text{s}}$ and $\boldsymbol{m}$ are the longitudinal and transverse components of $\boldsymbol{M}$, see Fig. 5.1. In this linearized approximation the L.L. relaxation term in (5.1) can be rewritten as

$$-\frac{\lambda}{\gamma}\left(\frac{\boldsymbol{m}}{\chi_{\perp}} - \boldsymbol{h}_{\text{eff}}\right), \tag{5.5}$$

where $\boldsymbol{h}_{\text{eff}}$ is the effective transverse rf field which is given by (5.2). The internal rf magnetic field has to be evaluated from Maxwell's equations using appropriate boundary conditions. For ultrathin films this field is in a very good approximation equal to the applied rf field and dipolar fields which have to be evaluated from Maxwell's equations, e.g. see the Damon Eschbach spin wave modes in [5.7]. The transverse dc susceptibility $\chi_{\perp}$ is given by

$$\chi_{\perp} = \frac{M_{\text{s}}}{H_{\text{eff}}}, \tag{5.6}$$

where H_{eff} is the effective dc field in equilibrium.

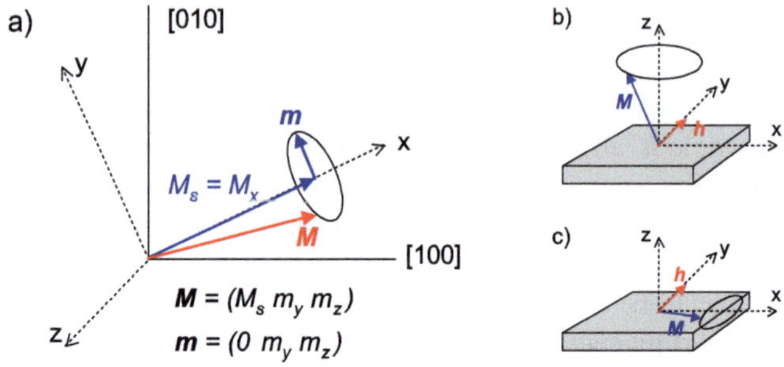

Fig. 5.1. (**a**) A schematic drawing of precessing magnetization $\boldsymbol{M}$, and the coordinate axes. The in-plane crystallographic axes correspond to a (001) oriented ferromagnetic film. For zero rf field the saturation magnetization M_s is oriented along the dc effective field $\boldsymbol{H}_{\text{eff}}$. The x-axis is chosen along the effective field $\boldsymbol{H}_{\text{eff}}$. For a small rf field $\boldsymbol{h}$ the precessional motion is small, $|\boldsymbol{m}| \ll M_s$, and the longitudinal component $M_x \simeq M_s$. (**b**) In the perpendicular configuration the saturation magnetization and applied dc field H are perpendicular to the sample plane. (**c**) In the parallel configuration the saturation magnetization and applied dc field are parallel to the sample plane. Note that in the parallel configuration the rf magnetization precesses with a pronounced elliptical polarization

Equation (5.5) is consistent with the Landau-Ginzburg treatment of the thermodynamics of irreversible processes. In this treatment the rate of approach to thermodynamic equilibrium is proportional to the appropriate thermodynamic force. For a magnetic moment $\boldsymbol{m}$ the thermodynamic force is proportional to the total transverse effective field. Since the transversal effective field is given by the derivative of the Gibbs energy with respect the magnetization vector, see (5.2), the Landau-Ginzburg relaxation corresponds to a steepest descent down the Gibbs energy. This effective field can be obtained from (5.2) using the partial derivatives with respect to the components of $\boldsymbol{m}$. In fact the first term in (5.5), $-\boldsymbol{m}/\chi_\perp$, corresponds to the Zeeman energy $-M_x H_{\text{eff}} \simeq -M_s H_{\text{eff}} + (H_{\text{eff}}/2M_s)(m_y^2 + m_z^2)$, see Fig. 5.1. It is worth noticing that the L.L. damping parameter λ is equivalent to the Onsager damping coefficient for the transverse magnetization $\boldsymbol{m}$ in neutron scattering studies [5.6]. However, in the neutron studies it is important to include the contribution from the exchange field, see (5.23), directly in the transversal susceptibility. Neglecting the anisotropy fields (elliptical polarization) the transversal susceptibility is given by

$$\chi_\perp(q) = \frac{M_s}{H + \frac{2A}{M_s}q^2} , \tag{5.7}$$

where A is strength of the exchange coupling and q is the wavevector of magnetic excitation studied by neutron scattering.

The magnetic relaxation can be expressed in other forms. One example is the modified Bloch-Bloembergen (M.B.B.) [5.7] relaxation term

$$-\frac{1}{\tau\gamma}(\boldsymbol{m} - \chi_\perp^{\text{BB}}\boldsymbol{h}) , \tag{5.8}$$

where $\boldsymbol{h}$ is the external rf driving field. The modified B.B. relaxation term is commonly used to describe magnetic insulators [5.8]. The M.B.B. relaxation term assumes that the amplitude of the rf eigenmodes (magnons) can be described by a single relaxation rate $1/\tau$, see further discussion in [5.9]. The dc effective transverse susceptibility $\chi_{\perp}^{\mathrm{BB}}$ is defined in such a way that the expressions in brackets of (5.8) and (5.5) lead to the same quasi static solution (for $\boldsymbol{h}$ slowly varying in time) for the transverse magnetization $\boldsymbol{m}$. The difference between (5.5) and (5.8) is that the relaxation parameters, λ and $1/\tau$, multiply different instantaneous deviations from the equilibrium, and consequently they have a different physical meaning. For the L.L. damping the total transverse effective field relaxes to zero; note that the Zeeman effective field has to be expressed in the transverse components. In the modified B.B. damping the rf magnetization relaxes towards the instantaneous equilibrium magnetization, $\chi_{\perp}^{\mathrm{BB}}\boldsymbol{h}$. Safonov and Bertram showed that one can formally rewrite the L.L. relaxation term (5.5) to the B.B. term (5.8) by a suitable matrix transformation [5.9]. However, this is only a formal transformation which does not apply in general. The particular type of damping has to be obtained from direct calculations involving the microscopic mechanism of damping, see Sect. 5.4, and experimentally by investigating the dependence of FMR linewidth on the microwave frequency and the configuration of applied magnetic field, see Sects. 5.3, 5.7.

The relaxation term in linear response theory can be treated more fundamentally using the theory of irreversible processes. Mori helped to develop the concept of irreversibility which is based on the correlation-function theory of kinetic coefficients [5.10]. The Mori theory involves in general a set of independent variables which represent measurable physical quantities. Using projection operators in Hilbert space one can separate out slowly varying parts in the equations of motion from fast random forces. For the further discussion of magnetic relaxation let us assume to have a system of N spins occupying volume V. In this case the right handed precession of the transverse spin moment $S^{+} = \sum_f (S_{f,y} + \mathrm{i}S_{f,z})$ (in units of $\hbar$, and using $\mathrm{e}^{\mathrm{i}\omega t}$ for the time dependence of precession) can be considered as a slowly variable parameter. f represents the lattice point, and y and z describe the transverse components of the local spin. The equation of motion can be written as, see (5.3) in [5.10],

$$\frac{\mathrm{d}}{\mathrm{d}t}S^{+} = \mathrm{i}\omega_{\mathrm{res}}S^{+} - \frac{1}{\tau}S^{+} + \frac{1}{\hbar}f(t)\ , \tag{5.9}$$

where $f(t)$ is a fast fluctuating random torque and ω_{res} is the eigenfrequency of a homogeneous precession described by S^{+}. The relaxation rate $1/\tau$, see (5.3) in [5.11], is given by

$$1/\tau = \frac{1}{2\hbar^2(S^{+}, S^{-})}\int_{-\infty}^{\infty} \varphi(t)\exp(-\mathrm{i}\omega t)\,\mathrm{d}t\ , \tag{5.10}$$

where $\varphi(t)$ is the scalar product (projection operator) of the fast fluctuating torque, $\varphi(t) = (f^{-}(t), f^{+}(0))$, and the scalar product (S^{+}, S^{-}) represents the transverse dc magnetic response. (S^{+}, S^{-}) can be expressed, using equations (3.6) and (3.8) in [5.11], in terms of the transverse susceptibility $\chi_{\perp} = M_{\mathrm{s}}/H$,

$$(S^+, S^-) = 2\frac{V}{(\mu_B g)^2}\frac{M_s}{H} \,. \tag{5.11}$$

Equation (5.10) is usually referred to as *the second fluctuation dissipation theorem*, see also [5.12–14].

It is useful to replace $\varphi(t) = (f^-(t), f^+(0))$ with the time correlation function of fluctuating torque $\langle f(t) f(0)\rangle$ which is a measurable quantity. By applying the fluctuation dissipation theorem for $kT > \hbar\omega$ one obtains

$$\varphi(t) = (f(t), f(0)) = \frac{1}{kT}\langle f(t) f(0)\rangle \,. \tag{5.12}$$

Equation (5.10) has a similar form as the L.L. damping term in (5.5), and therefore (5.10) can be converted into the expression which determines the L.L. damping coefficient λ from the correlation function of the fluctuating transverse field

$$\alpha = \frac{\lambda}{M_s\gamma} = \frac{\gamma M_s V}{2kT}\frac{1}{2}\int_{-\infty}^{\infty}\langle h(t)h(0)\rangle \exp(-\mathrm{i}\omega t)\,\mathrm{d}t \,. \tag{5.13}$$

The transverse fluctuating torque $f(t)$ was replaced by $M_s h(t)$, where the variable $h(t)$ represents a fast transverse fluctuating field inside the system. The real part (cos integral) of (5.13) corresponds to damping. The imaginary part (*sin* integral) of (5.13) contributes to a frequency shift of the resonant frequency ω_{res}. The time dependence $\langle h(t)h(0)\rangle$ can be approximated by a simple exponential decay $h^2 e^{-t/\tau_{fl}}$, where τ_{fl} represents the correlation time of fluctuations. For $\omega \ll 1/\tau_{fl}$ the L.L. damping is independent of frequency. However, that does not have to be always the case, the time response can be affected by slow fluctuating torques which results in the frequency dependent damping and susceptiblity, see the dynamics studies on Fe whiskers with the intersticial carbon [5.12, 15] and around the critical point T_c [5.16, 17].

The second fluctuation dissipation theorem, (5.13), allows one to determine the amplitude of the fluctuating field in an arbitrary cell ΔV which is a part of the magnetic sample. The fluctuations in different ΔV are statistically independent. In computer simulations an estimate for the fluctuating internal field (Langevin random noise) in dissipative systems is given by

$$h^2 = \frac{2kT}{\gamma M_s \Delta V}\frac{1}{\Delta t}\frac{\alpha}{1+\alpha^2} \,, \tag{5.14}$$

where Δt is the time increment in spin dynamics calculations, see [5.18, 19].

The above equations can be generalized for an arbitrary magnon with the wave number $\boldsymbol{q}$. Both the resonance frequency and transverse dc susceptibility can be determined by evaluating the rf transversal susceptibility tensor $\tilde{\chi}$ using the L.L. eq. of motion with a rf harmonic driving field $\boldsymbol{h}_0 \exp\{\mathrm{i}(\omega t - \boldsymbol{qr})\}$, where ω is the applied angular frequency, see further details in [5.3, 4]. The resonance frequency is given by the minimum value of the denominator of $\tilde{\chi}$, and the dc transversal susceptibilities are given by the (y, y) and (z, z) components of $\tilde{\chi}$ for $\omega = 0$.

Since the relaxation term in (5.10) is inversely proportional to the transverse static susceptibility it follows that *the L.L. damping is a more fundamental description of relaxation than the M.B.B. relaxation equation.* A nice discussion of the fluctuation dissipation theorem for Gilbert, L.L. and B.B. relaxation terms was carried out by Smith [5.20]. He has shown that only the Gilbert and L.L. damping provide correct correlations for the local fluctuating field components $\langle h_y(t)h_y(0)\rangle$ and $\langle h_z(t)h_z(0)\rangle$. On the other hand the M.B.B can be classified as intrinsically nonlocal damping.

It turns out that the L.L. equation (5.1) can be transformed to the L.L.G equation (5.4) by using the following substitutions

$$\frac{\lambda}{M_\mathrm{s}} = \alpha\gamma\,, \quad \gamma = \gamma_\mathrm{G}\frac{1}{1+\alpha^2}\,, \tag{5.15}$$

where γ_G is the Gilbert gyromagnetic ratio and α is the damping constant.

5.3 FMR Linewidth

FMR and rf susceptibilities were extensively described in Volume II of this book [5.2] and in [5.3]. For additional reading see [5.21]. Here a brief summary will be presented with emphasis on the FMR linewidth for parallel and perpendicular configurations. For a small rf magnetization $\boldsymbol{m}$ the L.L. equation of motion (5.1) can be solved using a linear expansion in small components of $\boldsymbol{m}$ and $\boldsymbol{h}$, see [5.3, 22]. Experimentally, the FMR linewidth is determined as the difference in the dc applied magnetic fields between the maximum and minimum of the field derivative of the FMR absorption peak, $\partial\chi''/\partial H$ [5.3, 22].

In ultrathin films the magnetic moments across the film are locked together by exchange coupling and they can be considered as giant magnetic molecules which have unique magnetic properties of their own [5.3]. For ultrathin films the role of the interface perpendicular uniaxial anisotropy and the dipole-dipole energy of an uniformly magnetized sample can be included in the effective field $4\pi M_\mathrm{eff} = 4\pi M_\mathrm{s} - 2K_\mathrm{u}^\mathrm{s}/dM_\mathrm{s}$, see [5.3, 22]. The K_u^s is the coefficient of the perpendicular interface uniaxial anisotropy and d is the film thickness. The following calculations were carried out in the absence of magnetic anisotropies except for the perpendicular uniaxial anisotropy allowing one to account for the ellipticity of the rf precession.

5.3.1 Gilbert Damping

Parallel Configuration. In the parallel FMR configuration the static magnetization and the applied field are in the film plane, see Fig. 5.1c; one can write for an ultrathin film the Gibbs energy density as

$$U_\mathrm{u}^\mathrm{s} = -\frac{K_\mathrm{u}^\mathrm{s}}{d}\left(\frac{m_z}{M_\mathrm{s}}\right)^2\,, \tag{5.16}$$

where K_u^s is the interface uniaxial perpendicular anisotropy in erg/cm^2 [5.3], and d is the film thickness. With the saturation magnetization M_s along the $\boldsymbol{x}$-axis and the rf field $\boldsymbol{h}$ applied in plane along the $\boldsymbol{y}$-axis, see Fig. 5.1, we have $\boldsymbol{h}_{\text{eff}} = (0, h, -4\pi D_{\text{eff}} m_z)$, where $4\pi D_{\text{eff}} = 4\pi - 2K_u^s/M_s^2 d$ is the effective demagnetizing factor. Solving the Gilbert equation of motion (5.3) results in the Gilbert

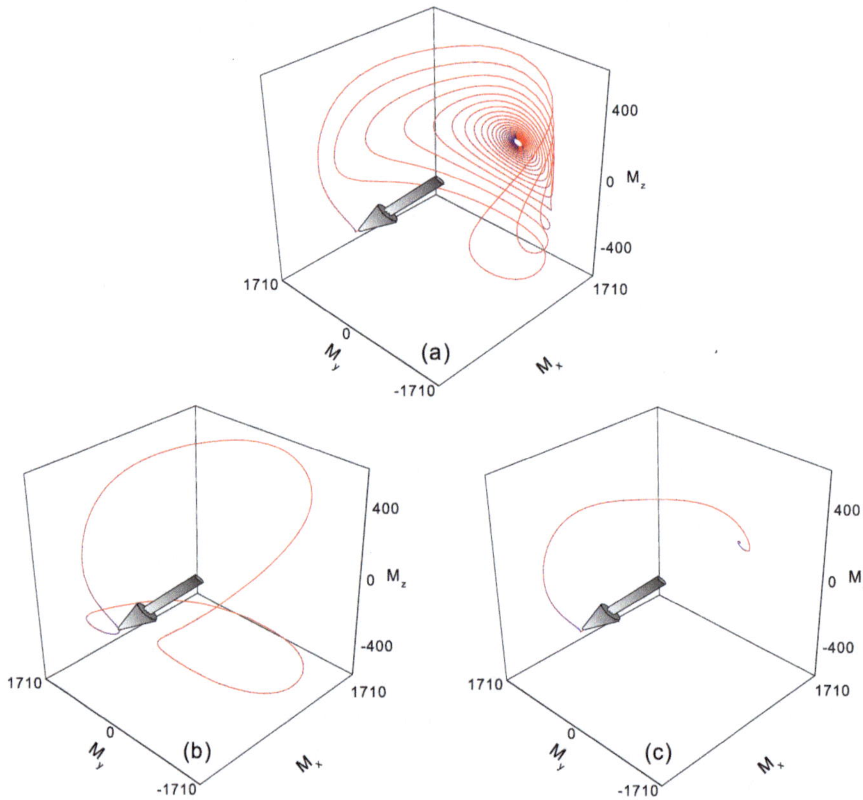

Fig. 5.2. Computer simulations of time evolution of the magnetization in an applied field $H = 0.5\,\text{kOe}$. The field is oriented along the magnetic easy axis [100], in the positive x-direction. The initial orientation of the saturation magnetization (represented by the arrow) is in plane, 5° away from the $[\bar{1}00]$ crystallographic axis (nearly antiparallel orientation with respect to $\boldsymbol{H}$). Simulations were carried out for the following Gilbert damping parameters α: (**a**) 0.01, (**b**) 0.0, and (**c**) 0.2 (the fastest approach to equilibrium). The magnetic parameters were taken from an Fe film of 2.3 nm (16 ML) in thickness grown onto a GaAs(001) substrate and covered by a 20 ML thick Au(001) cap layer [5.23]. The magnetic parameters are as follow: $4\pi M_s = 21.4\,\text{kG}$, $4\pi M_{\text{eff}} = 18\,\text{kOe}$, the in-plane four-fold anisotropy $K_1 = 3.5 \times 10^5\,\text{ergs/cm}^3$, the in-plane uniaxial anisotropy $K_u = -1.0 \times 10^5\,\text{erg/cm}^3$, and the g-factor $g = 2.09$. The in-plane uniaxial anisotropy is oriented 45° with respect to the cubic crystallographic axis [100]. $\alpha = 0.01$ corresponds to a realistic Gilbert damping ($G = 1.2 \times 10^8\,\text{s}^{-1}$). For $\alpha = 0$ the precession is not damped, and exhibits a *butterfly* like pattern. $\alpha = 0.2$ leads to a critical damping with no oscillations in the rotation

FMR linewidth

$$\Delta H_G = 1.16\left(\frac{G}{\gamma M_s}\right)\left(\frac{\omega}{\gamma}\right) = 1.16\alpha\frac{\omega}{\gamma}\,. \tag{5.17}$$

The factor 1.16 comes from using the inflection points of a Lorentzian line. Note, that the FMR linewidth is strictly linearly dependent on the microwave frequency ω and inversely proportional to the saturation magnetization M_s. This feature is a hallmark of the Gilbert damping. The FMR linewidth increases proportionally with the damping coefficient α. There is no explicit dependence on the applied field H and any magnetic anisotropies. With an increasing α the Gilbert torque eventually becomes dominant and $\partial \boldsymbol{M}/\partial t \to 0$. For $\alpha \to \infty$ the system behaves like molasse approaching its equilibrium infinitely slowly. For the in-plane oriented magnetization where the ellipticity of the precession is significant the fastest approach to equilibrium is achieved for $\alpha = 0.2$, see Fig. 5.2. The relaxation frequency Γ (the rate of approach of precessing magnetization to equilibrium) for $\alpha \ll 1$ is given by

$$\Gamma = \alpha\omega_0\,, \tag{5.18}$$

where ω_0 is the precessional frequency in appropriate effective fields. For a simple applied dc field H for which $\omega_0 = \gamma H$ (5.18) becomes

$$\Gamma = \frac{G}{\chi_\perp}\,. \tag{5.19}$$

Perpendicular Configuration. In the perpendicular FMR configuration the static magnetization and the applied field are perpendicular to the film plane, see Fig. 5.1b. The FMR linewidth is identical to that for the parallel configuration, see (5.17).

5.3.2 Landau Lifshitz Damping

Parallel Configuration. Gilbert and L.L. damping equations of motion are identical for small damping, $\alpha \ll 1$. For small α the FMR linewidth for L.L. damping is the same as for the Gilbert damping, see (5.15). The Gilbert damping parameter G is just replaced by the L.L. damping parameter λ. With an increasing λ the L.L. relaxation torque becomes dominant and $\boldsymbol{M} \times \boldsymbol{H}_{\text{eff}} \to 0$, see (5.1); i.e. the system is always in its quasi-equilibrium state. This behavior does not seem to be realistic. The system should get stuck in thick *molasses* never reaching equilibrium. With an increasing Gilbert damping the system slows down ($\partial \boldsymbol{M}/\partial t \to 0$). Obviously the Gilbert equation of motion is a more realistic description of damping in media with big losses, $\alpha > 1$.

Perpendicular Configuration. In the perpendicular configuration when the film is fully saturated ($H - 4\pi M_s + 2K_u^s/M_s d > 0$) the rf precession is circular, but the FMR linewidth is the same as in the parallel configuration. This means that for the Gilbert and L.L. damping the FMR linewidth does not depend on the elliptical polarization of the magnetization precession.

5.3.3 Modified Bloch-Bloembergen Relaxation

Parallel Configuration. The transverse susceptibilities in (5.5) are equal to $\chi_{y,y} = \chi_{z,z} = M_s/H$, and consequently $\chi_\perp^{\mathrm{MBB}} = M_s/H$. The transverse rf fields are $h_x = h$ and $h_z = -4\pi m_z$, where h is the internal field from Maxwell's equations. This results in the FMR linewidth

$$\Delta H_{\|}^{\mathrm{MBB}} = 1.16\left(\frac{1}{\tau\gamma}\right)\left(\frac{\omega}{\gamma}\right)\left(\frac{1}{H_{\mathrm{res}} + 2\pi M_{\mathrm{eff}}}\right), \tag{5.20}$$

where $(\omega/\gamma)^2 = H_{\mathrm{res}}(H_{\mathrm{res}} + 4\pi M_{\mathrm{eff}})$. At high microwave frequencies, $H_{\mathrm{res}} \to \omega/\gamma$, the FMR linewidth is frequency independent; at low microwave frequencies ($H_{\mathrm{res}} \ll 4\pi M_{\mathrm{eff}}$) the FMR linewidth is proportional to the microwave frequency ω. Clearly, the ellipticity of precession strongly affects the FMR linewidth.

Perpendicular Configuration. The effective field $\boldsymbol{H}_{\mathrm{eff}} = (0, 0, H - 4\pi M_{\mathrm{eff}})$. In the perpendicular configuration the rf precession is circularly polarized, and the FMR linewidth is

$$\Delta H_\perp^{\mathrm{BB}} = 1.16\,\frac{1}{\gamma\tau}\,. \tag{5.21}$$

In the perpendicular configuration the FMR linewidth has a constant value independent of the microwave frequency, saturation magnetization, and applied external field.

The FMR linewidths shown above are valid only for $\Delta H/H_{\mathrm{res}} \ll 1$. Otherwise one has to evaluate the full tensor susceptibility using equations of motion (5.3), see detailed calculations in [5.4, 21].

The L.L, Gilbert, and B.B. damping parameters λ, G, and $1/\tau_{\mathrm{eff}}$ have to satisfy the symmetry requirements of the system and, in principle, they can be different along non-equivalent axes. This could be particularly true for ultrathin films where one encounters tetragonal lattice distortions and interfaces. However, our FMR studies indicate that the dependence of damping on the in-plane orientation of the magnetic moment is *a smoking gun* indicating the presence of extrinsic damping due to sample defects, see Sect. 5.7.1.

5.4 Intrinsic Damping in Metals, Theory

First, let us discuss what one means by intrinsic damping. Magnetic damping is sample dependent, and therefore does not reflect the intrinsic properties of materials. However some relaxation processes are unavoidable. At finite temperatures one is not able to avoid phonons and magnons [5.24, 25]. In alloys one cannot avoid an inhomogeneous electron potential. In ultrathin films the electron scattering at the interfaces can be partly diffuse. The presence of these perturbations affects even true intrinsic interactions such as the intra and interlayer exchange coupling, dipole-dipole interaction and magnetic anisotropies. The magnetic relaxation processes which involve the electron scattering with phonons and thermally excited magnons can be called intrinsic because they are an integral part of the system. The contributions from structural defects and complex geometrical features can in principle be avoided and should be called extrinsic. The smallest measured damping (FMR linewidth) under well defined thermodynamic conditions is considered to be intrinsic to the system. Sadly, this is an imprecise definition and often leads to controversies, but in reality perhaps, it is the only criterion left to experimentalists.

5.4.1 Eddy Currents

In metallic films the magnetic damping can be affected by eddy currents. The role of eddy currents in thin films can be estimated by evaluating the effective Gilbert damping accompanying the magnetization precession in the presence of eddy currents. For thin films where the rf magnetization fully penetrates the film the contribution of eddy currents to the L.L.G. equations of motion (5.3) can be evaluated by integrating Maxwell's equations across the film thickness d. For a circularly polarized precession this results in the effective Gilbert damping, G_{eddy} (in Gauss's units)

$$\frac{G_{\mathrm{eddy}}}{(M_{\mathrm{s}}\gamma)^2} = \frac{1}{6}\left(\frac{4\pi}{c}\right) 2\sigma d^2 , \qquad (5.22)$$

where σ is the electrical conductivity and c is the velocity of light in free space. For Fe the measured intrinsic Gilbert damping parameter is $G_{\mathrm{Fe}} \sim 0.5 \times 10^8\ \mathrm{s}^{-1}$ [5.26, 27]. G_{eddy} becomes comparable to the intrinsic damping in Fe for a film thickness of 70 nm. Notice that G_{eddy} decreases rapidly with decreasing film thickness, $G_{\mathrm{eddy}} \sim d^2$. The ultrathin film limit in Fe is reasonably satisfied for $d < 10$ nm [5.28], and therefore the role of eddy currents is negligible in this case.

For thicker films one has to solve the L.L. and Maxwell's equations selfconsistently. The role of eddy currents in thin films was fully treated in [5.28–30]. The results of FMR linewidth as a function of the film thickness is shown in Fig. 5.3. Note that the role of eddy currents is insignificant below 100 nm for Py. However for Fe the damping is already affected by eddy currents above 25 nm.

In the thick film limit ($d >$ skin depth $\sim$ 100 nm at FMR) the FMR line broadening by eddy currents is proportional to $\sqrt{\sigma A}$ [5.31–33], where A is the exchange coupling coefficient defined by the exchange field [5.1, 3, 4]

$$\boldsymbol{H}_{\mathrm{exch}} = \frac{2A}{M_{\mathrm{s}}^2}\nabla^2\boldsymbol{m} = \frac{2A}{M_{\mathrm{s}}^2}(0, \nabla^2 m_y, \nabla^2 m_z) . \qquad (5.23)$$

∇ is the Nabla operator. Note that the rf exchange field has no contribution along the dc magnetization component in a linearized system, see Fig. 5.1. Within the skin depth the rf magnetization is inhomogeneous. The exponential decay of the rf magnetization can be described by a band of spin waves with an average $\boldsymbol{k}$ wave vector equal in magnitude to $1/\delta$ and a direction perpendicular to the sample surface. δ is the skin depth at FMR. The expression $\sqrt{\sigma A}$ reflects the fact that the FMR

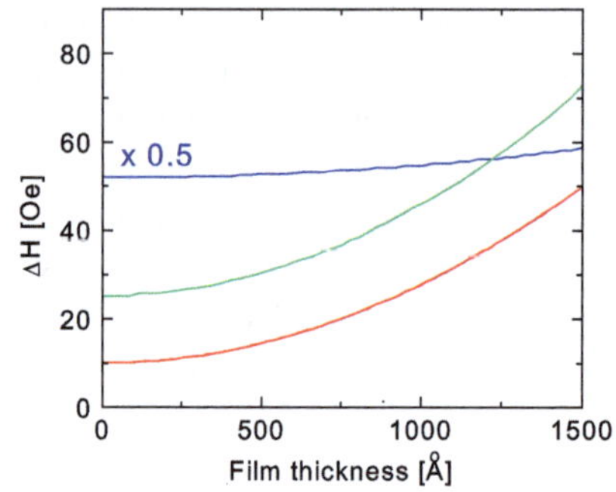

Fig. 5.3. The FMR linewidth in the parallel configuration for Fe (*red*), Py (*blue*), and Ni (*green*) films at RT temperature and at 10 GHz as a function of film thickness. In the limit of small thicknesses the FMR linewidth is given only by the Gilbert damping coefficient. The Gilbert damping parameters G were 0.8, 1.0, and $2.4 \times 10^8\ \mathrm{s}^{-1}$ for Fe, Py, and Ni, respectively. Calculations were carried out for a symmetric rf driving. This means the rf field components are same on both sides of the film

line-broadening originates from the rf exchange fields $(2A/M_s)k^2$(see (5.23)) inside the skin depth. The contribution of eddy currents and exchange coupling to the FMR linewidth was introduced by Ament and Rado [5.34] and it is usually referred to as the exchange-conductivity mechanism. The exchange-conductivity mechanism leads to a finite FMR linewidth even in the absence of intrinsic damping. The spatial distribution of the rf magnetization in the skin depth is altered by the surface magnetic anisotropy (surface pinning); and consequently the surface anisotropy strongly affects the FMR linewidth in the thick film limit [5.31–33]. E.g. for an Fe film of 200 nm thickness with an uniaxial perpendicular interface anisotropy $K_s = 0.3\ \mathrm{erg/cm^2}$ the FMR linewidth at RT is wider by 200 Oe at 25 GHz compared to the linewidth arising only from the intrinsic L.L. damping, see [5.28]. That represents a 7 fold increase in the FMR linewidth when compared to the intrinsic Gilbert damping contribution.

5.4.2 Phonon Drag

The magnetic relaxation by direct magnon-phonon scattering is another possible damping mechanism. Suhl recently investigated the role of phonon drag. His explicit results are limited to small geometries where the magnetization and lattice strain are homogeneous. The Gilbert phonon damping G_{ph} is [5.35],

$$\frac{G_{ph}}{\gamma^2} = 2\eta \left(\frac{B_2(1+\nu)}{E} \right)^2 , \tag{5.24}$$

where η is the phonon viscosity, B_2 is the magnetoelastic shear constant, E is Young's modulus, and ν is the Poisson ratio. All parameters can be readily obtained except the parameter for the phonon viscosity η. However, it turns out that the phonon viscosity parameter η in the microwave range of frequencies was experimentally determined by our microwave transmission experiments, see [5.36]. In these studies at 9.5 GHz a fast transversal elastic shear wave was generated by magnetoelastic coupling inside the skin depth of a thick Ni(001) X-tal slab 22 μm in thickness. The transverse elastic shear wave propagated across the slab, and was then re-radiated in a transmitted microwave power at the other side of the Ni slab. We called this effect *phonon assisted microwave transmission*. The experimental data were fitted using the L.L. and elastic wave equations of motion including magnetoelastic coupling. The elastic wave relaxation time was found to be $\tau_{ph} = 6.6 \times 10^{-10}\,s^{-1}$ at 9.5 GHz. The phonon viscosity as introduced in [5.35] is given by $\eta = c_{44}/\tau_{ph}\omega^2$, where c_{44} is the elastic modulus. For Ni $\eta = 3.4$ (in CGS units). Using (5.24) results in a phonon Gilbert damping $G_{ph} \sim 10^7\,s^{-1}$, that is $\sim$ 30 times smaller than the intrinsic Gilbert damping parameter of Ni, $G_{Ni} = 2.4 \times 10^8\,s^{-1}$. In Fe the intrinsic damping parameter $G_{Fe} = 0.7 \times 10^{8\;-1}$ is smaller than that in Ni, and consequently the phonon drag in Fe can be more important. However even in this case the phonon damping is 6 times smaller than G_{Fe}. Clearly, the direct magnon-phonon scattering in small geometries is not important.

It should be pointed out that Suhl's theory does not treat the magnon-phonon interaction selfconsistenly. As in the case of eddy currents one should carry out selfconsistent calculations using the L.L. and elastic wave equations of motion which include the magnetoelastic term of interaction. Such calculations were carried out by Kobayashi et al. [5.37]. They showed that the magnetoelasticity can have only an appreciable effect on the FMR linewidth if the excited elastic wave establishes a resonant mode across the film thickness at or near the FMR field. This effect was called Ferromagnetic Elastic Resonance (FMER). They calculated the frequency FMR linewidth, $\Delta\omega/\gamma$, as a function of the phonon quality parameter Q_{ph}. In our notation the phonon quality parameter $Q_{ph} = \omega\tau_{ph}$. The effect of elastic waves on the FMR lineshape is most pronounced when $Q_{ph} \gg Q_{mag} = 1/\alpha$. The quality parameters for Ni are $Q_{mag} = 10$ and $Q_{ph} = 35$. In Ni the FMER effect can be strong, and can even result in a split FMR peak, see details in [5.37]. The thickness range over which the effect is noticeable is given by $\Delta d/d = \Delta\omega/\omega$. After an extensive and systematic effort to observe FMER in films and perfect Ni platelets no convincing results were achieved. That indicates that the phonon resonance is hard to establish in real samples. The elastic wave wavelengths are $\sim$ 300 nm at 10 GHz. This means that for magnetic films thinner than 150 nm at and below 10 GHz the FMER effect is absent.

The above two sections lead to the following conclusions: Phonon drag is too small to explain the measured damping in magnetic metals. Eddy currents can be effective in thicker samples and at high frequencies. Eddy currents play a negligible role in ultrathin films.

5.4.3 Spin-orbit Relaxation in Metallic Ferromagnets

The literature on intrinsic damping in metals dates back to the late sixties and seventies. The purpose of this section is not to get involved in the details of the calculations but rather to outline the underlying physics.

We will start the description of intrinsic damping in metals by using a model introduced by Heinrich, Fraitová and Kamberský [5.26] which is based on the s-d exchange interaction. In this model the interaction of itinerant s-p like electrons with localized d-spins can be obtained by integrating the s-d exchange energy density functional [5.38]

$$H_{\mathrm{sd}} = \sum_{j} \int_{V} J(\boldsymbol{r}_j - \boldsymbol{r}) \boldsymbol{S}_{j,d} \cdot \boldsymbol{s}_s(\boldsymbol{r})\, \mathrm{d}r^3 \,, \tag{5.25}$$

where $J(\boldsymbol{r} - \boldsymbol{r}')$ is the s-d exchange interaction between the spin density $\boldsymbol{s}_s$ of s-p like itinerant electrons and the localized spins of d-electrons $\boldsymbol{S}_{j,d}$. j is the lattice site. One should point out that a strict distinction between s-p and d electrons is only historical. In fact one should consider two groups of electrons, those that are mostly localized (here denoted by d-electrons), and those that are itinerant (here denoted as s electrons). In reality itinerant electrons are hybridized states of s-p and d electrons. The transverse interaction Hamiltonian involving only the rf components of the magnetization vectors is described by three particle collision terms

$$E_{\mathrm{sd}} = \left(\frac{2S}{N}\right)^{\frac{1}{2}} \sum_{\boldsymbol{k}} J(\boldsymbol{q}) a_{\boldsymbol{k},+} a^{+}_{\boldsymbol{k}+\boldsymbol{q},-} b_{\boldsymbol{q}} + (\mathrm{h.c.}) \,, \tag{5.26}$$

where N is the number of atomic sites, a, a^+ annihilates and creates electrons with the appropriate momentum and spin, and b, b^+ annihilates and creates magnons with the wavevector $\boldsymbol{q}$. The $+$ and $-$ signs in the subscripts correspond to majority and minority spin electrons, respectively. Keep in mind that the electron spin momentum is oriented antiparallel to the magnetic moment. This equation provides a particle representation of the s-d exchange interaction. Magnons and itinerant electrons are coherently scattered by the s-d exchange interaction and that results in creation and annihilation (Hermitian conjugate term in (5.26)) of electron-hole pairs, see Fig. 5.4. The total angular momentum in the s-d exchange interaction is conserved, and consequently the itinerant electron spin flips appropriately during the scattering with magnons. The scattering shown in Fig. 5.4 alone does not lead to magnetic damping for magnons

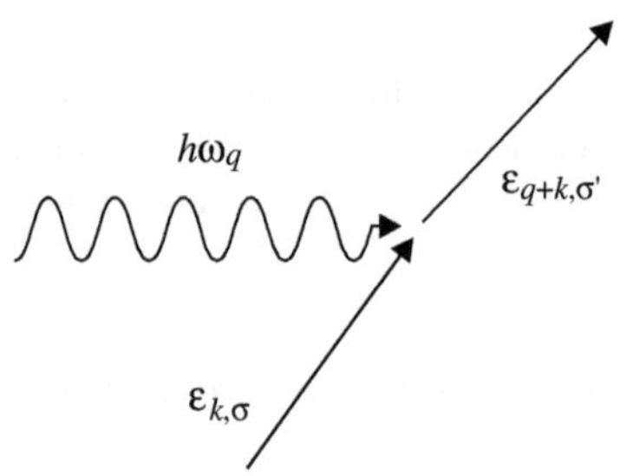

Fig. 5.4. The spin wave with the energy $\hbar\omega_{\boldsymbol{q}}$ collides with an itinerant electron with energy $\varepsilon_{\boldsymbol{k},\sigma}$ (σ represents the spin state), and creates an itinerant electron with the momentum $\boldsymbol{k} + \boldsymbol{q}$ and spin orientation σ'

with the momentum wave-vector $\boldsymbol{q} \sim 0$ (uniform mode observed at FMR). The s-d exchange interaction on its own only leads to a renormalized spectroscopic splitting factor γ, see [5.26]. The coherent scattering of magnons with itinerant electrons has to be disrupted by incoherent scattering with other excitations. The spin flip hole-electron pairs $(a_{\boldsymbol{k},+}a^{+}_{\boldsymbol{k}+\boldsymbol{q},-})$ can be incoherently scattered by thermally excited phonons and magnons. That results in a fast fluctuating torque resulting in magnetic relaxation, see (5.10). One can account for incoherent scattering by including a finite life time τ_{eff} for the electron-hole pair excitations [5.26]. In this case the spin-flip electron hole pair energy has an additional imaginary term $\mathrm{i}\hbar/\tau_{\mathrm{eff}}$:

$$\Delta\varepsilon_{\boldsymbol{k},\boldsymbol{k}+\boldsymbol{q}} = \varepsilon_{\boldsymbol{k}+\boldsymbol{q},-} - \varepsilon_{\boldsymbol{k},+} + \mathrm{i}\frac{\hbar}{\tau_{\mathrm{eff}}} \,. \tag{5.27}$$

The effective lifetime, τ_{eff}, is in this case given by the spin-flip electron-hole pair, τ_{sf}. τ_{sf} is enhanced compared to the momentum relaxation time τ_{m}, which enters in the conductivity. The reason is that one needs to invoke spin orbit interaction to flip the electron spin during the relaxation process of electron hole pairs by phonons. Elliot [5.39] showed that for simple normal metals $\tau_{\mathrm{sf}} = \tau_{\mathrm{m}}/\Delta g^2$, where Δg is the deviation of the g-factor from its purely electronic value, $g = 2$, due to spin orbit interaction. This is not directly applicable to our case. For normal metal electrons scattering by 3d impurities, the ratio between the spin flip scattering and non-spin flip cross section is around 10^{-2} [5.40]. One can make a better estimate of τ_{sf} from the spin diffusion length l_{sd}. The spin diffusion length is a part of the spin accumulation process which occurs in Current Perpendicular to Plane Giant Magnetoresistance (CPP GMR) studies. The spin diffusion length l_{sd} [5.41] is given by

$$l_{\mathrm{sd}} = \left(\frac{\lambda^m_{\mathrm{FM}} v_{\mathrm{F}} \tau_{\mathrm{sf}}}{6}\right)^{1/2} , \tag{5.28}$$

where v_{F} is the Fermi velocity of electrons participating in the spin accumulation process. The effective momentum mean free path in the ferromagnet λ^m_{FM} [5.41], assuming a simple free electron model, is

$$\frac{1}{\lambda^{\mathrm{m}}_{\mathrm{FM}}} = \frac{1}{2}\left(\frac{1}{\lambda_\uparrow} + \frac{1}{\lambda_\downarrow}\right) = \frac{ne^2\rho^\star}{mv_{\mathrm{F}}} , \tag{5.29}$$

where $\lambda_\downarrow$ and $\lambda_\uparrow$ are the momentum mean free paths for minority and majority electrons, and n is the total density of conduction electrons. $\rho^\star = \rho_{\mathrm{FM}}(1-\beta^2)$, where ρ_{FM} is the measured resistivity of FM, and β is a bulk spin asymmetry coefficient, see [5.40]. In Py (permalloy) $\rho_{\mathrm{Py}} = 12.3\,\mu\Omega$, $\beta_{\mathrm{Py}} = .73$, and one conduction electron per atom [5.41]. Assuming $v_{\mathrm{F}} = 1.5\times10^8$ cm/s that results in $\rho^\star_{\mathrm{Py}} = 26.3\,\mu\Omega$ and $\lambda^{\mathrm{m}}_{\mathrm{Py}} = 2.4$ nm [5.41]. In Py $l^{\mathrm{Py}}_{\mathrm{sd}} = 4.3$ nm [5.41]. From (5.28) $\tau^{\mathrm{Py}}_{\mathrm{sf}} = 3\times10^{-14}$ s. In Py the corresponding enhancement factor for the spin flip relaxation is only $\tau^{\mathrm{Py}}_{\mathrm{sf}}/\tau^{\mathrm{m}}_{\mathrm{Py}} \sim 20$. In Co $\rho^\star_{\mathrm{Co}} = 6.6\,\mu\Omega$, $\beta_{\mathrm{Co}} = 0.36$ [5.41], that results in $\lambda^{\mathrm{m}}_{\mathrm{Co}} = 10$ nm. Piraux et al GMR measurements led to $l^{\mathrm{Co}}_{\mathrm{sd}} = 59$ nm at LN_2 temperatures, and their data analysis resulted in $\tau^{\mathrm{Co}}_{\mathrm{sf}} = 3.8\times10^{-12}$ s. In Co the enhancement factor for the spin scattering time is $\tau^{\mathrm{Co}}_{\mathrm{sf}}/\tau^{\mathrm{m}}_{\mathrm{Co}} \sim 300$.

If one would blindly apply Elliot's formula then $\Delta g = 0.06$ and 0.22 for Co and Py respectively. These numbers are not entirely unreasonable, but they should be treated only as accidental results.

The rf susceptibility can be calculated by using the Kubo Green function formalism in the Random Phase Approximation (RPA) [5.42]. The imaginary part (damping) of the denominator of the circularly polarized rf susceptibility is usually expressed as an effective damping field [5.26]

$$\frac{2\langle S\rangle}{Ng\mu_{\mathrm{B}}}\sum_{\boldsymbol{k}}|J(\boldsymbol{q})|^2\left(n_{\boldsymbol{k}+\boldsymbol{q},-}-n_{\boldsymbol{k},+}\right)\delta(\hbar\omega_{\boldsymbol{q}}+\varepsilon_{\boldsymbol{k},+}-\varepsilon_{\boldsymbol{k}+\boldsymbol{q},-})\,, \tag{5.30}$$

where the summation is carried out over all available states around the Fermi surface, $\langle S\rangle$ is the reduced spin $M_{\mathrm{s}}(T)/M_{\mathrm{s}}(0)$. The factor $g\mu_{\mathrm{B}} = \gamma\hbar$ was used to convert the relaxation energy into an effective relaxation field. The incoherent scattering of electron-hole pair excitations broadens the delta function into a Lorenzian [5.26]

$$\delta(\hbar\omega_{\boldsymbol{q}}+\varepsilon_{\boldsymbol{k},+}-\varepsilon_{\boldsymbol{k}+\boldsymbol{q},-}) \Rightarrow \frac{\hbar/\tau_{\mathrm{sf}}}{(\hbar\omega_{\boldsymbol{q}}+\varepsilon_{\boldsymbol{k},+}-\varepsilon_{\boldsymbol{k}+\boldsymbol{q},-})^2+(\hbar/\tau_{\mathrm{sf}})^2}\,. \tag{5.31}$$

A brief explanation of (5.30) and (5.31) is needed. The difference in occupation numbers is $\Delta n = n_{\boldsymbol{k}+\boldsymbol{q},-} - n_{\boldsymbol{k},+} = \delta(\varepsilon_{\boldsymbol{k}} - \varepsilon_{\mathrm{F}})\hbar\omega_{\boldsymbol{q}}$, where δ is Dirac's delta function, and shows that the relaxations processes are limited to electrons at the Fermi level. This is a crucial equation. It shows that the energy $\hbar\omega_{\boldsymbol{q}} = \hbar\omega$ of a resonant magnon is the energy which participates during the scattering process. The Lorentzian function (5.31) should not be looked upon as a smeared conservation of energy, but rather as an expression for the probability of achieving a certain scattering event. In ferromagnetic metals one usually studies by means of FMR a nearly homogeneous mode, $q \sim 0$. For the spin-flip electron-hole pair excitations the difference in electron energy is dominated by the exchange energy, $\varepsilon_{\boldsymbol{k}+\boldsymbol{q},-} - \varepsilon_{\boldsymbol{k},+} = +2\langle S\rangle J(0)$. Using $N\langle S\rangle g\mu_{\mathrm{B}} = M_{\mathrm{s}}(T)$ (assuming a unit volume) one gets a damping field proportional to the frequency ω and inversely proportional to the saturation magnetization M_{s}. This is a typical feature of Gilbert damping, see (5.17). After integration over the Fermi surface one can extract the Gilbert damping parameter

$$G = \chi_{\mathrm{P}}\frac{1}{\tau_{\mathrm{sf}}}\,, \tag{5.32}$$

where χ_{P} is Pauli's susceptibility for itinerant electrons. Pauli's susceptibility was obtained from the following expression

$$\chi_{\mathrm{P}} = \left(\frac{\hbar\gamma}{2\pi}\right)^2\int k^2\,\mathrm{d}k\delta(\varepsilon_{\boldsymbol{k}}-\varepsilon_{\mathrm{F}}) = \mu_{\mathrm{B}}^2N(\varepsilon_{\mathrm{F}})\,, \tag{5.33}$$

where $N(\varepsilon_{\mathrm{F}})$ is the density of states at the Fermi level of itinerant electrons. Notice that the summation in (5.30) was replaced by integration $(1/2\pi)^3\int \mathrm{d}k_x\,\mathrm{d}k_y\,\mathrm{d}k_z$.

It is interesting to estimate the Gilbert damping using (5.32). The Pauli susceptibility for 3d transition metals can be expected to be in the range of of 3–9×10^{-6}

[5.43]. The question remains what is the relationship between τ_{sf} obtained in GMR measurements, see (5.28), and in magnetic relaxation, see (5.32). In GMR one studies the longitudinal spin accumulation, while in FMR one investigates the transverse motion. The corresponding spin relaxation times are related, but the coefficient of proportionality can change between 1 to 2. In ESR often the longitudinal relaxation is assumed to be 2 times slower ($1/\tau_{sf} = 1/T_1 = 1/2T_2$) than the transversal relaxation time [5.44]. However, some simulations using a random Brownian spin walk results in the coefficient of one [5.45]. It is not obvious if this relationship holds for itinerant electrons in ferromagnet. One needs to carry out some discussion to clarify this point. One should realize that the magnetic moments in ferromagnet are locked together in FMR (acoustic mode) by the exchange coupling and therefore their moments precess at low frequency compared to that of the optical mode where the precession is governed by the exchange coupling. Therefore the spin relaxation rates of itinerant electrons at FMR are similar to those expected in ESR type of measurements. Both points are summarized by (5.38). In my view this equation applies to both the transverse and longitudinal relaxations. The only uncertainty is the factor relating the transverse and longitudinal τ_{sf}. This factor is between 1 to 2, see above.

In order to get Gilbert damping by the s-d interaction mechanism, see (5.32), in the range of $1\times10^8\,\mathrm{s}^{-1}$ one needs to have τ_{sf} in the range of $5\times10^{-14}\,\mathrm{s}$. It is interesting to note that this is satisfied for Py, $\tau_{sf} = 3\times10^{-14}\,\mathrm{s}$. In fact recently Ingvarsson et al. [5.46] showed that the Gilbert damping parameter in Py($Ni_{81}Fe_{19}$) films scales proportionally with the sample resistivity. This is in good agreement with predictions by formula (5.32). The resistivity in Py samples used by Ingarsson et al. was changed as much as by a factor of four by changing the growth morphology of Py films. Py is in this respect a special case; in pure materials like Co and Fe the spin flip relaxation time is too long to contribute to this type of Gilbert damping.

The above calculations were carried out for circularly polarized magnons. It was shown that the ellipticity of magnons (parallel configuration) [5.47] does not change the intrinsic Gilbert damping which is based on scattering processes (shown in Fig. 5.4). That is not a surprising result considering that the FMR linewidth ΔH for circularly polarized magnons showed the explicit features of Gilbert damping. ΔH is proportional to ω and indirectly proportional to M_s, see (5.17).

Ignoring the presence of collisions with thermally excited magnons the s-d relaxation is indirectly proportional to τ_{FM}^{m}, and consequently scales with resistivity.

Kamberský [5.48] showed that the intrinsic damping in metallic ferromagnets can be treated more generally by using the spin-orbit interaction Hamiltonian. The spin-orbit Hamiltonian corresponding to the transverse spin and angular momentum components can be expressed in a three particle interaction Hamiltonian

$$\frac{1}{2}\left(\frac{2S}{N}\right)^{0.5}\xi\sum_{\boldsymbol{k}}\sum_{\alpha,\alpha',\sigma}\langle\beta|L^{+}|\alpha\rangle c^{+}_{\beta,\boldsymbol{k}+\boldsymbol{q},\sigma}c_{\alpha,\boldsymbol{k},\sigma}b_{\boldsymbol{q}} + \mathrm{h.c.}\,, \tag{5.34}$$

where ξ is the coefficient of spin-orbit interaction, $L^{+} = L_x + \mathrm{i}L_y$ is the right handed component of the atomic site transverse angular momentum $\boldsymbol{L}_{\mathrm{a,t}}$. $c_{\alpha,\boldsymbol{k},\sigma}$ and $c^{+}_{\beta,\boldsymbol{k}+\boldsymbol{q},\sigma}$ annihilate and create electrons in the appropriate Bloch states with the spin σ, and

b_q annihilates the spin wave with the wave-vector $\boldsymbol{q}$. The indices α, β represent the projected local orbitals of Bloch states, and are used to identify the individual electron bands. For simplicity no dependence of the matrix elements $\langle\beta|L^+|\alpha\rangle$ on the wave-vectors will be considered. The rf susceptibility can again be calculated by using the Kubo Green function formalism in the Random Phase Approximation (RPA). The imaginary part of the denominator of the circularly polarized rf susceptibility for a spin wave with the wave-vector $\boldsymbol{q}$ and energy $\hbar\omega$ can be expressed in a manner similar to that carried out for the s-d exchange interaction. The effective damping field is then given by

$$\frac{G}{\gamma M_s}\frac{\omega}{\gamma} = \frac{\langle S\rangle^2}{2M_s}\xi^2\left(\frac{1}{2\pi}\right)^3\int \mathrm{d}\boldsymbol{k}^3\sum_{\alpha,\beta,\sigma}\langle\beta|L^+|\alpha\rangle\langle\alpha|L^-|\beta\rangle$$
$$\times\,\delta(\varepsilon_{\alpha,\boldsymbol{k},\sigma}-\varepsilon_{\mathrm{F}})\hbar\omega\frac{\hbar/\tau_{\mathrm{m}}}{(\hbar\omega+\varepsilon_{\alpha,\boldsymbol{k},\sigma}-\varepsilon_{\beta,\boldsymbol{k}+\boldsymbol{q},\sigma})^2+(\hbar/\tau_{\mathrm{m}})^2}\,. \tag{5.35}$$

$\langle S\rangle$ is the reduced spin $M_s(T)/M_s(0)$. Since no spin flip is present during the scattering the relaxation time τ_{sf} in (5.31) is replaced by the momentum relaxation time τ_m which enters the conductivity of FM.

Intraband transitions, $\alpha=\beta$: For low frequency spin waves ($q \ll k_{\mathrm{F}}$) the electron energy balance $\hbar\omega+\varepsilon_{\alpha,\boldsymbol{k},\sigma}-\varepsilon_{\alpha,\boldsymbol{k}+\boldsymbol{q},\sigma}=\hbar\omega-(\hbar^2/2m)(2\boldsymbol{kq}+q^2)$ in the denominator of (5.31) can be significantly less than $\hbar/\tau_{\mathrm{m}}$. Even in good crystalline structures this limit is satisfied above cryogenic temperatures. After integration over the Fermi surface the Gilbert damping can be approximated by

$$G \simeq \langle S\rangle^2\left(\frac{\xi}{\hbar}\right)^2\left(\sum_{\alpha}\chi_{\mathrm{p}}^{\alpha}\langle\alpha|L^+|\alpha\rangle\langle\alpha|L^-|\alpha\rangle\right)\tau_{\mathrm{m}}\,, \tag{5.36}$$

where $\chi_{\mathrm{p}}^{\alpha}$ the the Pauli susceptibility of those states which participate in intraband electron transitions and satisfy $|\ \hbar\omega-\hbar^2/2m(2\boldsymbol{kq}+q^2)\ | \ll \hbar/\tau_{\mathrm{m}}$. The Gilbert damping in this limit is proportional to the relaxation time τ_{m}, and consequently scales with the conductivity.

Interband transitions, $\alpha\neq\beta$: Interband transitions are associated with energy gaps $\Delta\varepsilon_{\beta,\alpha}$. The electron hole pair energy can be dominated by these gaps. For the gaps larger than the relaxation frequency $\hbar/\tau_{\mathrm{m}}$ the Gilbert damping can be approximated by

$$G \simeq \langle S\rangle^2\sum_{\alpha}\chi_p^{\alpha}(\Delta g_{\alpha})^2\frac{1}{\tau_m}\,, \tag{5.37}$$

where $\chi_{\mathrm{p}}^{\alpha}$ are appropriate Pauli susceptibilities for Fermi sheets α, and Δg_{α} is the deviation of the g-factor from its purely electronic value $g=2$. $\Delta g_{\alpha} = 4\xi\sum_{\beta}\langle\alpha|L_x|\beta\rangle\langle\beta|L_x|\alpha\rangle/\Delta\varepsilon_{\beta,\alpha}$, see [5.49], determines the contribution of spin orbit interaction to the spectroscopic splitting factor g. Notice that in this limit the spin-orbit interaction results in a Gilbert damping coefficient similar to that found for the s-d exchange interaction, see (5.32), if one assumes Elliot's formula for the spin flip relaxation time τ_{sf}. χ_{p} only includes those electron states for which the change in energy during the interband transitions satisfy $\Delta\varepsilon \gg \hbar/\tau_{\mathrm{m}}$. In this approximation the Gilbert damping is proportional to $1/\tau_{\mathrm{m}}$, and consequently scales with resistivity. In reality a large distribution of energy gaps modifies the overall temperature dependence. The interband damping can be expected to be dependent on resistivity only at low temperatures. With an increasing temperature the relaxation rate $\hbar/\tau_{\mathrm{m}}$ becomes comparable to the energy gaps $\Delta\varepsilon_{\alpha,\beta}$, which results in a gradual saturation of the interband Gilbert damping with increasing temperature [5.48].

So far the treatment of intrinsic damping has been mostly formal. At this point it is useful to outline a simple physical description of intrinsic damping. Kamberský's model was originally based on the observation that the Fermi surface changes with the direction of the magnetization [5.50]. This model corresponds to intraband transitions. This is a relatively easy mechanism to describe by a classical picture. As the precession of the magnetization evolves in time and space the Fermi surface also distorts periodically in time and space. This is often referred to as a *breathing Fermi surface*. The effort of the electrons to repopulate the changing Fermi surface is delayed by a finite relaxation time τ_{m} of the electrons and this results in a phase lag between the Fermi surface distortions and the precessing magnetization. The interband transitions are connected with dynamic orbital polarization, i.e. changes of the electron wave functions beside changes of their energies.

The s-d exchange interaction can be viewed as two precessing magnetic moments corresponding to the d-localized and itinerant electrons which are mutually coupled by the s-d exchange field. In the absence of damping the low energy excitation (acoustic mode, FMR) corresponds to a parallel alignment of the magnetic moments precessing together in phase. However, due to a finite spin mean free path of the itinerant electrons, the eq. of motion for itinerant electrons has to include spin relaxation towards the instantaneous effective field,

$$-\frac{1}{\tau_{\mathrm{sf}}\gamma}(\boldsymbol{m} - \chi_{\mathrm{P}}\boldsymbol{h}_{\mathrm{eff}}) , \tag{5.38}$$

where $\boldsymbol{h}_{\mathrm{eff}}$ includes the exchange field between the localized and itinerant electrons That results in a phase lag between the two precessing magnetic moments [5.51], and consequently in magnetic damping. It is interesting to note that the magnetic damping using this classical approach is equivalent to that using the Kubo formalism.

The phase lag for the *breathing Fermi surface* and the s-d exchange interaction is proportional to the microwave frequency ω. Clearly, in both cases one gets a typical situation for "friction" like damping which is described in magnetism by the Gilbert relaxation term.

In Ni the Gilbert damping is significantly increased as the temperature approaches the cryogenic range of temperatures and saturates for temperatures less than 50 K [5.52]. The Gilbert damping parameter G was found to be $2.5 \times 10^8\ \mathrm{s}^{-1}$ at RT, and $14 \times 10^8\ \mathrm{s}^{-1}$ at Helium temperature. The saturation of G was explained by Korenman and Prange [5.53] using an equation similar to (5.35). For intraband electron-hole pair excitations $\varepsilon_{\boldsymbol{k}} - \varepsilon_{\boldsymbol{k}+\boldsymbol{q}} = (\boldsymbol{k}_{\mathrm{F}}\boldsymbol{q}/m + q^2/2m)\hbar^2$. With increasing τ_{m} the energy balance in the denominator of (5.35) becomes eventually comparable to $\hbar/\tau_m$. Momentum and energy conservation now play an important role in the sum over the Fermi surface in (5.35), and one finds

$$G \sim \frac{\tan^{-1}(qv_F\tau_m)}{qv_F}, \tag{5.39}$$

where q is the wave number of a resonant magnon. For $v_F\tau_m \gg 1/q$ the expression in (5.39) saturates and depends inversely on q. This behavior was already well known in connection with the anomalous skin depth where only electrons moving within the skin depth contribute to the effective conductivity. This effect is usually referred to as the concept of ineffective electrons [5.54]. The presence of ineffective electrons at low temperatures in the measured Gilbert damping shows very explicitly that the magnetic damping in metallic ferromagnets is caused by itinerant electrons. This is further supported by Ferromagnetic Antiresonance (FMAR) studies. By using microwave transmission at FMAR ($q \to 0$) we were able to avoid problems with ineffective electrons [5.55] and to get precise values of the intrinsic damping. We found that in high purity single crystal slabs of Ni the Gilbert damping below RT (keeping the role of thermal magnons low) was well described by the two terms which were equal in strength and proportional to the conductivity and resistivity [5.55]. This is in good agreement with the above predictions. At RT these two terms compensated each other and that resulted in a nearly temperature independent Gilbert damping at ambient temperatures. The FMAR measurements on Ni showed a nearly temperature independent L.L. parameter [5.56]. The saturation of the L.L. parameter above RT is quite well accounted for by the recent quantitative calculations by Kamberský [5.57]. It was explained by the interband contributions which saturate above RT, see above.

The conclusion: High quality crystalline Ni samples convincingly showed that the intrinsic damping in metals is caused by the itinerant nature of the electrons and the spin-orbit interaction. The models presented were intended to highlight in a simple way the underlying physics of magnetic relaxations in metals, and predict some basic trends of damping in metallic ferromagnets. Detailed calculations can be often complex and not easy to penetrate. Quantitative calculations [5.26, 48, 58–60] show that the spin-orbit interaction is indeed the leading mechanism underlying the intrinsic damping in ferromagnetic metals.

The control of damping allows one to speed up the approach to equilibrium in magnetization reversal processes. This is particularly important in Magnetic Random Access Memory (MRAM) applications. Memory pixels in MRAM should have a high damping. The shortest time of magnetization reversal is achieved by critical damping, see e.g. Fig.5.2c, and is given by one period of precession. Silva et al [5.61] have shown that one can control the damping in Py by adding rare earth elements. See further details in section Engineering High Frequency Dynamic Properties by Russek et al. [5.62] and Bailey et al. [5.63]. Rare earth elements have a large orbital momentum allowing one to significantly enhance the spin orbit contribution to the magnetic properties in 3d transition metals. It was found that a modest concentration of 2 percents of Tb in Py resulted in a 10 fold increase of the magnetic damping as measured by FMR [5.64]. This additional damping was found strongly temperature dependent. It increased by a factor of 6 going from RT to 150 K. The temperature dependence of resistance in Tb doped Py can be expected to be even weaker than in Py, and therefore this substantial increase in damping with decreasing temperature cannot be explained by the spin orbit contribution as discussed above. The FMR linewidth was found independent on the angle of the saturation magnetization with respect to the film surface suggesting that is caused by an intrinsic contribution. A precession of the Py magnetic moment is coupled by antiferromagnetic coupling to the Tb magnetic moments. The spin orbit interaction leads to a simultaneous rf motion of the Tb orbital momentum. This motion results in local lattice distortions (short wavelength phonons) which can provide an effective channel for dissipation of the magnetic energy to the lattice [5.64]. The FMR line broadening by two magnon scattering, due to magnetic inhomogeneities, can be excluded because it would lead to a substantial narrowing of the FMR linewidth when approaching the perpendicular configuration, see Sect. 5.7.1. However one should keep in mind that large scale lateral inhomogeneities (which are not described by two magnon scattering) can result in a significant FMR linewidth which in perpendicular configuration can even surpass that measured in the parallel configuration. Large scale inhomogeneities would result in a FMR linewidth independent on the microwave frequency, see Sect. 5.7.1. Measuring the dependence of the FMR linewidth on the microwave frequency would help to clarify the origin of magnetic damping in the Tb doped Py films.

5.4.4 Dynamic Studies

Numerous studies of magnetic relaxation in metallic bulk and thick film samples were already carried out in the sixties and seventies. The advent of MBE in the growth of magnetic ultrathin metallic films in the eighties brought back an appreciable interest in dynamics. However, it has been the recent requirement for fast switching MRAM and memory media bits that brought the study of magnetic relaxation to the forefront in the study of magnetic nanostructures.

5.4.5 Techniques for Dynamic Studies

The description of standard FMR techniques can be found in Chap. 3, Volume II of this book [5.2]. The situation has changed dramatically in experimental techniques during the last 8 years. FMR measurements can be carried out on micrometer sized samples. These techniques employ a variety of tunable Atomic Force Microscope (AFM) microcantilevers operating either on torque or calorimetric sensing. Magnetic Resonance Force Microscopy (MRFM) achieves high sensitivity by means of a mechanical resonator that detects the force between a small probe magnet (attached to the cantilever) and the longitudinal magnetization (perpendicular to the plane of precession) in the sample [5.65, 66]. The Caltech and Los Alamos groups [5.67] recently used the MRFM technique for imaging rf magnetostatic modes on micrometer scale samples of yttrium iron garnets. The spatial resolution of this technique presently is 10 μm. The NIST group in Boulder, Colorado developed FMR detection using micrometer sized thin film samples which are deposited onto a micromechanical cantilever detector [5.68–70]. The cantilever responds to FMR in one of three ways: (a) Measuring the change in torque on the sample in a uniform field using the changing longitudinal magnetic moment. (b) Measuring the FMR damping torque. (c) Measuring the absorbed energy in FMR using a bimaterial cantilever as a calorimetric sensor. Presently they are able to detect ferromagnetic samples having an effective magnetic volume of 2×10^{-11} cm^3. The spatial resolution is now in the micrometer range, but it is expected to eventually reach the nanometer scale.

Temporal (picosecond) resolution can be achieved by using a variety of microwave probe stations using coplanar wavequides. In this case the temporal detection of rotational motion of magnetization, is performed using a pulsed inductive microwave magnetometer (PIMM), see [5.61]. An excellent presentation of this technique can be found in [5.71]. The measurements were carried out by using a single coplanar wavequide. Magnetic field pulses of 50 ps rise time and 10 ns duration are created by a fast pulse generator on one side of the wavequide. The sample is placed (or deposited) on the coplanar waveguide. The time dependence of the magnetization rotation is inductively picked up by the coplanar transmission line. The resulting signal is detected by a 20 GHz bandwidth digital sampling oscilloscope. The pulse generator and oscilloscope are attached to the coplanar transmission line by impedance matched picoprobes [5.72]. One is able this way to investigate the free precession of magnetization. The angle of precession was adjusted by the amplitude of the pulsed field and was varied from 0.002 to 40 Degrees. For spin-valve samples the rotational motion of the magnetization can be detected by Giant Magneto-Resistance (GMR). In this case the measurements are carried out using two microwave strip lines. The test line creates short field pulses, and the read line is employed to monitor the GMR signal that depends on the instantaneous angle between the magnetic moments, see [5.73]. An excellent review of microwave testing probes and techniques for FMR studies of confined magnetic structures can be found in thesis by M. Bailleul [5.74].

Temporal (picosecond) and spatially (sub μm) resolved techniques are now employed to investigate the magnetization reversal and large angle precession in a variety of systems which are attractive for spintronics applications. This area of research is covered in this book (Volume III) in the chapter on *Nonequilibrium spin dynamics in laterally defined magnetic structures* by Byoung-Chul Choi and M. Freeman [5.75].

A significant effort has been carried out in computer simulations of spin dynamics. They provide a guide for researchers in identifying proper lateral geometries of memory pixels and show a detailed dynamic picture of magnetization reversal. The modes of magnetization reversal of small lateral geometries are addressed in this book (Volume IV) in the chapter on *An Introduction to Micromagnetics* by A.S. Arrott [5.76].

5.4.6 Intrinsic Damping, FMR Experiments

The discussion in this subsection will be limited to those experiments whose main purpose is to identify the intrinsic magnetic damping in metals. A long standing and persistent question concerns the applicability of L.L. and Gilbert relaxation terms in describing the intrinsic damping. This question has not been easy to answer. A simple proof can be done by measuring the FMR linewidth as a function of the microwave frequency and the angle of the saturation magnetization with respect to the sample normal. The L.L. and Gilbert damping models predict that the FMR linewidth in the parallel and perpendicular configuration have a strictly linear dependence on the microwave frequency, and the FMR linewidth is expected to be identical for the parallel and perpendicular configurations provided that the measurements are carried out along equivalent crystallographic directions, see Sects. 5.2–5.4. This seemingly simple set of experiments is obscured by a lack of perfect samples. One needs high quality single crystals with perfect surfaces to make a convincing interpretation of the measured FMR results.

Fe whiskers and single crystal platelets of Ni, Fe and Fe-Ni and Fe-Co alloys prepared by chemical vapor transport provided such perfect systems. In these samples the skin depth is smaller than the sample thickness, and the interpretation of experimental results requires complete calculations using either the L.L. or Gilbert equations of motion together with Maxwell's equations. Quach's et al. [5.77] FMR measurements were carried out at 9.5, 24.8, and 35.5 GHz using Ni-Co(100) platelets. Their results were well described by an angular independent L.L. damping parameter $\lambda = 2.25 \times 10^8\,\mathrm{s}^{-1}$. Anderson et al. [5.78] using single crystals of Ni(001) and Ni(110) showed that the FMR linewidth at 22 GHz as a function angle is again well described by a constant L.L. damping parameter $\lambda = 2.3 \times 10^8\,\mathrm{s}^{-1}$. Frait and Fraitová [5.27] in their extensive FMR studies of Fe whiskers over a wide range of microwave frequencies, 20–100 GHz, clearly showed that the intrinsic damping is well represented by L.L. damping, $\lambda_{\mathrm{Fe}} = 5.7 \times 10^7\,\mathrm{s}^{-1}$.

FMAR Measurements. Ferromagnetic Antiresonance (FMAR) for measurements of microwave properties was introduced by Heinrich and Mescheryakov [5.79, 80] in the late sixties. The measurements of the L.L. damping is significantly simplified at FMAR. At FMAR the real part of the rf permeability μ is zero, and the role of eddy-currents is substantially decreased, see [5.80]. In fact the microwave transmission is limited only by the imaginary part χ'' of the rf susceptibility at FMAR [5.80, 81]. FMAR is an anti-resonance effect ($\omega/\gamma = B_{\mathrm{int}}$), χ'' is very small resulting in an appreciable increase of the skin depth compared to the skin depth at $\mu = 1$. The skin depth at FMAR is $\delta \sim \left(\sqrt{4\pi M_{\mathrm{s}}/\Delta H_{\mathrm{FMR}}}\right)\delta_0$, where δ_0 is the skin depth for $\mu = 1$. The microwave transmission at FMAR increases by many orders of magnitude over that for an rf permeability $\mu = 1$, and results in a well defined peak in the microwave transmission measurements. Fitting of the microwave transmission peak at FMAR allows one to determine very accurately the magnetic damping [5.79, 80]. In addition, the FMAR linewidth is unaffected by extrinsic relaxation mechanisms, see Sect. 5.7.1. The measurements of the microwave transmission at FMAR provide accurate results of the intrinsic damping in metals. FMAR was extensively used for Py (permalloy) [5.80, 81], amorphous Fe rich ribbons [5.82, 83], Ni [5.55, 56, 84] and Fe [5.85, 86] single crystal samples. In these measurements the rf response was well described by the L.L.G. equation of motion.

A significant change in the rf permeability at FMR and FMAR results in large variations of the absorbed microwave power as a function of applied field in thick samples where the sample thickness exceeds the skin depth. In this limit the microwave absorption is given by the surface impedance which is proportional to the inverse value of the skin depth [5.87]. Consequently at FMR and FMAR one gets maximum and minimum absorption of microwave power, respectively. The difference in microwave absorption can vary as much as by a factor of a hundred in the microwave range of frequencies. Large variations in the surface impedance as a function of applied field is called giant magnetoimpedance (GMI) effect [5.88, 89].

In Conclusion: *Careful FMR measurements using high quality crystalline metallic bulk materials and the FMAR microwave transmission studies strongly support the conclusion that the intrinsic relaxation term in most ferromagnetic metals can be described by either L.L. or Gilbert damping terms.*

5.4.7 Relaxation at Large q Wave-numbers, Dipole-dipole Damping

The relaxation processes discussed so far were limited to small q wave-numbers (wave-vectors). Small q-vectors (< 0.01 Å^{-1}) are definitely most important in a wide range of studies including submicrometer laterally defined structures. However, in some cases the magnetic excitations having large q-wave vectors can play a decisive role in spin dynamics. For example a magnetic pixel can be brought above its critical point T_c, and then it can be cooled down rapidly to a desired ferromagnetic state. In this case the magnetic response involves thermally excited magnons with a wide range of q wave-vectors. The purpose of this section is to provide some insight into relaxation processes which include fluctuating torques in the proximity of the critical point and magnetic excitations characterized by large q wave-vectors.

The dipole-dipole interaction in metals usually represents a dominant contribution to the effective field. A typical example is the demagnetizing field perpendicular to the surface. In Fe $4\pi M_s \sim 20$ kOe at RT. One would expect that temperature fluctuations of such dipolar fields can also result in a significant contribution to the magnetic damping. The role of fluctuating dipolar fields in relaxation processes around the critical point T_c has been investigated in a direct and systematic way by inelastic neutron scattering studies. A good review of this field can be found in the article by Frey and Schwabl [5.90]. Recent experimental studies have been concentrated mostly on classic Heisenberg ferromagnets such as EuS and EuO, see papers by Goerlitz and Koetzler [5.6] and Boeni et al. [5.91].

Before we discuss the data from neutron diffraction experiments it is useful to briefly describe the inelastic neutron scattering technique with an emphasis on magnetic excitations. The intensity of inelastic neutron scattering is proportional to the imaginary part of the magnetic susceptibility $\chi''(\boldsymbol{q}, \omega)$. In this respect nothing new has to be introduced. The susceptibility is determined by solving (5.1) which was fully described in Sects. 5.1, 5.2. The neutron scattering allows one to investigate both the spin wave (transversal susceptibility) and longitudinal (longitudinal susceptibility) excitations by using spin-flip and non-spin-flip intensities with the incident polarization σ and wave vector $\boldsymbol{Q}$, see further details in [5.90, 91]. For completeness let us summarize the expressions for the imaginary parts of transverse and longitudinal susceptibilities. The imaginary part of the transverse susceptibility is given by

$$\chi''_{\perp}(\boldsymbol{q},\omega) = \frac{\chi_{\perp}(q)\omega\Gamma_{\perp}(q)}{(\omega-\omega(q))^2+\Gamma_{\perp}(q)^2} + \frac{\chi_{\perp}(q)\omega\Gamma_{\perp}(q)}{(\omega+\omega(q))^2+\Gamma_{\perp}(q)^2}, \tag{5.40}$$

where the relaxation frequency $\Gamma_{\perp}(q) = \lambda_{\perp}(q)/\chi_{\perp}(q)$. $\lambda_{\perp}(q)$ and $\chi_{\perp}(q)$ are the transverse L.L. damping parameter and the transverse susceptibility corresponding to the wave vector q. Note that the first and second term in (5.40) correspond to the right (resonant) and left handed (non-resonant) precession of rf magnetization, respectively. The expression in (5.40) assumes a circularly polarized precession. The wave vector $\boldsymbol{q}$ was replaced by its magnitude q, no directional dependence will be discussed. The longitudinal susceptibility is described by the same expression as in (5.40) only the transverse susceptibility and L.L. damping are replaced by their longitudinal counterparts and the resonant frequency $\omega(q) = 0$. In neutron studies the L.L. coefficient is usually referred to as the Onsager coefficient of spin dynamics. The neutron scattering technique is able to measure all 5 parameters, $\chi_{\perp}(q)$, $\chi_{||}(q)$, $\Gamma_{\perp}(q)$, $\Gamma_{||}(q)$, and $\omega(q)$.

Small q Wave-vectors. Mezei in 1982 [5.92] carried out a pioneering work on Fe for temperatures above T_c using time of flight (TOF) and neutron spin echo (NSE) techniques. The TOF and NSE techniques allowed him to investigate relaxation rates for small scattering q-vectors down to $0.006\,\text{Å}^{-1} = 6\times10^5\,\text{cm}^{-1}$ which for the first time overlapped with those used in FMR measurements for the bulk metallic ferromagnets, see Sect. 5.4.6. As expected for small q wave-vectors Mezei observed a spin diffusion like scattering ($\omega(q) = 0$), with the relaxation frequency, $\Gamma = \lambda/\chi$, independent of q. Its strength was found to be proportional to $\sqrt{\chi}$, where χ is the critical susceptibility around T_c. For Fe $\chi = (\Delta T/1.45)^{-1.33}$ [5.12], where $\Delta T = T - T_c$. It is instructive to look at some of Mezei's quantitative results. The temperature $\Delta T = T - T_c = 1.4\,\text{K}$ was the closest to T_c. The relaxation frequency Γ (energy width) was found to be of $1.8\times10^9\,\text{s}^{-1}$. The Onsager (L.L.) damping parameter is given by $\Gamma\times\chi$, see (5.19), and that results in a L.L. parameter of $2\times10^9\,\text{s}^{-1}$. For the highest measured temperature, $\Delta T = 51\,\text{K}$, the relaxation parameter was found to be $\Gamma = 2.3\times10^{10}\,\text{s}^{-1}$. The corresponding susceptibility is $\chi = 0.0088$, and results in a L.L. damping of $1.9\times10^8\,\text{sec}^{-1}1$. It is interesting to note that at $\Delta T = 41\,\text{K}$ the measured L.L. damping is close to that measured by FMR at moderately high temperatures where the spin-orbit interaction determines the L.L. damping, e.g. at $T\sim 600\,\text{K}$ the L.L. parameter from FMR and FMAR measurements was found to be equal to $0.9\times10^8\,\text{s}^{-1}$ and $0.8\times10^8\,\text{s}^{-1}$, respectively. A significant increase of the L.L. damping close to T_c is caused by the dipole-dipole interaction [5.93]. Mezei's measurements were carried out above T_c. It is fair to ask how relevant these results are for temperatures below T_c. Boeni et al. [5.91] showed that passing through T_c from above, the dipolar dynamic universality class for the transverse (critical) fluctuations is not changed, and it is therefore reasonable to take the L.L. damping from above T_c as a good measure of the damping below T_c.

Large q wave-vectors. For large q-vectors ($q > 0.04\,\text{Å}^{-1}$) the dipolar relaxation frequency reaches the universal behavior as in Mezei's measurements. It scales as q^z with $z = 5/2$ which corresponds to the exchange dynamic universality region where the magnetic fluctuations are dominated by the exchange interaction [5.92]. In the exchange region the relaxation frequency Γ for $T > T_c$ was found to be independent of temperature [5.92]. For $q = 0.06\,\text{Å}^{-1}$ the relaxation frequency was found to be of $1.6 \times 10^{11}\,\text{s}^{-1}$ [5.92]. The susceptibility for large q-wave vectors can be estimated by using the Ornstein-Zernike formula $\chi(q) = q_d^2/(\kappa^2+q^2)$, where κ is the inverse value of the correlation length and q_d is the dipolar wave-number. In Fe $q_d = 0.045\,\text{Å}^{-1}$ and $\kappa = 0.09\,\text{Å}^{-1}$ at $T - T_c = 1.4\,\text{K}$ [5.93]. The corresponding $\chi(\Delta T = 1.4) = 0.16$ and results in a L.L. damping parameter of $3 \times 10^{10}\,\text{s}^{-1}$.

Lynn and Mook [5.94, 95] showed that the neutron excitations in Ni and Fe for large q wave-vectors ($\sim 0.3\,\text{Å}^{-1}$) exhibit a well defined resonance peak even above T_c. In itinerant ferromagnets only the long range order is lost; a significant degree of the short range order exists even well above T_c. Lynn [5.94] measured the relaxation parameter Γ in Fe for $q = 0.5\,\text{Å}^{-1}$ from 4 to 1200 K. For $T > T_c$ the relaxation parameter was, as in Mezei's measurements, temperature independent $\Gamma(T > T_c) = 2.7 \times 10^{13}\,\text{s}^{-1}$, but it rapidly decreased for temperatures less than T_c. In the ferromagnetic state the transversal susceptibility for large q wave-vectors is given by $\chi_\perp(q) = M_s/H_q$, where $H_q = (2A/M_s)q^2$ is the exchange field. In RPA the transverse susceptibility $\chi_\perp(q)$ is independent of temperature below T_c [5.42]. For $q = 0.5\,\text{Å}^{-1}$ the transversal susceptibility is $\chi_\perp = 3 \times 10^{-4}$ and the corresponding Onsager coefficient $\lambda(q = 0.5\,\text{Å}, T \sim T_c) = 8 \times 10^9\,\text{s}^{-1}$. Well below T_c at $T \sim 800\,\text{K}$ the relaxation frequency is smaller by a factor of 10 resulting in a 10 times smaller L.L. damping compared to that at T_c, $\lambda(q = 0.5\,\text{Å}, T = 800\,\text{K}) = 8 \times 10^8\,\text{s}^{-1}$.

In itinerant ferromagnets one also has to consider simple Stoner type excitations where an electronic spin flip excitation across the Fermi surface creates a single particle electron-hole pair with opposite spins. The spin waves can decay into Stoner excitations when the energies of the spin wave is comparable to single particle excitations of an equivalent q wave-vector. This means the spin-wave excitations are not present across the whole Brillouin zone. The disappearance of spin waves in Fe and Ni due to heavy damping caused by Stoner type excitations occurs above $\sim 100\,\text{meV}$ ($q > 0.5\,\text{Å}^{-1}$). First principles numerical calculations of the dynamic properties of Ni and Fe using fully itinerant electron band calculations were found to be in excellent agreement with the neutron scattering experiments [5.96].

In conclusion: The neutron studies showed that the L.L. damping parameter is significantly enhanced around the critical point T_c. Even well below T_c there is a significant enhancement of the L.L. damping parameter λ at large q wave-vectors compared to that at $q \sim 0$. The enhancement of the L.L. parameter is caused by dipole-dipole interaction.

Note, that close to T_c the relaxation rates Γ are around 300 fs. One should point out that the above neutron studies were obtained on bulk materials. In ultrathin films the fluctuations and magnon electron interactions are constrained to a two dimensional space, and therefore the bulk results can be used only as rough estimates. Recently Vollmer et al. [5.97] were able to study large wave vector spin waves in an 8ML thick ultrathin film of fcc Co(001) (grown by MBE on Cu(001) at RT) by Spin Polarized Electron Energy Loss (SPEEL) spectrometry. For ultrathin films the neutron studies are not possible. The SPEEL spectroscopy allowed them to investigate the spin waves up to the surface Brilloin zone boundary with $q_{\max} \sim 1.2\,\text{Å}^{-1}$ and the spin wave energy $E \sim 220\,\text{meV}$. The spin wave energy approached the surface Brilloin zone with zero slope as expected for the Heissenberg model. This is indeed a remarkable result. In the neutron studies on the bulk Ni and Co samples the spin waves disappeared into the single particle excitation already at $\sim 100\,\text{meV}$. This difference between Co and Ni, Fe is not clear at this point, but is most likely caused by much weaker itinerant effects in ultrathin Co films. The spin wave peaks in the SPEEL spectra were significantly broadened. After removing the instrumental resolution the energy width ranged from 40 to 75 meV. This life time would be only of 10 fs. This is significantly faster relaxation than that observed by the neutron studies in Ni and Fe at RT, see above. However one should realize that the neutron studies were only able to investigate the spin waves up to the energy of $\sim 100\,\text{meV}$ and $q_{\max} \sim 0.5\,\text{Å}^{-1}$. The short life time of spin waves at large q-vectors in the Co ultrathin film is most likely caused by the enhanced decay of the spin wave into Stoner excitations.

5.4.8 Magnetic Relaxation at Large Precessional Angles

FMR measurements are usually carried out using small precession angle. One can ask whether the magnetic damping in samples undergoing a large angle of precession (encountered in magnetization reversal processes) is comparable to that studied by FMR. In domain wall mobility measurements the magnetization angle changes by 180°, and therefore these studies allow one to estimate the effective Gilbert (L.L.) damping for large precessional angles. The domain wall motion in Ni and Py thin films (less than several 100 nm in thickness) is mostly limited by the magnetic damping. The L.L. damping creates a viscous force on the domain wall $f_{\text{visc}} = \beta v$, where v is the domain wall velocity. The viscous force opposes the applied magnetic pressure which is proportional to the applied field. The parameter β is proportional to the damping parameter α, and is indirectly proportional to the effective domain wall width $\Delta W_{\text{eff}} = \epsilon_{\text{dw}}/2A$, where ϵ_{dw} is the domain wall energy density and A is the exchange coupling coefficient [5.99, 100]. The measurements of the domain wall velocity were usually carried out using a constant applied field. The velocity per unit magnetic field is called the domain wall mobility. Regrettably the dependence of β on the effective domain wall width makes the interpretation of domain wall mobility measurements a non trivial problem.

Patton and Humphrey [5.98] showed that the domain wall mobilities in permalloy (Py) films are significantly larger than those expected for a Bloch wall with the Gilbert damping obtained from FMR. Significantly wider domain walls were used to explain the measured data. This was in agreement with the results of static wall-shape measurements by Fuchs [5.101] and Suzuki et al. [5.102] using defocused mode Lorentz microscopy.

The study of the domain wall ac response by Leaver and Vojdani [5.100] was carried out using nearly perfect single crystal Ni platelets, 200 nm thick which were prepared by chemical vapor transport. In their measurements the coefficient of the viscous force β was found to be in good agreement with predictions using a Neel-like wall and the intrinsic Gilbert damping obtained from Rodbell's FMR measurements on Ni platelets [5.103]. The Leaver-Vojdani studies are unique because they were carried out on perfect crystalline samples. Their measurements can be used for a reliable estimate of the Gilbert damping at large angles of precession. In this respect it is prudent to critically assess their conclusions.

Leaver and Vojdani assumed that the domain wall in Ni platelets has a fully developed vortex, completely free of magnetic charge, see [5.104, 105]. This is strictly valid for systems with a negligible crystalline anisotropy. The importance of anisotropy is gauged by the reduced anisotropy coefficient (reduced quantity) Q which is given by the ratio of the crystalline and magnetostatic energy [5.99]. In Ni $Q = 0.2$; this means that the magnetostatic energy cannot be entirely neglected. In the other extreme limit assuming a simple Bloch wall (no vortex structure around the interfaces) one can calculate β using the wall energy of Ni, see Fig. 3.67 in [5.99]. The effective Bloch wall width in Ni is 350 nm, which would result in a Bloch coefficient β_{Bloch} that is by a factor of 2.2 larger than that measured by Leaver and Vojdani. This is similar to the situation in Patton's measurements. The Bloch domain wall is too thin to explain the parameter β with the Gilbert damping parameter obtained from FMR. The domain wall width calculations by Rave and Hubert [5.106] using $Q = 0.1$ is more realistic for the Leaver-Vojdani studies. For a Ni platelet of 200 nm thickness the calculated wall width around the surface was by a factor of two larger than that in the interior, see Fig. 7 in [5.106]. This means that the effective domain wall width in the Ni platelets can be as much as a factor of two bigger than that in a Bloch wall. Clearly, no appreciable change in the value of the Gilbert damping from that obtained by Rodbell using FMR is needed to explain Vojdani and Leavers's results.

In high quality samples, there is no compelling evidence for having the Gilbert damping at large angles of precession considerably different from that obtained by standard FMR.

It is an interesting point that the effective angular frequencies in domain wall mobility measurements are rather low compared to standard FMR measurements. In Patton's experiments the domain wall velocity reached 5000 cm/s, in the Leaver-Vojdani ac experiments the amplitude of the domain wall oscillations reached a maximum of 10 μm at 1 MHz oscillatory frequency. Assuming that the domain wall width is approximately 300 nm in Py and 80 nm in Ni one obtains an angular frequency in the range of $\sim 2\pi \times 10^8\ \mathrm{s}^{-1}$ in the Py and Ni measurements.

The angular independent damping in Py films (10, 25 and 50 nm in thickness) was reported in the work by Nibarger, Lopusnik and Silva [5.71]. By measuring the time resolved precessional motion of the magnetization after applying a variable transverse pulse field to the applied dc field they found that the damping was independent on the angle of precession from 0.002 to 40 Degrees. The frequency of precession was in the range of 700 MHz to 3 GHz. These Py samples did not exhibit a simple Gilbert damping. The effective α was increased by decreasing the frequency of precession. It reached a constant value for $f > 1.7$ GHz, and therefore some degree of damping by inhomogeneities can be expected, see Sect. 5.7. The measurements were carried out on films where the sample edges were not important. This result has shown clearly that in the absence of edge scattering the rotational motion even at large angles is not noticeably affected by higher order spin wave scattering which could be in principle generated by nonlinearities at high angles of precession [5.107]. In measurements using a high rf power the large angle of precessional motion is strongly affected by multi-magnon scattering processes. The multi-magnon scattering can lead to a substantial line broadening of FMR (Suhl-Damon-Bloembergen-Wang four magnon scattering process) and can even result in subsidiary resonance (three magnon scattering process) [5.108]. The Suhl-Damon-Bloembergen-Wang main resonance saturation is important in the performance of microwave devices employing FMR [5.109]. The results by Nibarger et al. [5.71] have shown that the Suhl instabilities in films with a large lateral scale do not appear easily during a free rotation even at high angles of rotation.

The conclusion of this section is that the Gilbert damping from FMR measurements can be also applied for large angles of precession. However a word of caution is needed. In FMR one usually studies a homogeneous mode of precession. In experiments with a large angle of rotation that does not always have to be the case. One can start with a nearly homogeneous rotation, but due to scattering by sample inhomogeneities (particularly for small lateral geometries where the sample edges play a crucial role) this rotation can evolve into complex spin wave and soliton patterns. In this case one has to carefully assess the situation and decide what one means by damping. The intrinsic damping measured by FMR determines the time scale of dissipation of the magnetic energy into the lattice. It is a common error to replace the time of de-phasing of homogeneous rotational mode with the damping parameter. De-phasing leads to the spacial incoherence, but the magnetic energy has to be removed by the intrinsic damping which in principle can be measured by FMR. See further detailed discussion in Sect. 5.5.2.

5.5 Magnetic Relaxations in Multilayers

5.5.1 Current Induced Torque

Magnetic multilayers provide a special case where dynamic interactions between the itinerant electrons and the magnetic moments in ultrathin films offer new exciting possibilities. The non-local spin dynamics in metallic multilayers is one of the main research topics of magnetic nanostructures, and has much promise for spintronics applications. It has been shown in a number of recent experiments using either pillar shape nanoscopic samples [5.110], or point contact geometries [5.111], that the magnetization reversal can be driven by a current flowing perpendicularly to

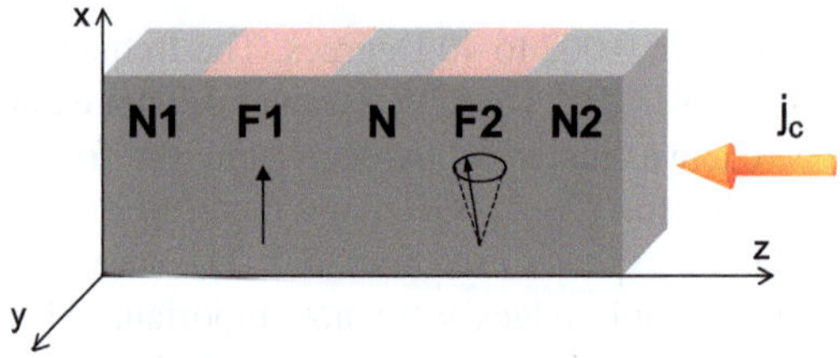

Fig. 5.5. A five layer structure. F1 is assumed to have a stationary magnetic moment, and F2 is allowed to precess. N1, N, N2 are non-magnetic layers. The coordinate system is identical to that in Fig. 5.1. The thick arrow represents the current density j_c applied along the z-axis (perpendicular to the film plane)

the magnetic layers. Slonczewski [5.112] showed that the transfer of vector spin momentum accompanying an electric current flowing through the interfaces of two magnetic films separated by a non-magnetic metallic spacer (magnetic double layer) can result in *negative Gilbert-like* torques. This torque leads, for sufficiently high current densities, to spontaneous magnetization precession and switching phenomena [5.113].

Assuming ballistic electron transport and using a five layer system in Fig. 5.5, Slonczewski [5.112] evaluated the spin transport across the interfaces using Wentzel-Kramers-Brillouin (WKB) Hartree-Fock spinor wave functions. The torque generated by a current perpendicular to the interfaces has a L.L. like form

$$\frac{\partial \boldsymbol{S}_{1,2}}{\partial t} = \frac{j_c g}{e} \boldsymbol{s}_{1,2} \times (\boldsymbol{s}_1 \times \boldsymbol{s}_2) , \tag{5.41}$$

where $\boldsymbol{S}_{1,2}$ are the total spins per unit area in F1 and F2, $\boldsymbol{s}_{1,2}$ are the unit vectors along the total spin vectors, e is the electron charge, and j_c is the electron current density. The scalar function g is given by

$$g = \frac{4\sqrt{P^3}}{-16\sqrt{P^3} + (1+P)^3(3 + \boldsymbol{s}_1 \cdot \boldsymbol{s}_2)} , \tag{5.42}$$

where $P = (n_+ - n_-)/(n_+ + n_-)$ is the spin polarizing factor. Note that the torques in (5.41) do not cancel each other. They are perpendicular to their respective magnetic moments maintaining the same circulatory pattern. Assuming that F1 has a stationary magnetic moment oriented along the unit vector $\boldsymbol{u}$ the current induced torque can be described by an effective field acting on F2

$$\hbar \frac{j_c g}{e d_2 M_{s,2}} \boldsymbol{u} \times \boldsymbol{s}_2 , \tag{5.43}$$

where $M_{s,2}$ is the saturation magnetization in F2. Note, that the current induced effective field is inversely proportional to the film thickness, and therefore the current induced effective torque has an interface character. Assuming a circularly polarized magnetization m^+ with a small angle of precession around the $\boldsymbol{u}$ direction, and adding the effective field in (5.43) to the L.L. eq. of motion, see (5.5) results in the magnetic relaxation term

$$-\left(\frac{\lambda}{\gamma}\frac{1}{\chi_\perp} - \hbar\frac{j_c g}{edM_s}\right)m^+ . \tag{5.44}$$

The current induced relaxation depends linearly on the current density and its sign can be reversed by changing the current direction. For current densities larger than the critical value

$$j_{c,crit} = \frac{\lambda}{\gamma\chi_\perp}\frac{edM_s}{\hbar g} . \tag{5.45}$$

The damping term is negative, and F2 is allowed to precess away from its equilibrium. With a further increasing current density the negative damping can reach a threshold value at which the magnetic moment in F2 undergoes a complete magnetization reversal.

Recently Fert et al. [5.114] derived more accurate expression for the g parameter in (5.41) for Co/Cu/Co systems. Using the standard diffusive transport equations of the CPP-GMR theory and assuming the spin momentum transport for small angles between the magnetic moments they have shown that the torque equation (5.41) has contributions originating in the Cu spacer and at the Cu/Co interface. These equations were used to discuss the results of magnetization reversal by the current injection and transfer of spins in their Co/Cu/Co trilayer pillars.

Berger evaluated the role of non-local spin momentum transport in magnetic double layers using a somewhat different approach than Slonczewski. F1 was again assumed to be static, and the direction of its magnetic moment determined the axis of the static equilibrium. Magnons were introduced by allowing the magnetic moment of the second (thinner) layer to precess around the equilibrium direction. Itinerant electrons entering the thin ferromagnetic layer through a sharp interface cannot immediately accommodate the direction of the precessing magnetization. Berger showed that this leads to an additional exchange torque which is directed towards the equilibrium axis, and represents an additional relaxation term. This relaxation torque is confined to the vicinity of the N/F2 interface. The resulting relaxation torque in a magnetic double layer structure contributes to an additional magnetic relaxation torque which can be expressed in an additional FMR linewidth [5.115], ΔH_{add}. The additional FMR linewidth ΔH_{add} is proportional to

$$\Delta H_{add} \sim (\Delta\mu + \hbar\omega) , \tag{5.46}$$

where $\Delta\mu = \Delta\mu_\uparrow - \Delta\mu_\downarrow$ is the difference in the spin up and spin down Fermi level shifts, and ω is the microwave angular frequency. Berger [5.115] showed that the spin momentum transfer between F1 and F2 is mostly confined to a very thin region around the N/F2 interface. Its depth is given by

$$L_0 = \frac{\pi}{k_\uparrow - k_\downarrow} . \tag{5.47}$$

The majority and minority electrons entering F2 from N come with their spins polarized by F1. The majority and minority electrons entering F2 contribute with opposite torques. However, the time to reach a given distance d in F2 away from the N/F2 interface depends on the perpendicular k_z component. Keep in mind that the parallel component of $\boldsymbol{k}$ wave-vector is preserved in ballistic transport. Thus for an electron coming in at oblique incidence the phase of precession is bigger than that for an electron entering the interface along the film normal. That results in a fast de-phasing leading to a rapid decay of the overall torque with distance d from the N/F2 interface. The effective depth is given by L_0, and it is expected in to be less than 1 nm. Therefore, the additional FMR linewidth in (5.46) represents an interface effect.

The second term in (5.46) will be discussed in Sect. 5.6. In this section the discussion will be limited to the spin transport which is induced by an applied perpendicular current.

The layer F1 acts as a spin polarizer of the current I (assuming that the layer F1 is thicker than the spin diffusion length). The majority and minority spin channels have different conductivities and consequently the current densities $j_c^{\uparrow\downarrow}$ for majority and minority spin carriers, are different. In a traditional treatment this results in a shift of the Fermi levels for the spin up and down electrons. Assuming that the layer N is much thinner than the spin diffusion length the appropriate shifts in the electron $\boldsymbol{k}$ space in N are [5.115]

$$\Delta k_z^{\uparrow\downarrow} = -\frac{2m}{en\hbar} j_c^{\uparrow\downarrow} , \tag{5.48}$$

where n is the total number of conduction electrons per unit volume in N, m and e are the electron mass and charge, and the z coordinate is perpendicular to the interfaces, see Fig. 5.5. These shifts produce displacements in the local majority and minority electron Fermi levels

$$\Delta\mu^{\uparrow\downarrow} = \hbar \Delta k_z^{\uparrow\downarrow} v_z^{\mathrm{N}} , \tag{5.49}$$

where v_z^{N} is the z-component of the electron velocity in N. Using the parameters $\beta = j_c^{\uparrow}/j_c^{\downarrow} = \sigma_{\mathrm{F1}}^{\uparrow}/\sigma_{\mathrm{F1}}^{\downarrow}$ and $j_c = j_c^{\uparrow} + j_c^{\downarrow}$ results in

$$\Delta\mu = \Delta\mu^{\uparrow} - \Delta\mu^{\downarrow} = -2\left(\frac{\beta-1}{\beta+1}\right)\frac{\hbar k_z}{en} j_c . \tag{5.50}$$

$\sigma_{\mathrm{F1}}^{\uparrow}, \sigma_{\mathrm{F1}}^{\downarrow}$ are spin up spin down conductivities in F1 far from any interface. The total spin current (current driven torque) across the N/F2 interface is obtained by integration over the displaced Fermi surfaces. The integration also incorporates the wave function transmission coefficients t at the N/F2 interface. The current driven torque is proportional to $\Delta\mu\sin(\theta)$, where θ is the angle between the magnetic moments of F1 and F2. The pre-factor accompanying $\Delta\mu\sin(\theta)$ depends on the coefficient β, transmission coefficients t, the Fermi k wave-vector in N, and the Fermi $k^{\uparrow,\downarrow}$ wave-vectors in F2, see details in [5.116].

This calculation assume purely ballistic transport across the N/F2 interface. One also has to consider thick layers (compared to spin diffusion lengths). The spin imbalance in this case is associated with isotropic expansion and contraction of the Fermi surface. This is commonly used in CPP GMR theory, see [5.40], and was originally introduced by Aronov [5.117]. Berger [5.116] showed by solving diffusion equations for thick layers F1 and N2 that the difference in the Fermi levels, $\Delta\mu_{\text{diff}} = \Delta\mu^{\uparrow} - \Delta\mu^{\downarrow}$, for the contracted and expanded Fermi surface and the displaced Fermi surfaces $\Delta\mu$ in (5.50) satisfy

$$\frac{\Delta\mu}{\Delta\mu_{\text{diff}}} = \left(\frac{\Lambda_{\downarrow}}{l_{\text{sd}}}\right)_{\text{F1}} + \left(\frac{\Lambda_{\downarrow}}{l_{\text{sd}}}\right)_{\text{N2}} , \tag{5.51}$$

where $\Lambda_{\downarrow}$ and l_{sd} are the momentum mean free path for spin down electrons and the spin diffusion length in F1 and N2, respectively. In most materials the ratio $\Lambda_{\downarrow}/l_{\text{sd}} \gg 1$. It means that the isotropic expansion or contraction of the Fermi surface is significantly bigger than $\Delta\mu$. The pre-factors (accompanying $\Delta\mu$ and $\Delta\mu_{\text{diff}}$) for the current driven torque are nearly the same. It follows that the critical current requirement is appreciably dropped in thick layers where the electron transport obeys Aronov's diffusion equations. However, this advantage is lost rapidly for samples where the leads N1 and N2 spread out immediately into a cross-sectional area much wider than that of the F1/N2/F2 multilayer. The current flow in this case is not anymore one dimensional which results in a significant drop of $\Delta\mu_{\text{diff}}$ [5.116].

Stiles and Zangwill [5.118] carried out excellent model calculations describing the spin current transport at a non-ferromagnet/ferromagnet interface for a non-collinear orientation of spins. The paper has a great educational character and for that reason the main results will be briefly highlighted.

For the spin degree of freedom the spin density is

$$\boldsymbol{m}(\boldsymbol{r}) = \Sigma_{i\sigma\sigma'} \psi^{*}_{i\sigma} \boldsymbol{s}_{\sigma,\sigma'} \psi_{i\sigma'}(\boldsymbol{r}) , \tag{5.52}$$

and the spin current density is expressed as

$$\mathfrak{R} = \Sigma_{i\sigma\sigma'} Re[\psi^{*}_{i\sigma}(\boldsymbol{r}) \boldsymbol{s}_{\sigma,\sigma'} \otimes \boldsymbol{v} \psi_{i\sigma'}(\boldsymbol{r})] , \tag{5.53}$$

where $\boldsymbol{s} = (\hbar/2)\boldsymbol{\sigma}$, and $\boldsymbol{\sigma}$ is a vector whose components are the three Pauli matrices. $\otimes$ represents the matrix product between the Pauli matrix and the electron velocity operator $\boldsymbol{v} = -\mathrm{i}(\hbar/m)\nabla$ (which is in the direction of the applied current) of an appropriate electron state $\psi_{i\sigma}$. The spin current density $\mathfrak{R}$ is a tensor with its indices describing the spin and real space directions. The spin balance in space is given by

$$\frac{\partial \boldsymbol{m}}{\partial t} = -\nabla\mathfrak{R} - \frac{\delta \boldsymbol{m}}{\tau_{\text{sf}}} + \boldsymbol{T}_{\text{ext}} , \tag{5.54}$$

where the first term on the right hand side accounts for a non local origin of the spin current density, the second term describes the spin relaxation term of the spin accumulation density $\delta\boldsymbol{m} = \boldsymbol{m} - \boldsymbol{m}_{\text{eq}}$, and the last term is the Landau-Lifshitz-Gilbert torque density, see (5.1) and (5.5). Stiles and Zangwill evaluated the spin currents in the NM/FM double layer system assuming that the incident current flows from the NM layer with the spin polarization having the spin moment oriented with an arbitrary polar and azimuthal angle with respect to the magnetic moment in the FM layer. For a stationary incident current the total flow of the spin current through a pill box which just straddles the interface can be used (using the divergence theorem for (5.54)) to calculate the current induced spin momentum transfer torque at the NM/FM interface. The current induced spin transfer torque $\boldsymbol{N}$ per unit area is given by

$$\boldsymbol{N} = -\boldsymbol{\mathfrak{R}}_{\mathrm{inc}} + \boldsymbol{\mathfrak{R}}_{\mathrm{t}} + \boldsymbol{\mathfrak{R}}_{\mathrm{r}} \,, \tag{5.55}$$

where indexes inc, t, r stand for incidence, transmission and reflection components of $\boldsymbol{\mathfrak{R}}$, respectively. Calculations were carried out using a semiclassical Boltzmann transport equation which included interface spin dependent reflection and transmission coefficients. The reflection $R_\uparrow$, $R_\downarrow$ and transmission $T_\uparrow$, $T_\downarrow$ coefficients were calculated from the usual quantum mechanical matching conditions. The longitudinal (parallel to the magnetization) incident spin current was found to be equal to the sum of the reflected and transmitted spin currents. This means no interface torque parallel to the magnetization is present. The situation is different for the transverse components. The reflection and transmission coefficients are spin dependent and that results in the discontinuity of the transmitted and reflected spin density currents. This effect is usually referred to as spin filtering. In addition the reflection coefficients are generally complex which leads to an appreciable spin rotation after reflection. The rotation angles are widely spread with the transverse electron momentum $\boldsymbol{k}_\parallel$ (parallel to the interface), which results in an extensive cancellation of the total transverse reflected spin current. Finally, the spin precession for the transmitted spin current leads to a third source of the spin transfer torque. The transmitted component decreases rapidly to zero away from the interface. In the asymptotic limit the dependence on the distance x away from the interface is given by

$$\frac{\sin[(k_{\mathrm{F}}^{\uparrow} - k_{\mathrm{F}}^{\downarrow})x]}{(k_{\mathrm{F}}^{\uparrow} - k_{\mathrm{F}}^{\downarrow})x} \,, \tag{5.56}$$

where $k_{\mathrm{F}}^{\uparrow}$ and $k_{\mathrm{F}}^{\downarrow}$ are the Fermi k_{F} wavevectors for minority and majority spins in the ferromagnet. This is in agreement with Berger's calculations, see (5.47). The electron transient time in the ferromagnet is different for the majority and minority spins, $k_\downarrow \neq k_\uparrow$, leading to the phase difference $(k_z^{\uparrow} - k_z^{\downarrow})z$ for the transmitted transversal spin current. The electrons entering the interface under different angles spread out the phase difference and that results in a rapid decrease of the oscillatory amplitude of the total transmitted transversal spin current with an increasing distance from the interface. The decay length of the transmitted transversal spin current is of a few ML given by (5.47).

The incident spin current is fully absorbed by the interface, and consequently the incident transversal spin angular momentum delivers an interface transverse torque. This is in agreement with Slonczewski's and Berger's model of interface torque, see (5.41).

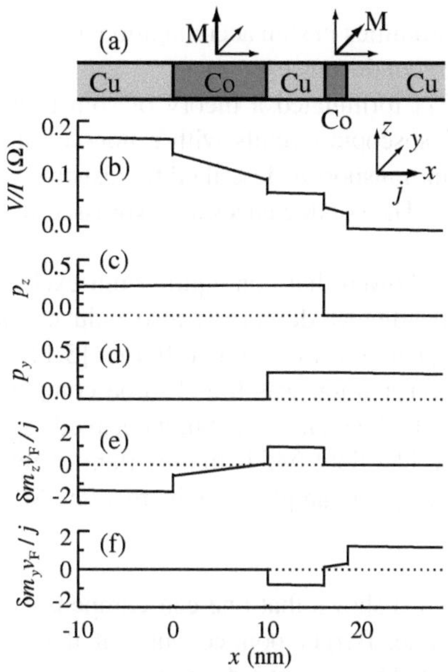

Fig. 5.6. Voltage, current polarization, and spin accumulation for a trilayer. (**a**) A heterostructure with two Co layers, with a Cu spacer, and two semi-infinite Cu leads. The electron current flows in the x-direction and the left magnetization is in the z-direction and the right magnetization is in the y-direction. (**b**) The voltage drop through the structure. (**c**) and (**d**) z and y components of the current polarization, respectively. (**e**) and (**f**) z and y components of the spin accumulation, respectively. This figure was provided by courtesy of M. Stiles and A. Zangwill

The rapid de-phasing of the transverse spin current on the length scale of a few ML is a purely quantum mechanical effect which has no real equivalent in classical equations for spin transport. A similar situation was found by Šimanek [5.119] for the spin transport through a Bloch wall where the magnetization gradually rotates in space.

An excellent theoretical extension of non-collinear spin current transfer in a Cu/Co/Cu/Co/Cu multilayer can be found in the paper by Stiles and Zangwill [5.120]. The numerical results were carried out for each part of the heterostructure (leads, ferromagnets, and the spacer Cu layer) and are summarized in Fig. 5.6, which was provided by courtesy of M. Stiles and A. Zangwill.

The transverse spin currents are discontinuous at the Co/Cu and Cu/Co interfaces, which results in transverse interface torques. Figure 5.6e,f shows the spin accumulation along the corresponding magnetization directions that is required for the calculated current polarizations. One should point out that the spin transport in Cu is carried out by diffusion, consequently the spin accumulation $\delta \boldsymbol{m}$ in Cu reverses its sign around the adjacent Cu/Co/Cu interfaces. The spacial variation of the spin accumulation as a function of x is not explicitly visible due to a long spin diffusion length in Cu (2000 nm). The drop in the voltage at the Cu/Co and Co/Cu interfaces, see Fig.5.6a, and discontinuities in spin accumulation across the interfaces are due to the presence of interface conductances.

Brataas et al. [5.121] formulated a theory of spin transport in hybrid normal metal-ferromagnetic mesoscopic circuits with a non-collinear orientation of magnetic moments. The spin transport is described by resistive elements using interface conductance parameters. The conductances are expressed in terms of scattering matrices.

Heide et al. [5.122] showed that non-equilibrium exchange interaction (NEXI) in the presence of perpendicular dc current also induces interlayer like exchange fields $b\boldsymbol{M}$ from one magnetic layer to another. Berger [5.116] evaluated this field for ballistic transport. Its contribution was found to be comparable to the spin torque contribution for the film thickness $d < L_0$, but it decayed rapidly to zero for $d > L_0$. The interlayer exchange like field $b\boldsymbol{M}$ is unimportant in reasonably thick ultrathin films; it can be present only for samples which are a few ML thick where $d < L_0$ is satisfied.

SWASER. Equation (5.44) shows that one can compensate the relaxation term by the current induced torque. Berger pointed out that there is an analogy between magnon emission generated by current induced torque and a semiconductor injection laser. The Fermi level difference $\Delta\mu < 0$ pumps electrons up from the majority to minority states. The spin relaxation back to majority states creates a magnon in F2 to preserve the total spin in the s-d interaction. The number of magnons is proportional to the deviation of the longitudinal magnetization component from its equilibrium $\sim (1 - \cos(\theta))$, for small θ it is $\sim \theta^2$, where θ is the angle between the instantaneous magnetization direction and the equilibrium direction. The rate of change of the transversal magnetization component, $\partial m_\perp/\partial t$, by the current induced torque is proportional to $\sin(\theta)$, see (5.41). For small angles of precession the rate of change of the longitudinal magnetization component, $\partial M_{||}/\partial t$ is proportional to $(m_\perp/M_{||})(\partial m_\perp/\partial t)$. Since $m_\perp/M_{||}$ and $\partial m_\perp/\partial t$ are proportional to θ, the longitudinal component $\partial M_{||}/\partial t \sim \theta^2$. Therefore the rate in which magnons are generated ($\sim \partial M_{||}/\partial t$) and the number of magnons have the same dependence on the angle θ. This means the rate of creation of magnons is proportional to the number of magnons, and that is called stimulated emission. The energy corresponding to precessing magnetization (magnon) plays the role of the laser semiconductor band gap. Berger used the term *spin wave amplification by stimulated emission of radiation (SWASER)* to describe a device which operates on the current induced negative damping term.

Devices based on spin instability can operate either on complete magnetization reversal, see [5.123], or on the stationary precession of the magnetization [5.124]. One can switch the magnetization of a memory pixel by applying a perpendicular current above its critical threshold. The reversal of magnetization is achieved simply by reversing the direction of current. Albert et al. [5.123] studied magnetization reversal using a 2.5 nm thick Co thin film nanomagnet in nanopillar ($60 \times 130\,\mathrm{nm}^2$) devices based on multilayer Cu(80 nm)/Co(40 nm)/Cu(6 nm)/Co(2.5 nm)/Au(10 nm) structures deposited on an oxidized Si wafer. The saturation magnetization was oriented in the plane of the sample. The switching currents were 2.3 mA and − 3.3 mA for the forward and backward switching, respectively. The corresponding critical current density was $\sim 2.5 \times 10^7$ A/cm^2. The obvious advantage of this approach is in having a local field with no effect on surrounding memory pixels. Since Maxwell's magnetic field is inversely proportional to $1/r$ (the radius of the memory pixel) and the critical current is inversely proportional to $1/r^2$, the current-induced exchange torque is more favorable for switching very small memory pixels.

Tsoi et al. [5.124] reported generation and detection of phase coherent magnons driven by the current in [Co(1.5 nm)/Cu(2–2.2 nm)]$_N$ multilayer structures employing a point contact configuration for the dc electrodes. In their studies they needed relatively high critical currents, 1×10^9 A/cm^2. By using an external microwave source (40–50 GHz) they were able to detect an increase in the dc voltage across the point contact junction when the applied microwave frequency was either equal to or greater than the resonance frequency of the uniform precession mode. A possible explanation lies in a rf non-linear mixing of two microwave signals. One originates in the applied dc current and other is supplied by an external microwave source. Pufall et al. [5.125] extensively investigated the current induced magnetic excitations in multilayer samples grown using a wide range of magnetic materials. The measurements were carried out by using a mechanical point contact measurements. The tip material was Ag. The multilayer structure was Fe(5 nm)/Cu(0.9 nm)/FM(1.2 nm) $\times$10/Cu(1. nm) in which FM = Co,$Co_{90}Fe_{10}$, $Ni_{80}Fe_{20}$, Fe and $Ni_{40}Fe_{10}Co_{50}$ sputtered deposited onto oxidized Si. They studied the critical current required to induce the step in resistance, dI/dV, that is generally believed to be given by the onset of magnetization precessional mo-

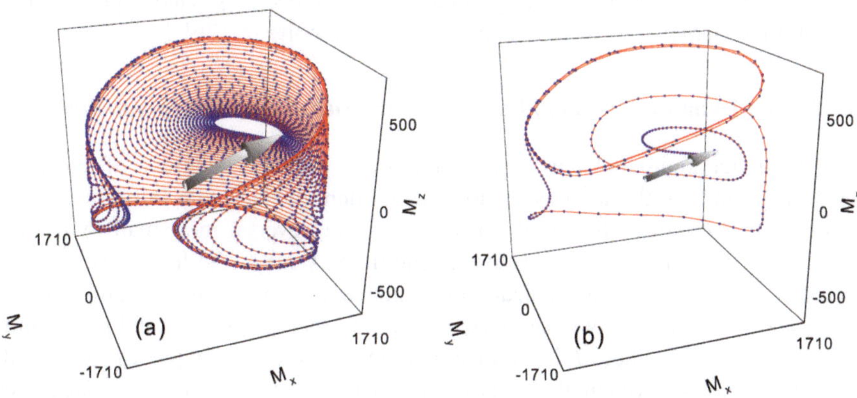

Fig. 5.7. Time evolution of the magnetic moment in an external field $H = 0.6$ kOe which is applied along the hard magnetic axes $[1\bar{1}0]$. The starting orientation of the magnetic moment was a few degrees away from the easy axis [100], see the arrow. Computer simulations were carried out with the following current densities: (**a**) 3.5×10^7 A/cm^2 and (**b**) 2.5×10^8 A/cm^{-2}. The magnetic properties of layer F2 are described in Fig. 5.2. Maxwell's field in this calculation was neglected. Note, that one is able to establish complex stationary precessions having large amplitudes. For the low current density in (**a**) the tip of the magnetization follows the edge of a *bracelet*. With an increasing current the bracelet closes and eventually evolves into a *lasso shape* orbit, see (**b**)

tion. In these measurements the magnetization and the exchange coupling was varied by a factor of 4 and 30, respectively. They found that their results are consistent with with Slonczewski's theory for describing the current induced magnetic effects [5.113].

Fig. 5.7 shows a stationary precession of the magnetization under very simplifying assumptions using equation (5.41) and assuming that the ferromagnetic layer F1 is stationary and the second layer F2 is allowed to process as a single macroscopic spin. In realistic experiment one would have to use small size magnetic pixels to avoid the dominant role of Maxwell's field. In this case one has to include the inhomogeneous demagnetizing field from the ferromagnetic pixels.

In the recent experiments carried out by Kiselev et al. [5.126] using a single pillar (130 nm $\times$ 70 nm) of 80 nm Cu/40 nm Co/10 nm Cu/3 nm Co/2 nm Cu/30 nm Pt on an oxidized Si wafer and Rippard et al. [5.127] using lithographically defined point contacts of 40 nm diameter circles made on spin-valve mesas of Cu(50 nm)/$Co_{90}Fe_{10}$(20 nm)/Cu(5 nm)/$Ni_{80}Fe_{20}$(5 nm)/Cu(1.5 nm)/Au(2.5 nm) it has been shown for the first time that a dc current can result in a coherent large angle microwave precession. Microwave probes attached to the structure allowed them to detect the microwave signal using a spectrum analyzer. The critical current density was in the range of $3 \times 10^8\ \mathrm{A/cm^2}$. The onset of microwave emission was found to be strongly correlated with the peak in the dc $\mathrm{d}V/\mathrm{d}I$ curve. However, significant microwave signals were observed for the currents somewhat larger than that corresponding to the peak of $\mathrm{d}V/\mathrm{d}I$. By applying the in-plane and out of plane magnetic field the microwave signal was observed in 6 to 36 GHz range of frequencies [5.127]. The maximum microwave power generated by the precessing nanomagnet devices was 40 times larger than room temperature Johnson noise [5.126].

5.5.2 Spin Dynamics in Small Lateral Geometries, Computer Simulations

The computer simulations in this section were provided courtesy of Professor M. Scheinfein [5.129], and this section was written together with Prof. Scheinfein. In Fig. 5.8 micromagnetic simulations for the time reversal in a small Py sample $640 \times 320 \times 5\ \mathrm{nm}^3$ (x, y, z) are shown. The time extent of the field pulse was selected so that the initial magnetization oriented along $+x$ was rotated and oriented along $-x$ when the field was set back to zero. In other words, the field pulse duration was $\Delta t = T/2$ where T is the macroscopic uniform rotation precession period. Note that the total moment along the x axis reached its nearly stationary value after 1 ns. However, a small transverse in-plane y component is still present and keeps further oscillating. An ongoing precession of the magnetic moment for $t > 1$ ns is very pronounced in Fig. 5.8b. In Fig. 5.8b the averaged local magnetic moment in a $20 \times 20 \times 5\ \mathrm{nm}^3$ patch of the total structure center shows very pronounced oscillatory behavior in the y direction. This means that the system is not yet close to a true static equilibrium. This marked difference between the global (total) and local behavior is a direct consequence of mode-mode coupling. The long wavelength mode gets nearly stationary very quickly, but the magnetic energy of the system is not yet fully dissipated because the intrinsic relaxation parameter $\alpha = 0.01$ is not sufficient

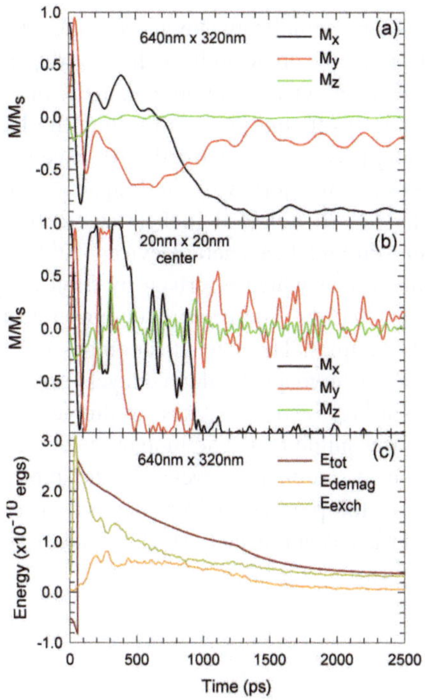

Fig. 5.8. The time reversal of the magnetic moment in a submicron Py square sample of $640 \times 340 \times 5\,\text{nm}^3$. The magnetic parameters for Py are $A = 1.05 \times 10^{-6}$ ergs/cm, $M_s = 800\,\text{G}$, $K_u = 1000\,\text{erg/cm}^3$. The time reversal of the magnetic moment was carried out in the following way. The sample has its initial magnetic state in a s-state (see Arrott [5.76] and Shi [5.128] in this book) with the magnetic moment along the $+x$ magnetic easy direction. A magnetic field of 500 Oe was applied along the y hard magnetic axis for a duration of 57 ps. (**a**) The time dependence of the components of the total magnetic moment. (**b**) The time dependence of the components of the magnetic moment in the center of the sample. (**c**) The time dependence of the total, demagnetizing and exchange energies

to dissipate energy that quickly. In fact, one can estimate the true relaxation time by using (5.18). The period of oscillations can be obtained from the red line in Fig. 5.8b, and that results in a relaxation time of 40 ns. Obviously this is an appreciably longer period than 1 ns which is required to reach near equilibrium for the total moment along the $-x$ direction. In Fig. 5.8c one can follow the approach to equilibrium. The maximum energy occurs right after the field was switched off. Note that almost half of the energy is left in the system when the total magnetic moment was reversed. The demagnetizing and exchange energies are out of phase. They drive each other into equilibrium. The initial low frequency oscillations produce magnetostatic modes which collide with boundaries and couple to higher frequency waves (spin waves and solitons). These waves carry more energy (higher exchange internal fields) and have a large state density which can lead to higher damping rates. The mode-mode coupling effectively moves the magnetic energy into short wavelength modes where it is dissipated. They become completely attenuated on the time scale of 40 ns. Note that if the relaxation parameter is obtained from the time dependence of the total moment then one would get an effective damping parameter of $\alpha = 0.4$. This is clearly wrong. In small magnetic pixels one needs to follow the local moments in order to determine the damping parameter α correctly. An excellent review of switching processes in spin sensors and memory pixels in Magnetic Random Access Memory (MRAM) is reviewed by S.E. Russek et al. in their chapter on High Speed Switching and Rotational Dynamics in Small Magnetic Devices [5.62].

The magnetization reversal which is driven by Slonczewski' spin torque, see (5.41), are shown in Figs. 5.9–5.12. These figures demonstrate the complexities which can occur in finite sized samples. The magnetization reversal is simulated gradually from a simple to more realistic magnetic double layers separated by a nonmagnetic spacer. In the left frame of Fig. 5.9 one assumes the simplest case. The magnetization reversal is carried out for a single magnetic moment where no lateral variations are present. Note that the magnetization time reversal by the spin current follows a simple precessional mode where the precession amplitude in the perpendicular direction (blue line) to the film surface is appreciably smaller than that in the in-plane. This is

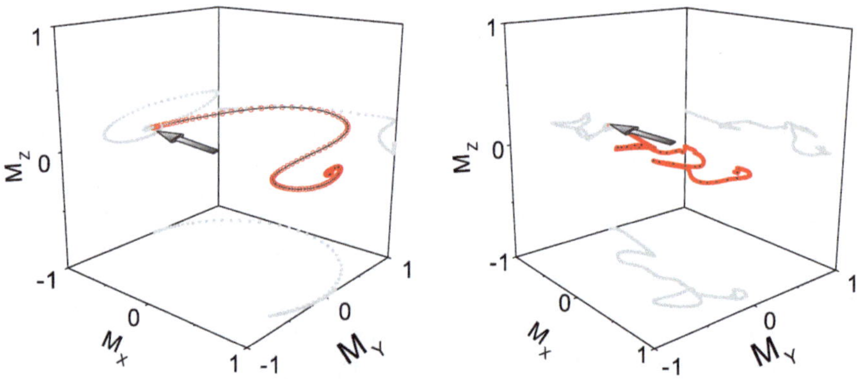

Fig. 5.9. The time evolution of the average magnetic moment in zero external field using Slonczewski's induced spin current torque, see (5.41). The density of current is 6×10^8 A/cm^2. The magnetic double structure consists of two Fe(001) pixels of 60×60 nm^2 $\times$ 2.2 nm (layer F2) and 60×60 nm^2 $\times$ 220 nm (layer F1) which are separated by a 2.2 nm thick NM spacer. The magnetic properties of the Fe films correspond to those of the Fe bulk. The relaxation parameter $\alpha = 0.01$. The thick layer F1 is pinned along its perimetry with Dirichlet boundary conditions. The thickness of F1 and Dirichlet boundary conditions assures that this layer remains stationary, and creates no demagnetizing field acting on F2. The left figure is calculated using a single spin approximation. This means the lateral dimensions of the soft magnetic layer F2 are ignored. There is no contribution of Maxwell's field from the passing dc current. The initial direction of the magnetic moment in F2 is a few degrees away from the negative x-direction (red line) corresponding to an easy magnetic crystallographic direction [100] in Fe. In the right frame of the figure one uses the full lateral dimension of F2. This means that the Maxwell's field and the dipolar field in F2 are included self consistently. The layer F1 is still fully pinned with Dirichlet boundary conditions

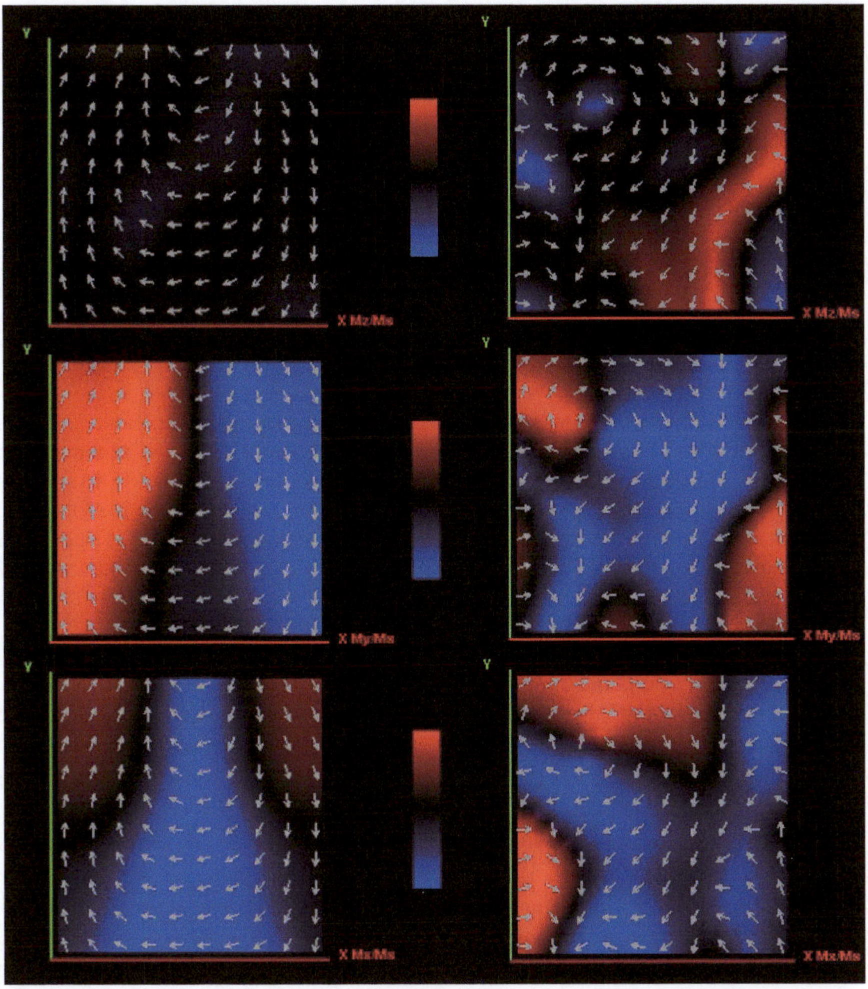

Fig. 5.10. The magnetization reversal in the thin film which is described in the right frame of Fig. 5.9. The magnetization direction cosines (M_z/M_s – *top*, M_y/M_s – *middle* and M_x/M_s – *bottom*) are shown at times $t = 20$ ps (*left*) and $t = 60$ ps (*right*). At $t = 0$ the magnetic moment in the layer was oriented everywhere in $-x$ direction (*full red color*). Evident at $t = 20$ ps the magnetization is trying to create a vortex, see the left part of the figure above. At $t = 60$ ps the mode mode-coupling results in a full vortex in the left upper corner above, but the rest of the sample has a complex magnetization pattern which is a consequence of all competing internal fields, see the right part of the figure above. Eventually, $t > 200$ ps, the magnetic moment is fully reversed in the $+x$ direction. The reversal of the magnetic moment is caused by Slonczewski's spin current induced torque which is able to overcome Maxwell's field created by a passing dc current

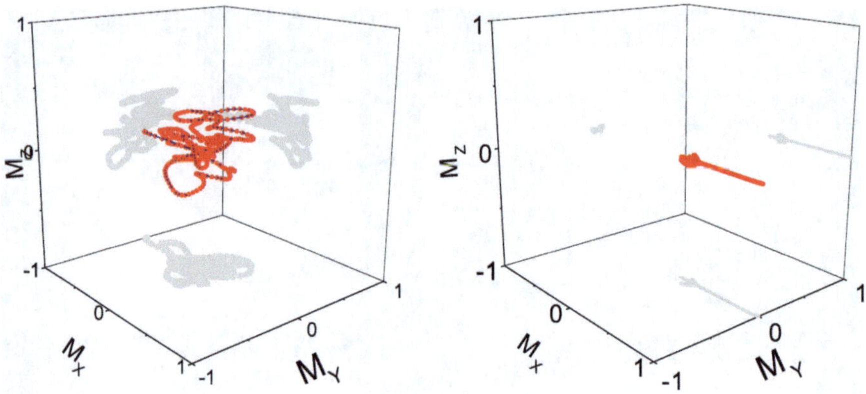

Fig. 5.11. The time dependence of the total (average) magnetic moment. Both Co layers are allowed to precess, and they are now coupled by the dipolar fields. No constrains are imposed. The dimensions of the thick F1 and thin F2 layers are $10 \times (130 \times 130)\,\text{nm}^3$ and $2.5 \times (130 \times 130)\,\text{nm}^3$, respectively. The left and right frames correspond to the F1 and F2 layers, respectively. At $t = 0$ one assumes that the magnetic moments are oriented antiparallel (due to mutual dipolar field). For $t > 0$ one applies a dc current with the current density of $6 \times 10^8\,\text{A/cm}^2$ to drive the system into magnetization reversal state

caused by a strong perpendicular demagnetizing field. The time of reversal is 200 ps. In the right frame of Fig. 5.9 the constraint of a single spin in layer F2 is released. The magnetic moments inside layer F2 (1000 pixels) are allowed to rotate independently and Maxwell's field from the applied dc current is included self consistently. The time reversal of the total magnetic moment is clearly more complicated. It does not follow a simple precession, the local magnetic moments create domains and vortices as shown at $t = 60\,\text{ps}$ in Fig. 5.10. However, in this case where $F1$ is pinned the magnetic moment in F2 still eventually reaches a reversed steady state. Figures 5.11 and 5.12 correspond to the situation. where all constrains are removed, i.e. F1 is no longer pinned and can be subjected to the torques from F2 due to the spin current and Maxwell's fields. With no constrains the magnetic moments in both layers are allowed to rotate. The simulations are meant to replicate experimental conditions encountered for spin-torque measurements in pillar-like structures. The polarizing layer, composed of Co with bulk like properties (except that the anisotropy is small) is 10 nm thick. The interlayer paramagnet is 6 nm thick, and the free layer, also composed of Co is 2.5 nm thick. When the constrains on the polarizing layer are relaxed, the magnetic field from the injected current and the spin torque from the free layer destabilize the magnetization in the polarizing layer. In general, the spin torque is competing with that produced by Maxwell's fields from the current, and symmetry is necessarily broken. The resulting magnetization configuration consists of persistent spin wave excitations that destabilize the magnetization in both layers. Once the symmetry is broken, chaotic magnetization states result. The magnetization as a function of time is shown for the free layer (right) in Fig. 5.11. That the magnetization follows a complex path in time is evident by the tangled path of the trajectories. One magnetization state

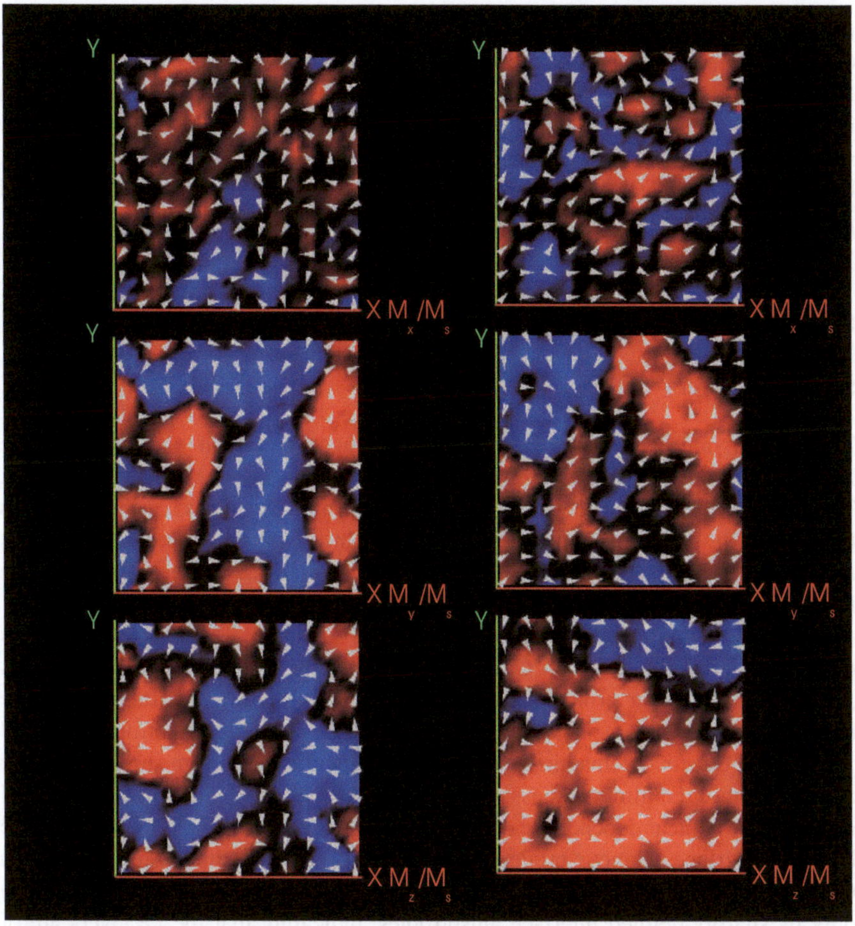

Fig. 5.12. The magnetization direction cosines (M_z/M_s – *top*, M_y/M_s – *middle* and M_x/M_s – *bottom*) are shown at times $t = 1$ ns for the thin free layer (*left*) and the thick polarization layer (*right*, see details in text). The symmetry breaking through the competition of the spin-torque and Maxwell's field introduces persistent spin wave excitations which destabilize the magnetization in both layers (continuously)

snap-shot is shown in Fig. 5.12 ($t = 1$ ns). The left panels are the direction cosines for the free layer and the right panels those for the polarizing layer. When the current is (suddenly) turned off at $t = 1$ ns, the magnetization relaxes into ferromagnetically coupled antisymmetric C-states. However, depending up the chaotic magnetization state present when the current is removed, the layers may form ferromagnetically (metastable state) or antiferromagnetically coupled C-state or vortices.

The above simulations have shown that the full reversal of magnetic moment by current requires a good pinning of the thick FM layer. The results by Myers et al. and Albert et al. [5.130, 131] have shown that one is able to carry out reliably magnetic reversal by using a nano-pillar structure of Cu(80 nm)/Co(40 nm)/Cu(6 nm)/Co(3 nm)/Au(10 nm) with the cross section ranging from 50×50 to $130 \times 60\,nm^2$. Using a laterally extended film for the polarization Co layer (40 nm) resulted in its effective magnetic pinning. Its coercive field was of $\sim$ 1 kOe. The computer simulations in right frame of Fig. 5.9 are relevant to this experiment. Myers et al. have shown also that magnetic reversal by spin polarized current exhibit statistical properties of thermal activation near room temperature [5.130].

5.6 Non-local Damping: Experiment

5.6.1 Multilayers

The role of interface damping has been investigated in high quality crystalline Au/Fe/Au/Fe(001) structures grown on GaAs(001) substrates, see details in [5.23]. The in-plane FMR experiments were carried out using 10, 24, and 36 GHz systems [5.22]. The in-plane resonance fields and resonance linewidths were measured as a function of the azimuthal angle φ between the external dc magnetic field and the Fe in-plane cubic axes.

Single Fe ultrathin films with thicknesses of 8, 11, 16, 21, and 31 monolayers (ML) were grown directly on GaAs(001). They were covered by a 20 ML thick Au(001) cap layer for protection in ambient conditions. FMR measurements were used to determine the in-plane four-fold and uniaxial magnetic anisotropies, K_1 and K_u, and the effective demagnetizing field perpendicular to the film surface, $4\pi M_{eff}$, as a function of the film thickness d [5.23, 24]. The magnetic anisotropies were well described by a linear dependence on $1/d$. The constant and linear terms represent the bulk and interface magnetic properties, respectively. The ultrathin Fe films grown on GaAs(001) have their magnetic properties nearly equal to those of bulk Fe, modified only by sharply defined interface anisotropies, indicating that the Fe layers are of a high crystalline quality with well defined interfaces. The lineshape of the FMR peaks is Lorentzian and the FMR linewidths are small (under 100 Oe for our microwave frequencies) and only weakly dependent on the film thickness. The reproducible magnetic anisotropies and small FMR linewidths provided an excellent opportunity for the investigation of non-local relaxation processes in magnetic multilayer films.

The thin Fe films which were studied in the single layer structures were regrown as a part of magnetic double layer structures. The thin Fe film (F2) was separated from the second thick layer (F1) of 40 ML thickness by a 40 ML thick Au spacer (N). The magnetic double layers were covered by a 20 ML Au(001) layer (N1) for protection under ambient conditions. Note that the layer notation from Fig. 5.5 was used. In this case N2 corresponds to GaAs(001). The thickness of the Au spacer layer was much smaller than the electron mean free path (38 nm) [5.132], and hence ballistic spin transfer between the magnetic layers is allowed.

The interface magnetic anisotropies separated the FMR fields of F1 and F2 by a big margin ($\sim$ 1 kOe), see [5.23]. That allowed us to carry out FMR measurements in F2 with F1 possessing a negligible angle of precession compared to that in F2. The

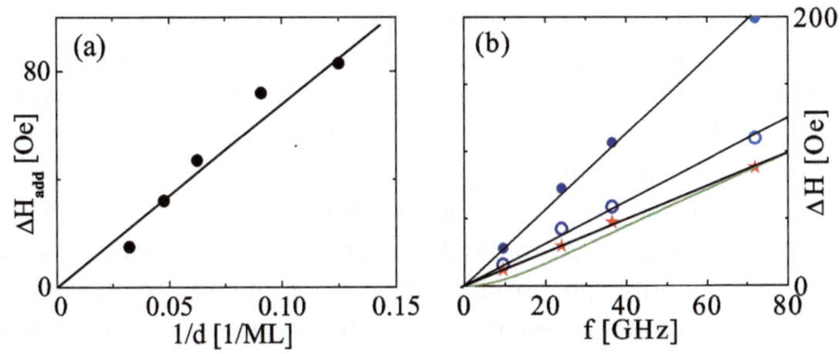

Fig. 5.13. The FMR linewidth measurements (half width half maximum (HWHM)) in the parallel configuration. (**a**) The dependence of the additional FMR linewidth $\Delta H_{\text{add}} = \Delta H_{\text{d}} - \Delta H_{\text{s}}$ on $1/d$ at $f = 36\,\text{GHz}$. d is the thickness of F2 (16 ML Fe layer in a GaAs/16Fe/40 Au/40Fe/20Au(001) structure). ΔH_{d} and ΔH_{s} represent the FMR linewidths for the Fe films in the double and single layer magnetic structures, respectively. (**b**) The frequency dependence of FMR linewidth in the 16 ML Fe layer: ΔH_d (*blue filled circles*) in a GaAs/16Fe/40Au/40Fe/20Au(001) structure; ΔH_{s} (*open blue circles*) in a single magnetic layer GaAs/16 Fe/20Au(001) structure; and the additional FMR linewidth ΔH_{add} is shown in red stars. The magnetic properties of 16 Fe were listed in Fig. 5.2. The solid line in olive green shows the fit to the data using the non-local damping mechanism which is based on (5.57). The integers represent the number of ML

FMR linewidths in single and double layer structures were only weakly dependent on the azimuthal angle φ of the saturation magnetization with respect to the in-plane crystallographic axes.

The thin Fe film in the single and double layer structures had the same FMR field showing that the interlayer exchange coupling [5.3] through the 40 ML thick Au spacer was negligible, and the magnetic properties of the Fe films grown by MBE on well prepared GaAs(001) substrates were fully reproducible.

The FMR linewidth in the thin films always increased in the presence of a thick layer F1. The additional FMR linewidth, ΔH_{add}, followed an inverse dependence on the thin film thickness d, see Fig. 5.13a. ΔH_{add} deviates from the straight line mainly for the thinnest Fe layer, $d = 8\,\text{ML}$ indicating that the thinnest film approaches the length $L_0 = \pi/(k_\uparrow - k_\downarrow)$, see (5.47). The non-local damping originates at the film interface (F2/N). The linear dependence of ΔH_{add} on the microwave frequency for both the parallel and perpendicular configuration with negligible zero-frequency offset, see Fig. 5.13b is equally important. This means that the additional contribution to the FMR linewidth can be described by an interface Gilbert damping. The additional Gilbert damping for the 16 ML thick film was found to be weakly dependent on the crystallographic direction, $G_{\text{add}} = 1.2 \times 10^8\,\text{s}^{-1}$ along the cubic axis. Note that its strength is comparable to the intrinsic Gilbert damping in the single Fe film, $1.45 \times 10^8\,\text{s}^{-1}$.

Discussion of the Interface Torque. Berger showed that in magnetic multilayers one can expect non-local interface damping, see the second term in (5.46). The additional relaxation term is proportional to $\hbar\omega$ and does not require the presence of a dc current.

It is based solely on the conservation of the total angular momentum; this means that the electrons in the non-magnetic spacer have to flip from spin down to up as a magnon is annihilated in the thin magnetic layer, and vice versa. The relaxation equations for spin up and spin down electrons in the spacer layer N were obtained using Fermi's golden rule which includes the change in energy of an electron when emitting or absorbing a magnon, and the spin up and spin down Fermi level shifts. The spin flip transitions in the N spacer lead to the displacement of the chemical potentials $\Delta\mu$ for majority and minority spins from their common thermal equilibrium value (in absence of magnetization precession). $\Delta\mu$ reaches its stationary value when the rate of the spin transitions due to spin-orbit interaction (see Sect. 5.3.3) is equal to the rate of spin flip transitions caused by the precessing magnetization. The system then reaches its dynamic equilibrium. This means that the additional interface damping is always accompanied by a non-zero $\Delta\mu$ even in the absence of an applied current. The sign of this shift is always negative and its amplitude is dependent on the precessional angle (number of magnons). Berger showed that $\Delta\mu$ is negligible for small angles of precession, but it approaches asymptotically $-\hbar\omega$ for large angles of precession [5.133]. It follows, that the interface damping is dependent on the angle of precession and gradually disappears at large precessional angles. Numerical estimates showed that a precessional angle of 15° is sufficient to remove the interface damping.

Berger's expression for the FMR linewidth, (5.46), was derived for a circularly polarized precession. One has to ask, what can be expected for the parallel FMR configuration where the demagnetizing effect leads to a strong ellipticity in the precession. Equation (5.46) suggests that the interface damping and the transport of the angular momentum by an applied dc current should be included in the same torque. A tempting possibility is to include these contributions in (5.41). In this case the effective field for F2 can be written as

$$\mathrm{coef}(\Delta\mu + \hbar\omega)\boldsymbol{c} \times \frac{\boldsymbol{M}}{M_s}\,, \tag{5.57}$$

where $\boldsymbol{c}$ is the direction of the magnetization in the stationary layer F1, and coef is a common pre-factor for the contributions by the interface damping, $\hbar\omega$, and spin transport by perpendicular current, $\Delta\mu$. Equation (5.57) results in Gilbert damping for the perpendicular configuration, but it leads to B.B. like damping for the parallel configuration which scales with the microwave frequency ω. For the parallel configuration the FMR linewidth would be given by $\Delta H_{\mathrm{add}} \sim (\omega/\gamma)^2/(B+H)$. The frequency dependence of the FMR linewidth using the damping effective field in (5.57) is shown by the solid curve in olive color, see Fig. 5.13b. It clearly does not fit the measured data at low microwave frequencies. The non-local FMR linewidth was found to be strictly linearly dependent on the microwave frequency as expected for Gilbert damping [5.25, 134]. Clearly, the mechanism of non-local damping has to be formulated differently.

Tserkovnyak, Brataas and Bauer [5.135] showed that the interface damping can be generated by a spin current from a ferromagnet into adjacent normal metallic layers. The spin current is generated by a precessing magnetic moment in F2. The spin current was calculated by using Brouwer's scattering matrix [5.136] which evolves under a time dependent parameter (phase angle of precession). N metal layers surrounding a magnetic layer were viewed as reservoirs in common thermal equilibrium in contact with an infinite thermal bath. The calculations were carried out by assuming that the reservoirs acted as ideal spin sinks. This approximation is valid when the injected spin momentum into a normal metal decays or leave the interface sufficiently fast to avoid the flow of spin current back into the ferromagnet. The resulting spin current is

$$\boldsymbol{j}_{\text{spin}} = \frac{\hbar}{4\pi} A_{\text{r}} \boldsymbol{n} \times \frac{\partial \boldsymbol{n}}{\partial t} , \tag{5.58}$$

where $\boldsymbol{n}$ is the unit vector along the magnetic moment $\boldsymbol{M}$, and A_{r} is the scattering parameter

$$A_{\text{r}} = \sum_{m,n} \left(|r^{\uparrow}_{m,n} - r^{\downarrow}_{m,n}|^2 + |t^{\uparrow}_{m,n} - t^{\downarrow}_{m,n}|^2 \right) , \tag{5.59}$$

where $r^{\uparrow\downarrow}_{m,n}$ are the reflection parameters for spin up and spin down electrons in the N reservoirs, and $t^{\uparrow\downarrow}_{m,n}$ are the transmission parameters into the reservoirs. The index m and n in (5.59) labels the modes (channels) of $\boldsymbol{k}_{\|,\perp}$ wave-vectors (parallel and perpendicular to the interface) at the Fermi energy. Note that, in contrast to Berger's theory, the contribution to damping disappears in the absence of s-d exchange splitting. The spin pump effect is not present in the absence of a long range ferromagnetic state. the coefficient. A_{r} can be rewritten as

$$A_{\text{r}} = \sum_{m,n} (\delta_{m,n} - 2r^{\uparrow}_{m,n} r^{\downarrow}_{m,n} - 2t^{\uparrow}_{m,n} t^{\downarrow}_{m,n}) . \tag{5.60}$$

It can be shown that the coefficient A_{r} is proportional to the interface mixing conductance, $A_{\text{r}} = \frac{h}{e^2} g_{\uparrow\downarrow}$, see [5.121, 137]. For ferromagnetic layers which are thicker than the lateral coherence length L_0, see (5.47), and the electron scattering at interfaces is partly diffuse the coefficient A_{r} is close to the number of transverse channels in the normal metal N, $\sum_{m,n} \delta_{m,n}$, see [5.121, 137, 138]. In simple metals this sum is given by

$$\frac{A_{\text{r}}}{S} = \frac{k_{\text{F}}^2}{4\pi} = 0.85 n^{2/3} , \tag{5.61}$$

where S is the area of the interface, k_{F} is the Fermi wavevector and n is the density of electrons per spin in the normal metal N. The spin current has the form of Gilbert damping. The Gilbert damping is given by conservation of the total spin momentum

$$\boldsymbol{j}_{\text{spin}} - \frac{1}{\gamma} \frac{\partial \boldsymbol{M}_{\text{tot}}}{\partial t} = 0 , \tag{5.62}$$

where $\boldsymbol{M}_{\text{tot}}$ is the total magnetic moment of F1. After simple algebraic steps one obtains the expression for the dimensionless damping parameter

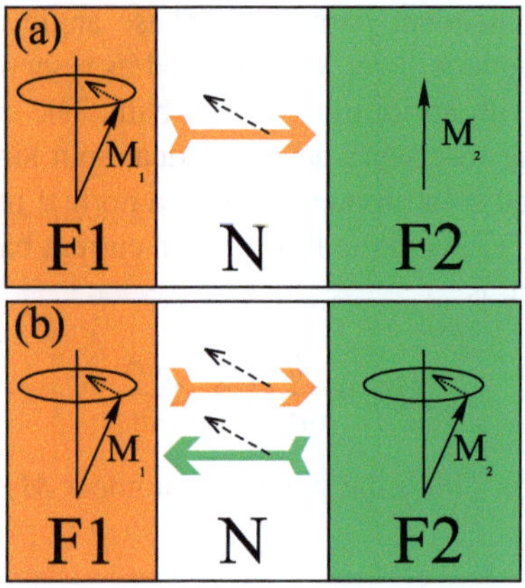

Fig. 5.14. A cartoon representing the dynamic coupling between two magnetic layers which are separated by a non-magnetic spacer. (**a**) represents two magnetic layers with different FMR fields. F1 is at resonance,and F2 is nearly stationary. A bow like arrow in the normal spacer describes the direction of the spin current. The dashed line represents the instantaneous direction of the spin momentum. F1 acts as a spin pump, F2 acts as a spin sink, and consequently F1 acquires an additional Gilbert damping. (**b**) represents a situation when F1 and F2 resonate at the same field. Both layers act as spin pumps and spin sinks. In this case the net spin momentum transfer across each interface is zero. No additional damping is present as the precession is in phase

$$\alpha = \frac{G}{\gamma}\frac{1}{M_s} = g\mu_B \frac{A_r}{4\pi M_s}\frac{1}{d_1} , \tag{5.63}$$

where d_1 is the thickness of the ferromagnetic layer F1, and A_r is now evaluated for a unit interfacial area. The inverse dependence of the Gilbert damping on the film thickness clearly testifies to its interfacial origin. The layer F2 acts as a spin pump. Now another important point has to be answered: How is the generated spin current dissipated? This answer can be found in the recent article by Stiles and Zangwill on *Anatomy of spin transfer torque* [5.118], see Sect. 5.5.1. They showed that the transverse component of the spin current in a normal layer N is entirely absorbed at the N/F interface. For small precessional angles the spin current $\boldsymbol{j}_{spin}$ is almost entirely transverse. It means that the N/F2 interface acts as an ideal spin sink, and provides an effective spin brake for F1, see Fig. 5.14a.

The layers F1 and F2 act as mutual spin pumps and spin sinks. The equation of motion for F2 can be written as

$$\begin{aligned}\frac{1}{\gamma}\frac{\partial \boldsymbol{M}_2}{\partial t} =& -[\boldsymbol{M}_2 \times \boldsymbol{H}_{\mathrm{eff},2}] + \frac{G_2}{\gamma^2 M_s^2}\left[\boldsymbol{M}_2 \times \frac{\partial \boldsymbol{M}_2}{\partial t}\right] \\ &+ \frac{\hbar}{4\pi d_2} g_{\uparrow\downarrow,2}\boldsymbol{n}_2 \times \frac{\partial \boldsymbol{n}_2}{\partial t} - \frac{\hbar}{4\pi d_2} g_{\uparrow\downarrow,1}\boldsymbol{n}_1 \times \frac{\partial \boldsymbol{n}_1}{\partial t} ,\end{aligned} \tag{5.64}$$

where $\boldsymbol{M}_2$ is the magnetization vector of F2, $\boldsymbol{n}_{1,2}$ are the unit vectors along $\boldsymbol{M}_{1,2}$, and d_1, d_2 are the thicknesses of layers F1 and F2. The exchange of spin currents is a symmetric concept and the eq.uation of motion for the layer F1 is obtained by interchanging the indices 2 ⇔ 1.

Equation (5.64) can be tested by investigating the FMR linewidth around an accidental crossover of the resonance fields for F1 and F2 [5.139]. In this case the resonant field of F2 approaches the resonant field of F1. When they reach the same resonant field the rf magnetization components of F1 and F2 are parallel to each other. Each precessing magnetization creates its own spin current which is pumped across the N spacer. The electron mean free path in Au thick films is 38 nm [5.132], and consequently the spin transport even in a 80 ML thick Au spacer is purely ballistic. At the same time both interfaces F1/N and N/F2 act as spin sinks. It follows that the net flow of the spin current through each interface can be zero, and the additional FMR linewidth can disappear, see Fig. 5.14b. A word of caution is needed. The spin currents compensate each other if the bulk relaxation rates in F1 and F2 are close to each other. The compensation of the spin currents is well demonstrated in our measurements. We observed an accidental crossover in the measured FMR fields due to a large in-plane uniaxial anisotropy in the 16ML thick Fe layer. The bulk Gilbert damping in F1 and F2 were very close to each other which resulted in the marked disappearance of the additional FMR linewidth at the crossover of the resonance fields, see Fig. 5.15. The good agreement between theory and experiment clearly shows the validity of the spin pumping theory based on (5.64). The magnetic layers even in the absence of static interlayer exchange coupling, are coupled by dynamic interlayer exchange. The increase in the additional FMR linewidth for 16Fe in Fig. 5.15 which appears close to the accidental crossover in the resonance fields shows that the spin pumping effect can be enhanced with the rf magnetic moments partly out of phase. This is

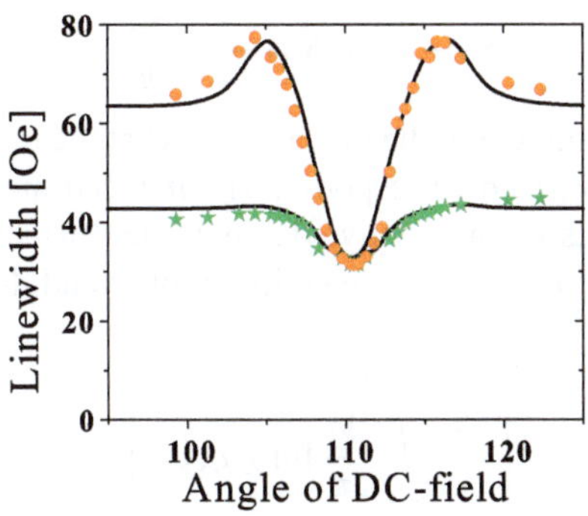

Fig. 5.15. The FMR linewidth at 24 GHz as a function of the angle φ around the crossover of the FMR fields for 20Au/40Fe/14Au/16Fe/GaAs(001). The measured and calculated FMR signals were analyzed using two Lorenzian lineshapes. The Lorenzian peaks were characterized by their amplitudes, resonance fields and linewidths. The solid lines were obtained from calculations using (5.64). Orange filled circles correspond to F1 (16Fe) and green stars correspond to F2(40ML). Note that the FMR linewidth for the thinner sample, F1, first increases before it reaches its minimum value corresponding to its single 20Au/16Fe/GaAs(001) layer structure. Note also that the additional line broadening scales with the film thickness

particularly present in ferromagnetic films which are exchange coupled. In this case the optical mode exhibits a larger linewidth than that expected from a simple spin pumping in which spin sink has negligible precessional amplitude, see [5.140].

The quantitative comparison with spin pumping theory is very good [5.140]. First principles electron band calculations [5.138] resulted in $g_{\uparrow\downarrow} \approx 1.4\times10^{15}\,\mathrm{cm}^{-2}$ for Cu/Co(111) interface. By scaling this value to Au using (5.61) one obtains $G_{\mathrm{sp,cal}} = 1.4\times10^{8}\,\mathrm{s}^{-1}$ for a 16 ML thick Fe film which is very close to the experimental value $G_{\mathrm{sp,exp}} = 1.2\times10^{8}\,\mathrm{s}^{-1}$ measured by FMR at RT. This is amazing agreement considering the fact that calculations of the intrinsic damping in bulk metals have been carried out over the last three decades and have not been able to produce a comparable agreement with experiment, see Sect. 5.4.3.

Spin pumping allows a new look at spintronics. One can in principle move information by spin current in the GHz range of frequencies which does not involve directly the net transport of electron charge. This represents potentially a truly different approach to electronics than that employed in semiconductors.

The spin pump model is a rather exotic theory to those who are used to magnetic studies. One would expect that there is a direct connection to a more common concept which is applicable to magnetic multilayers. The obvious choice is interlayer exchange coupling. One would expect that its dynamic part could create magnetic damping. A ferromagnetic sheet surrounded by a normal metal can be investigated by using a contact exchange interaction between the ferromagnetic spins and the electrons in the normal metal. A similar model was used by Yafet [5.141] for calculating the static interlayer coupling. One can expand Kubo's linear response theory [5.142] for slow precessional motion using a linear approximation for a retarded magnetic moment,

$$\boldsymbol{S}(t-\tau) \cong \boldsymbol{S}(t) - \tau\frac{\partial \boldsymbol{S}(t)}{\partial t}\,, \tag{5.65}$$

where $\boldsymbol{S}(t)$ is the spin moment of the magnetic sheet at the instantaneous time t and τ is the time delay of the retarded response. The induced moment in the N metal at the F/N interface results in an effective damping field which is proportional to the imaginary part of the rf transverse susceptibility of N and the time derivative of the magnetic moment

$$\boldsymbol{H}^{\mathrm{sd}}_{\mathrm{damp}} \sim \left[\frac{\partial}{\partial\omega}\int_{-\infty}^{\infty}\frac{\mathrm{d}q}{2\pi}\,\mathrm{Im}\,\chi(q,\omega)\right]_{\omega\to 0}\frac{\mathrm{d}\boldsymbol{M}(t)}{\mathrm{d}t}\,. \tag{5.66}$$

This damping term again satisfies the Gilbert phenomenology. By using the same interaction potential it was shown [5.143] that the Gilbert damping in dynamic interlayer exchange coupling, G_{s-d}, is similar to that using the spin-pumping theory [5.135] combined with a perfect spin sink. This leads to an important conclusion: *The spin pumping theory is directly related to the dynamic response of the interlayer exchange coupling.*

The spin pump effect can be also observed in single layer F films surrounded by N metal layers provided that the spin pump current diffuses away from the FM/N interface. Interface damping was observed for N/Py/N sandwiches [5.144], where N = Pt, Pd and Ta were 5 nm thick non-magnetic layers surrounding a permalloy (PY) magnetic layer. However, no interface damping was observed with Cu layers. Tserkovnyak et al. explained the lack of interface damping in Cu/Py/Cu structures by a long spin relaxation time in Cu compared to those in Pt, Pd, and Ta. The 5 nm thick Cu does not provide an effective spin sink. A non-equilibrium spin accumulation in the 5 nm thick Cu layer opposes the pumped spin current. A detailed semiclassical theory of spin transport in magnetic multilayer films which are based on the L.L.G. equation of motion for ferromagnets and spin diffusive equations for paramagnets is presented in paper by Urban et al. [5.145].

5.7 Extrinsic Damping

In realistic samples structural inhomogeneities and defects can play a major role in the magnetic relaxation processes.

5.7.1 Two Magnon Scattering

Inhomogeneous magnetic properties can result in scattering of magnons. The uniform mode ($k \sim 0$) can get scattered to nonuniform modes ($\boldsymbol{k} \neq 0$ magnons). This scattering process is usually referred to as *two magnon scattering*. The two magnon scattering process has been used extensively to describe extrinsic damping in ferrites [5.8, 146–149]. A more general treatment which also includes the scattering among non-uniform modes was carried out by Schloemann [5.150]. Patton and co-workers pioneered the use of two magnon scattering in metallic films [5.151].

In ultrathin film structures the k-vectors of magnons are held within the film plane [5.3], which constrains two magnon scattering to the two-dimensional (2D) spin-wave manifold. Let us briefly outline the main features of two magnon scattering. A simple example can be demonstrated by assuming an ultrathin film with an inhomogeneous interface perpendicular uniaxial anisotropy K_u. For simplicity, let us assume that K_u is inhomogeneous along the in-plane x-direction. In the parallel configuration the inhomogeneous part of the energy density is equal to $(\Delta K_u(x)/d)m_z^2(x)/M_s^2$, where $\Delta K_u(x)$ describes deviations of the uniaxial interface anisotropy from its average value, $m_z(x)$ is the perpendicular component of the magnetization $\boldsymbol{M}$, and d is the film thickness, see the coordinate system in Fig. 5.1. The $1/d$ term in the energy density $K_u(x)/d$ is a consequence of averaging of the interface energy K_u over the film thickness, see [5.3]. The perpendicular component m_z can be expressed using magnon creation and annihilation operator notation, see e.g. [5.152]

$$m^- = \left(\frac{2M_s\gamma\hbar}{V}\right)^{1/2} \sum_{\boldsymbol{k}} e^{i\boldsymbol{k}\cdot\boldsymbol{r}} b_{\boldsymbol{k}}$$

$$m^+ = \left(\frac{2M_s\gamma\hbar}{V}\right)^{1/2} \sum_{\boldsymbol{k}} e^{-i\boldsymbol{k}\cdot\boldsymbol{r}} b_{\boldsymbol{k}}^+ , \qquad (5.67)$$

where $m^+ = m_y + \mathrm{i}m_z$ and $m^- = m_y - \mathrm{i}m_z$ are the transverse right and left handed magnetization density operators, b_k^+ and b_k are magnon creation and annihilation operators, and V is the system volume. The contribution of the inhomogeneous interface anisotropy to the total magnetic energy is

$$E_{\mathrm{inh}} = \frac{\gamma\hbar}{M_{\mathrm{s}}Vd} \int_V \mathrm{d}^3\boldsymbol{r}\, \Delta K_{\mathrm{u}}(x) \sum_{\boldsymbol{k},\boldsymbol{k}'} (b_{\boldsymbol{k}}^+ \mathrm{e}^{-\mathrm{i}\boldsymbol{k}\cdot\boldsymbol{r}} + b_{\boldsymbol{k}'} \mathrm{e}^{\mathrm{i}\boldsymbol{k}'\cdot\boldsymbol{r}})^2 , \tag{5.68}$$

where the integration is carried out over the sample volume $V = \mathrm{d}S$, where S the surface area of the film. The summation is over all $\boldsymbol{k}, \boldsymbol{k}'$ wave vectors of the magnons. Equation (5.68) provides one with a simple physical picture of the two-magnon scattering process. The scattering of magnons is given by the cross-product of the bracket in (5.68). In FMR the uniform precession (magnon) $k = 0$ is scattered into magnons having non-zero $\boldsymbol{k}$ with a scattering matrix proportional to $(1/Vd)\int \mathrm{d}\boldsymbol{r}\, \Delta K_{\mathrm{u}}(x)\, \mathrm{e}^{-\mathrm{i}\boldsymbol{k}\cdot\boldsymbol{r}}$. Note that a scattering matrix element is proportional to the Fourier transform of the sample inhomogeneities. In ultrathin films the magnon $\boldsymbol{k}$ vectors are confined to the film plane, see below. The integration in (5.68) can be then replaced by two dimensional integration along the film surface. The integration perpendicular to the surface results in factor d. Consequently, the strength of this two magnon scattering is proportional to the scattering matrix $(1/V)\int \mathrm{d}\boldsymbol{r}\, \Delta K_{\mathrm{u}}(x)\, \mathrm{e}^{-\mathrm{i}\boldsymbol{k}\cdot\boldsymbol{r}}$ per unit surface area of the film, and inversely proportional to the film thickness d. The corresponding relaxation term can be calculated by evaluating the effective rf susceptibility using the Kubo formalism [5.42, 153].

One should emphasize that this type of relaxation is just a mode conversion. In FMR a homogeneous mode (magnon with $k \simeq 0$) relaxes by scattering into magnons with $\boldsymbol{k} \neq 0$. The magnetic excitations do not disappear by two magnon scattering. Two magnon scattering pumps the magnetic energy into other modes which leads to de-phasing of the transversal rf magnetization of the resonant mode. In fact even the longitudinal component of the magnetic moment is not changed by two magnon scattering. In order to reach equilibrium the magnetic energy has to be damped to the lattice by intrinsic damping. In this respect the two magnon scattering mechanism is similar to the mode-mode coupling in small lateral geometries described in Sect. 5.5.2.

Recently Arias and Mills [5.5] addressed the role of two magnon scattering in ultrathin film structures in the parallel configuration. They showed that lateral variations in the perpendicular uniaxial interface anisotropy field (due to the interface roughness) are a leading source of the two magnon scattering in ultrathin films. The imaginary part of the rf susceptibility denominator is given by [5.5]

$$\left(\frac{1}{d}\right)^2 \frac{1}{\omega} \sum_{\boldsymbol{k}} \mid A(k,\omega) \mid^2 \delta(\omega - \omega_{\boldsymbol{k}}) , \tag{5.69}$$

where $A(k)$ is the effective interaction matrix (includes the ellipticity of precession) and $\delta(\omega - \omega_{\boldsymbol{k}})$ is the delta function expressing the conservation of energy in the two magnon scattering. The dependence on the ellipticity of precession is different from that for the Gilbert damping $(2H + 4\pi M_{\mathrm{eff}})$, and has to be evaluated for a particular type of defects. Note, that the magnon momentum is not conserved in two magnon scattering due to sample inhomogeneities (loss of translational invariance), but the energy is conserved. The number of degenerate magnons is proportional to ω^2 when $\omega \to 0$, and consequently the two magnon scattering eventually decreases linearly to zero with decreasing microwave frequency. This treatment is based on perturbation theory. The two magnon scattering is relatively simple mechanism. It couples magnons together (mode-mode coupling) by the scattering matrix which is given by components of the Fourier transform of magnetic defects, see above. This results in coupled linear equations which can be solved exactly by evaluating eigenmodes of the system. The resonant mode is then given by superposition of the eigenmodes, and the FMR linewidth is due to the spread of the eigenfrequencies (dispersion relationship). The details of this approach can be found in [5.154]. An example of this treatment is shown in Fig. 5.23.

In order to evaluate (5.69) one has to calculate the magnon energy spectrum. In ultrathin films (no dependence on the direction perpendicular to the film surface) the $\boldsymbol{k}$ wave vectors are confined to the film plane. For a simple film which is described by the effective magnetization $4\pi M_{\text{eff}}$ the magnon energies are given by simplified Damon Eshbach modes (including the exchange coupling effective field) which are described for an arbitrary angle of the saturation magnetization with respect to the film surface in [5.5, 155],

$$\begin{aligned}\left(\frac{\omega_k}{\gamma}\right)^2 = & \left[H_{\mathrm{i}} + \frac{2A}{M_{\mathrm{s}}}k^2 + 4\pi M_{\mathrm{s}}(1 - N_k)\sin^2(\psi_k)\right]\left[H_{\mathrm{i}} + \frac{2A}{M_{\mathrm{s}}}k^2\right. \\ & \left. - \frac{2K_{\mathrm{u}}^s}{M_{\mathrm{s}}d} + 4\pi M_{\mathrm{s}} N_k \cos^2(\varphi) + 4\pi M_{\mathrm{s}}(1 - N_k)\sin^2(\varphi)\cos^2(\psi_k)\right] \\ & - \left[4\pi M_{\mathrm{s}}(1 - N_k)\cos(\psi_k)\sin(\psi_k)\sin^2(\varphi)\right]^2 , \end{aligned} \tag{5.70}$$

where $N_k = (1 - \exp(-kd))/kd$, H_{i} is the internal effective field, φ is the angle of the saturation magnetization $\boldsymbol{M}_{\mathrm{s}}$ with respect to the film surface, and ψ_k is the angle between $\boldsymbol{k}$ and the projection of the saturation magnetization $\boldsymbol{M}_{\mathrm{s}}$ into the sample plane. Note, that for the in-plane orientation the magnon energy with $\boldsymbol{k}$ parallel to the saturation magnetization decreases its energy with increasing k, and eventually at $k = k_0$ crosses the frequency of the homogeneous mode f, see Fig. 5.16a. This means that the magnon with $\boldsymbol{k}_0$ is degenerate with the homogeneous mode and can be involved in the two magnon scattering relaxation process. The value of $k = k_0$ decreases with increasing angle ψ, see Fig. 5.16. No degenerate modes are available above $\psi_{\max}$; $\sin^2(\psi_{\max}) = H_{\mathrm{i}}/(2H_{\mathrm{i}} + 4\pi M_{\text{eff}})$, no in-plane anisotropies are included in this expression. The number of degenerate modes further decreases with increasing angle φ between the magnetization $\boldsymbol{M}_{\mathrm{s}}$ and the film surface. k_0 as a function of the angle ψ_k and θ is shown in Fig. 5.16b. θ is the angle of the dc field with respect to the sample surface.

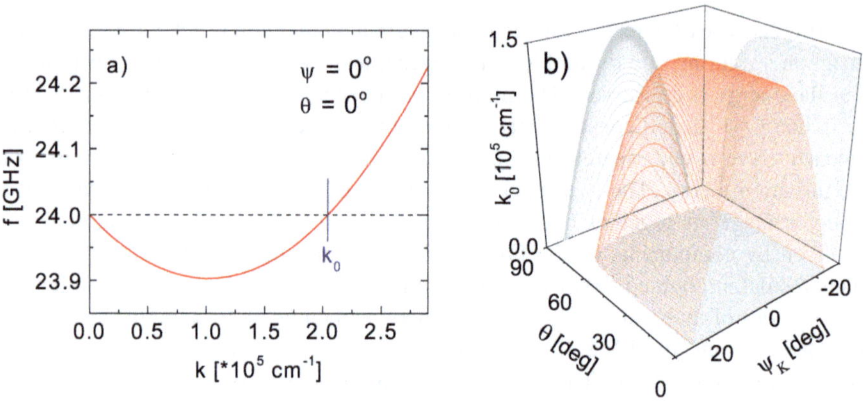

Fig. 5.16. The *solid red line* in (**a**) shows the magnon energy (in units of frequency) as a function of k wave-number for $\psi_k = 0$ and $\theta = \phi = 0$. The angle ψ_k is the in plane angle between $\boldsymbol{k}$ and the in-plane component of the saturation magnetization $\boldsymbol{M}_s$, and θ is the angle between the applied magnetic field $\boldsymbol{H}$ and the sample plane. The intersect between the *dashed line* (the frequency of uniform mode) and the solid line determines the wave-number k_0 of a degenerate magnon. (**b**) shows the dependence of k_0 on the angle ψ_k and θ. Calculations were carried out for $f = 24$ GHz and $4\pi M_{\text{eff}} = 16$ kG. All other anisotropies were set to zero. Notice that $\psi_{\max}$ as a function of θ is nearly constant except in vicinity of $\varphi \sim \pi/4$ (corresponding to $\theta \sim 70°$) where the contribution of two magnon scattering to FMR linewidth rapidly decreases to zero

Expression (5.69) can be rewritten as

$$2 \int_{-\psi_{\max}}^{\psi_{\max}} I(k_0, \psi) \frac{k_0 d\psi}{\frac{\partial \omega}{\partial k}(k_0, \psi)}, \tag{5.71}$$

where $I(k_0, \psi)$ represents the effective two magnon scattering matrix. Notice that $\frac{\partial \omega}{\partial k}(k_0, \psi)$ is the magnon group velocity. It depends on the strength of the dipolar and exchange interaction, and leads to effective narrowing of ΔH. The expression $\frac{k_0}{(\partial \omega/\partial k)(k_0, \psi)}$ in (5.71) accompanying $I(k_0, \psi)$ describes a weighting factor of magnon scattering. The line given by the expression $k_0(\psi)$ follows a lobe in the reciprocal k space of magnons, see Fig. 5.17. It is interesting to note that for a given microwave frequency the weighting factor is only weakly dependent on ψ. It is even weakly dependent on the angle φ up to vicinity of $\varphi \sim \pi/4$ where the contribution of two magnon scattering disappears. Consequenly the contributions to two magnon scattering in (5.71) along the lobe have to be included with an equal weight. However, this is not true for long wavelength magnons (close to the origin of the reciprocal space) where the two magnon scattering model is not applicable, see below.

For both the 2D and 3D spin-wave manifold there are no magnons degenerate with the FMR mode in the perpendicular configuration, hence the FMR linewidth, $\Delta H_\perp$, should be smaller than that in the parallel configuration, $\Delta H_\parallel$.

One expects that variations sample in magnetic properties which are far apart have to lead to a simple superposition of the local FMR lines. The FMR linewidth in

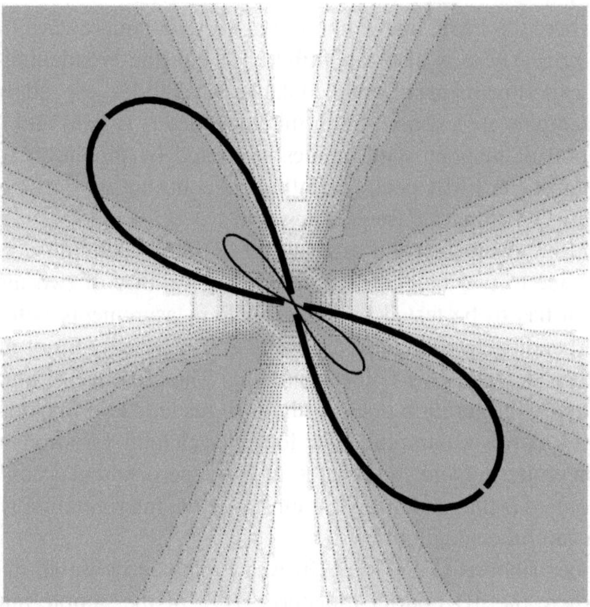

Fig. 5.17. The two magnon scattering lobes at 24 and 73 GHz in the q-space of the magnetic scattering matrix intensity $I(\boldsymbol{q}) \sim \cos^4(2\varphi_q)$, where φ_q is the angle of the $\boldsymbol{q}$ vector with respect to the [100] axis. This Figure represents the two magnon scattering in Au/Pd/Fe/GaAs(001) systems which are affected by misfit dislocation network, see below. The dashed lines are contour lines of $I(\boldsymbol{q})$. Note that the orientation of lobes (magnetization) affects the angular dependence of FMR linewidth caused by the misfit dislocation network. When the lobes are oriented close to $\langle 110 \rangle_{\text{Fe}}$ they have a weaker contribution than those oriented close to $\langle 100 \rangle_{\text{Fe}}$

this case merely reflects large length scale sample inhomogeneities and has nothing to do with the two magnon treatment of damping. McMichael et al. [5.154] recently addressed this point. They found that the FMR linewidth is given by superposition of local resonances when *characteristic inhomogeneity field is larger than interaction field* [5.156]. In the range of long wavelength defects the important part of the interaction field is magnetostatic contribution $2\pi M_s kd$ in (5.70). The FMR measurement is given by a simple superposition of local FMR peaks when

$$H_p D \geq 3\pi M_s d \, , \tag{5.72}$$

where H_p is the root mean square value of random variations of local anisotropy field satisfying a Gaussian distribution, and D is the corresponding average grain size, see Fig. 5.4 in [5.154]. The local inhomogeneous fields simply add to the applied field and one can get indeed a genuine zero frequency offset $\Delta H(0)$, see [5.154]. This offset $\Delta H(0)$ ought to increase in the perpendicular configuration where inhomogeneous magnetic anisotropy fields contribute more effectively to the FMR field due to the absence of a strong elliptical polarization accompanying the parallel configuration. This is perhaps the case of the FMR measurements on Py films by Patton et al.

[5.157, 158] where the FMR linewidth even at very low microwave frequencies did not approach zero. There is also a possibility to get line broadening using certain equipment. In experiments using small coplanar wavequides (especially for μm scale systems) the samples are exposed to an inhomogeneous rf field and that can result in the generation of magnetostatic modes separated by dipolar rf fields resulting in a broadening of the FMR peak [5.159]. This contribution becomes particularly important at low GHz range of frequencies.

One should be aware that the extrinsic FMR linebroadening does not have to be necessarily due to two magnon scattering. The validity of two magnon scattering contribution has to be tested by using FMR measurements with the magnetic moment in perpendicular configuration where ΔH as a function of microwave frequency should be described by Gilbert damping. The presence of non zero $H(0)$ in the perpendicular configuration does not have to exclude two magnon scattering in interpretation of extrinsic damping. The long wavelength variations are not a part of two magnon scattering and can be present in the perpendicular configuration, see above. The problem is that under this circumstance the interpretation is not as simple and requires a careful evaluation.

In amorphous ribbons [5.153] $\Delta H(0)$ was explained by a two magnon scattering process where the 3D Fourier transform, $I(k)$, of the sample inhomogeneities were represented by a step function in the k-wave vector with an upper cutoff less than the maximum wave vector $\boldsymbol{k}_{\max}$ allowed by the spin-wave manifold. In these calculations the two magnon scattering did not have to affect the slope of (5.75) in the measured microwave frequency range. In fact, in this case the Gilbert damping from the FMR measurements (obtained from the slope of (5.75)) was found to be equal to that obtained from the microwave transmission peaks at Ferromagnetic Antiresonance (FMAR) [5.83]. At FMAR ($\omega/\gamma = H + 4\pi M_{\mathrm{eff}}$) the spin manifold is shifted downwards compared to that at FMR by several kOe (5–10 kOe). In this case the intersection with the spin manifold starts at large k wave-numbers. The Fourier components of the sample inhomogeneities at these large k wave-vectors were found negligible, and consequently the FMAR measurements allowed one to determine the intrinsic Gilbert damping in thick metallic samples, see Sect. 5.4.6.

The structural defects in ultrathin film multilayers are often caused by the lattice misfit between individual layers. A typical case are crystalline Au/Pd/Fe and Au/Fe/Pd/Fe epitaxial layers grown on GaAs(001) templates. Pd has a lattice mismatch of 4.4% with respect to Fe and 4.9% with respect to Au, and therefore samples with a sufficient thickness of Pd are affected by the relaxation of lattice strain [5.160]. In the structures with a sufficiently thick Pd layer a self-assembled network of misfit dislocation half loops was observed by plan view Transmission Electron Microscopy (TEM) [5.160]. The network of misfit dislocations was found to be oriented along the $\langle 100 \rangle$ crystallographic axes of Fe. The onset of a rectangular network of misfit dislocations was accompanied by the presence of a strong extrinsic damping. This system provides also a nice example where all basic features of two magnon scattering are well represented. In the Pd/Fe/GaAs(001) structures with the number of Pd atomic layers $n \geq 130$ the FMR linewidth was strongly dependent on the angle φ_M between the magnetization and the crystallographic $\langle 100 \rangle_{\mathrm{Fe}}$ axes, showing a dis-

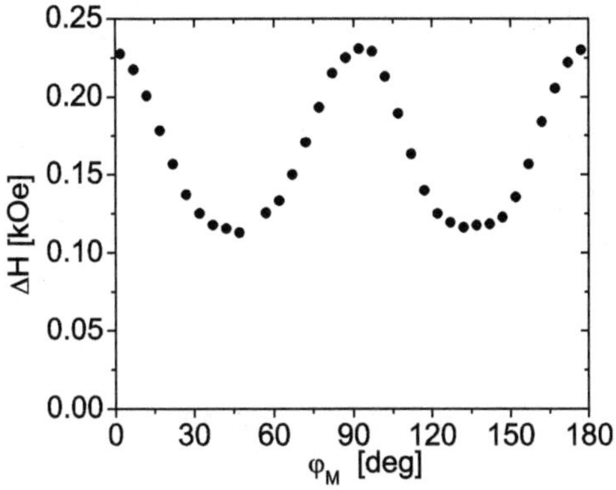

Fig. 5.18. The ferromagnetic resonance linewidth for the 200Pd/**30Fe**/GaAs(001) film (the integers represent the number of atomic layers) at 36 GHz as a function of the in-plane angle φ_M between the magnetization and the $[100]_{Fe}$ axis

tinct four fold symmetry, see Fig. 5.18. This angular dependence reflects the overall in-plane symmetry of the defects. The minima and maxima in ΔH are along the $\langle 110 \rangle_{Fe}$ and $\langle 100 \rangle_{Fe}$ crystallographic directions, respectively. The FMR linewidth as a function of microwave frequency also changed in a very pronounced way.

The frequency dependence of the FMR linewidth, $\Delta H(f)$, along the $\langle 100 \rangle_{Fe}$ (in-plane easy magnetic axes) and $\langle 110 \rangle_{Fe}$ (in-plane hard magnetic axes) directions is shown in Fig. 5.19. Along $\langle 110 \rangle_{Fe}$ the FMR linewidth between 10 and 73 GHz is nearly linearly dependent on the microwave frequency, but is accompanied by a small zero frequency offset $\Delta H(0) = 50$ Oe. However, the slope is close to that corresponding to the intrinsic Gilbert damping obtained in samples without extrinsic damping. Several observations for the $\langle 100 \rangle_{Fe}$ orientations can be made. Firstly, ΔH has clearly non linear dependence on the microwave frequency contrary to expectations for Gilbert damping. Secondly, the slope of the FMR linewidth is close to that expected for the intrinsic damping only between 36 and 73 GHz, but $\Delta H(0) =$ 160 Oe is significantly increased compared to the $\langle 110 \rangle_{Fe}$ orientations. Thirdly, below 36 GHz the frequency dependence of ΔH shows a clear downturn indicating that ΔH approaches zero at low microwave frequencies. In fact the frequency dependence for the $\langle 100 \rangle_{Fe}$ orientations resembles the calculations of two magnon scattering in ultrathin films by Arias and Mills [5.161]. A similar frequency dependence of ΔH was found also by Lindner et al. [5.162] in Fe/V superlattices.

Notice, that the extrapolated slope of $\Delta H(f)$ at low microwave frequencies in Fig. 5.19 is noticeably larger than that corresponding to the intrinsic damping. The rotational magnetization motion during the magnetization reversal of memory pixels is usually in the range of a few GHz. In this case the measured effective damping parameter α can be strongly affected by the slope of two magnon contribution. However it is worthwhile of pointing out that in many systems the slope of the FMR linewidth as a function of microwave frequency is close to that expected from the

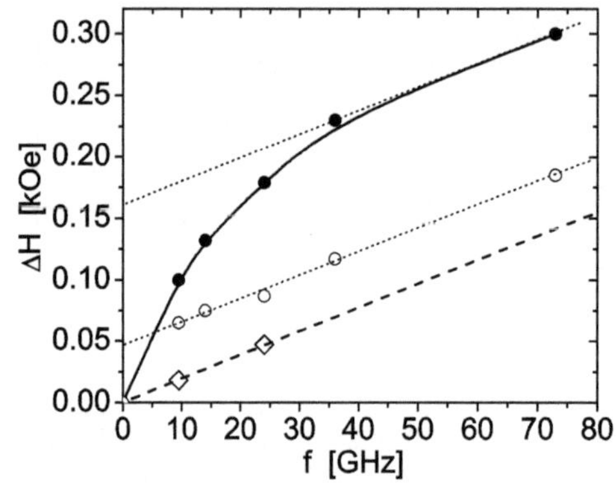

Fig. 5.19. The frequency dependence of the FMR linewidth for the 200Pd/**30Fe**/GaAs(001) structure (the integers are in number of atomic layers) along the $\langle 110 \rangle_{\mathrm{Fe}}$ (○) and $\langle 100 \rangle_{\mathrm{Fe}}$ (•) directions, respectively. The purpose of the solid line spline fit is to guide the reader's eye. The dashed line shows the frequency dependence of the intrinsic FMR linewidth (Gilbert damping) obtained by using the 100Pd/**30Fe**/GaAs(001) sample with no magnetic defects in Fe. Note that (◇) points are right on the dashed line indicating that in the perpendicular configuration the FMR linewidth ΔH is only given by the Gilbert damping ($\alpha = 0.006$) with no zero frequency offset ($\Delta H(0)_{\perp} = 0$). The dotted lines indicate the range of microwave frequencies where the slope of $\Delta H(f)$ is close to that expected from the Gilbert damping. Note that the dotted lines have zero frequency offsets

intrinsic Gilbert damping if the microwave frequency is held in a certain range (often between 10 to 36 GHz but not always).

In order to give convincing evidence for two magnon scattering mechanism one has to investigate the FMR linewidth as a function of the angle θ_{M} between the magnetization $\boldsymbol{M}$ and the sample surface by applying the dc field away from the surface. The dependence of the FMR linewidth on the angle θ_H between the dc magnetic field and the sample plane is shown in Fig. 5.20. These results have shown that the damping decreases significantly when the magnetization is inclined significantly from the film surface. In fact, the measured ΔH in the perpendicular configuration at 10 and 24 GHz was given exactly by the intrinsic damping. The appreciable FMR peak narrowing shown in Fig. 5.20, and the absence of $\Delta H(0)$ in the perpendicular configuration provide a strong support for the presence of two magnon scattering in the Pd/Fe/GaAs(001) structures.

This is further supported by some additional features of the out of plane FMR measurements. Fig. 5.20 shows (dashed line) the calculated dependence of the FMR linewidth as a function of θ_H taking the intrinsic value of the Gilbert damping $G_{\mathrm{int}} = 1.4 \times 10^8\,\mathrm{s}^{-1}$. The calculated increase for the intermediate angles is caused by dragging of the magnetic moment behind the external applied field $\boldsymbol{H}$. The difference between the measured FMR linewidth and that expected for the intrinsic damping ΔH_{ext} is caused by two magnon scattering. The two magnon scattering physics as a function of the angle θ_H is better represented by the adjusted frequency linewidth, $\Delta\omega$, see Hurben and Patton [5.163],

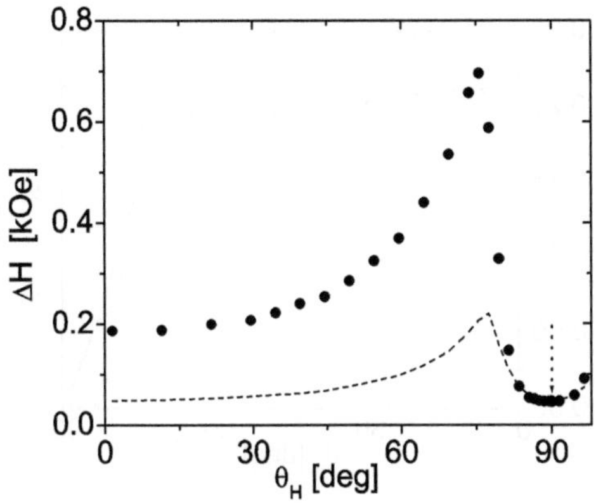

Fig. 5.20. Measured ferromagnetic resonance linewidth ΔH for the sample 200Pd/**30Fe**/GaAs(001) as a function of θ_H at 24 GHz. The dots represent the measured data and the dashed line represents the FMR linewidth $\Delta H_{\mathrm{G}}(\theta)$ that was calculated using the Gilbert damping from the perpendicular configuration. The peak in the FMR linewidth for $\theta = 78°$ is caused by dragging the magnetization behind the applied field. The dotted vertical line identifies the perpendicular configuration

$$\Delta\omega(\theta) = \left(\frac{\mathrm{d}\omega}{\mathrm{d}H}\right)_{\text{at FMR}} \Delta H_{\text{ext}} \,, \tag{5.73}$$

where $\mathrm{d}\omega/\mathrm{d}H = [\omega(H+\Delta H, \theta+\Delta\theta) - \omega(H,\theta)]/\Delta H$ using the FMR condition for the resonance frequency which includes the in plane and out of plane magnetic anisotropies. It is easier to calculate $\mathrm{d}\omega/\mathrm{d}H$ by picking $\Delta\omega$ and evaluating appropriate change in ΔH and $\Delta\theta$ satisfying the resonance condition for the out of plane configuration.

The values of $\Delta\omega/\gamma$ calculated from the measured FMR linewidth data are shown in Fig. 5.21. Remarkably, the extrinsic contribution to the frequency linewidth $\Delta\omega/\gamma$ is nearly independent of θ_H but decreases abruptly when the magnetization angle $\theta_M \sim \pi/4$, see the small vertical dotted line. $\Delta\omega/\gamma$ and $\psi_{\max}$ follow well each other as a function of θ_H, see the dashed line in Fig. 5.21. The proportionality between $\Delta\omega/\gamma$ and $\psi_{\max}(\theta_H)$ suggests that this extrensic damping is caused by two magnon scattering and the effective scattering matrix $I(\boldsymbol{q}, \theta_{\mathrm{M}})$ is weakly dependent on θ_{M}, see (5.71).

It is interesting to address the origin of a strong in-plane angular dependence of the two magnon scattering in self assembled network of misfit dislocations, see Fig. 5.18. The lattice defects decrease the local symmetry and create inhomogeneous magnetic anisotropies. Their angular dependence is given by the symmetry of defects. This means that the intensity of two magnon scattering can have an explicit dependence on the direction of the magnetization with respect to the axes of magnetic defects. This case was addressed by Lindner et al. [5.162]. They observed an anisotropic extrinsic damping (measured along the ⟨100⟩ and ⟨110⟩ axes) for FeV superlattices. It was assumed that defects were caused by surface steps. Their argument is based on

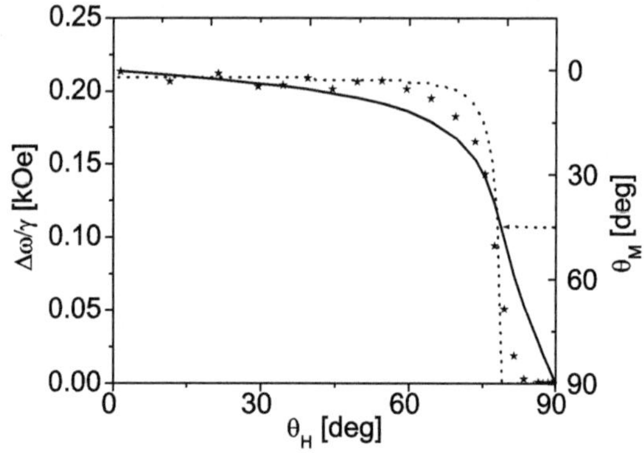

Fig. 5.21. (⋆) represent the adjusted frequency FMR linewidth $\Delta\omega/\gamma$ from the extrinsic contribution as a function θ_H at 24 GHz. The solid line shows the angle of magnetization θ_M as function of θ_H. The dashed line shows the critical angle ψ_{max} as a function θ_H. Notice that ψ_{max} describes the angular dependence of $\Delta\omega/\gamma$ quite well. ψ_{max} was scaled in order to compare it with $\Delta\omega/\gamma$

the behavior of uniaxial anisotropy. The line defects are the source of local uniaxial anisotropy. However the uniaxial internal effective field disappears for the magnetization oriented close to 45 Degrees away from the uniaxial axis. In this direction the uniaxial anisotropy does not affect the internal field and two magnon scattering is switched off. In systems with a rectangular network of defects the contribution of two magnon scattering is explicitly dependent on the angle φ_M (with respect to the defect lines) as $\cos^2(2\varphi_M)$. This argument can be applied to Pd/Fe(001) systems where the rectangular network of misfit dislocations creates local uniaxial anisotropies with the magnetic axis aligned along the dislocation lines. Since the defect lines of the network of misfit dislocations are oriented along the crystallographic axis $\langle 100\rangle$ of Fe the observed anisotropy in ΔH could be misinterpreted as a genuine anisotropic intrinsic damping. The Fourier components of the scattering intensity have to include two parts. One is explicitly dependent on the angle φ_M and the other on the wave vector $\boldsymbol{q}$,

$$I(\boldsymbol{q}, \varphi_M) = Q(\boldsymbol{q}) \cos^2(2\varphi_M) . \tag{5.74}$$

The term $Q(\boldsymbol{q})$ has to satisfy the symmetry of defects. It is needed to explain the frequency and angular dependence of ΔH. Further discussion can be found in [5.164].

The zero frequency offset $\Delta H(0)$ in Fig. 5.19 for the measurement along the $\langle 110\rangle$ orientation is often observed in many systems affected by extrinsic damping. In a wide range of metallic amorphous ribbons [5.83], films [5.157], and metallic ultrathin film multilayers [5.3, 165, 166] the FMR linewidth, ΔH, between 10 to 36 GHz is described by a linear dependence on microwave frequency with a zero frequency offset $\Delta H(0)$,

$$\Delta H(\omega) = \Delta H(0) + 1.16 \frac{\omega}{\gamma} \frac{G_{eff}}{\gamma M_s} . \tag{5.75}$$

$\Delta H(0)$ originates in the extrinsic contribution to the FMR linewidth. The subscript eff in the Gilbert damping parameter G_{eff} was used to express the fact that the slope of the FMR linewidth in (5.75) can in general include both the intrinsic, G_{int}, and extrinsic, G_{ext}, contributions. One should point out that (5.75) does not imply that the contribution of two magnon scattering approaches $\Delta H(0)$ when the microwave frequency approaches zero. Two magnon scattering eventually has to approach zero with decreasing microwave frequency. This point was well established in [5.153]. In our view the zero frequency offset represents a simple test for identifying the presence of extrinsic FMR linebroadening.

A typical example of a linear dependence of the FMR linewidth on the microwave frequency with a zero frequency offset can be seen on Au/Cr/Fe/GaAs(001) systems [5.167], see Fig. 5.22. It is appealing to find out whether the extrapolated $\Delta H(0)$ from limited range of microwave frequencies can be caused by two magnon scattering. The main test was done by investigating the FMR linewidth as a function of the angle θ_{M} with respect to the film surface using out of plane FMR measurements. The FMR linewidth ΔH for a 20Cr/15Fe sample as a function of the angle θ_H (between the applied external field and the sample plane) behaved in a very similar manner to that of Pd/Fe/GaAs(001) structures, see Figs. 5.20 and 5.21. The perpendicular FMR measurements at 9.5 and 24 GHz, see Fig. 5.22, showed no measurable $\Delta H(0)$, and the measured slope led to the intrinsic Gilbert damping parameter $G_{\mathrm{int}} = 1.4 \times 10^8\ \mathrm{s}^{-1}$. The appreciable FMR peak narrowing and the absence of $\Delta H(0)$ in the perpendicular configuration provide strong support for the presence of two magnon scattering in the Cr/Fe/GaAs(001) structures. Clearly, the zero frequency offset $\Delta H(0)$ for the Cr/Fe/GaAs(001) films is compatible with two magnon scattering. This means that the contribution of two magnon scattering can be constant across this frequency range. In crystalline Fe films the lowest microwave frequency is limited to 10 GHz due to the presence of crystalline in-plane anisotropies, and therefore we were not

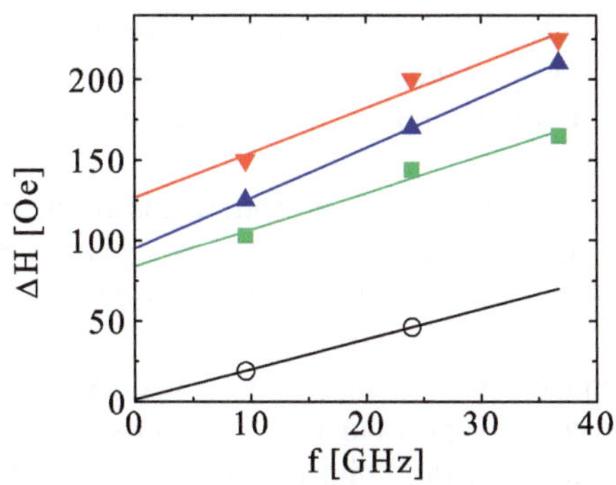

Fig. 5.22. The FMR linewidth for the parallel configuration for 20Cr/15Fe/GaAs(001) as a function of the microwave frequency along the cubic axes (*red*), hard in-plane uniaxial axis $[1\bar{1}0]$ (*blue*), and the direction [110] (*green*). The black line represents the perpendicular FMR linewidth. Note that $\Delta H(0) = 0$ for the perpendicular FMR measurements. The slope of the black solid line determines the intrinsic Gilbert damping $G_{\mathrm{int}} = 1.5 \times 10^8\ \mathrm{s}^{-1}$

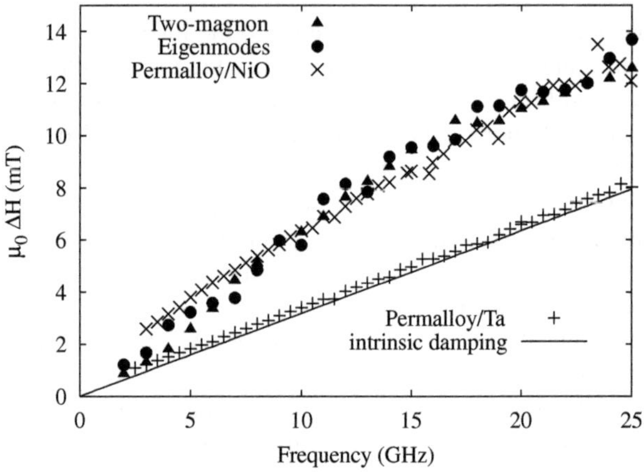

Fig. 5.23. FMR linewidth as a function of microwave frequency. The measurements were carried out on Py films grown by sputtering on a NiO template (× *points*) and a Ta seed layer (*crosses*). The two magnon scattering is present only on the Py films grown on a NiO template. Calculations of the inhomogeneous FMR line broadening was carried out by subdividing the sample into 1000 × 1000 grains. Each grain is 20 × 20 nm^2 and 100 nm in thickness. The grains have internal fluctuating fields. The FMR linewidths are calculated using a simple two magnon scattering theory (filled triangles) and a full two magnon scattering process which involves the scattering between the magnons with all non zero $\boldsymbol{k}$ wave-vectors (filled circles) corresponding to a full mode-mode coupling, see (5.68). The upper cut-off for the magnon k wave-vectors is given by the grain size. Both calculations trace each other and fit well the FMR linewidth for the Py films grown on NiO templates. Note, that the two magnon scattering decreases appreciably when the microwave frequency is smaller than 10 GHz. See further details in [5.168]. This Figure was obtained by courtesy of R. McMichael

able at this point to comment on what happens at very low frequencies where the damping should approach zero. The FMR measurements can be easily carried out below 10 GHz on Py. Two magnon scattering contribution to ΔH in the Py films grown by sputtering on NiO templates decreased gradually to zero below 10 GHz, see Fig. 5.23 and [5.168].

One should realize that the two magnon scattering is a dynamic effect affecting not only damping but also the frequency of spin waves. This means that the anisotropy in the angular dependence of the FMR linewidth is dependent on the microwave frequency, see further details in [5.164].

The angular dependent extrinsic damping created by a rectangular network of defects appears to be a common phenomenon. It was observed in our previous studies using the metastable bcc Ni/Fe(001) bilayers grown on Ag(001) substrates [5.169], and Fe(001) films grown on bcc Cu(001) [5.170]. In the Ni/Fe bilayers after depositing three ML of Ni the structure went through a major structural change going towards the stable fcc phase of Ni(001). That resulted in a network of rectangular lattice defects perhaps similar to those observed by Wulfhekel et al. [5.171]. In this case not only the magnetic damping developed a large anisotropy, but even the in-plane four fold anisotropy field was enhanced to several kOe which is significantly above that corresponding to cubic bulk Fe (0.5 kOe). Coercive fields of several hundred Oe were reached due to the presence of the enhanced anisotropy and lattice defects [5.172]. The enhancement was dependent on the Ni film thickness. The angular dependence of the FMR linewidth indicated that the defect lines were oriented along the ⟨100⟩ axes of Fe(001). Bcc Cu(001) layers grown on Ag(001) substrates are another example of such behavior. In this case the bcc Cu(001) layer went through a lattice transformation after the thickness of the Cu layer was larger than 10 ML. Again a strong anisotropy in ΔH was observed for the Fe(001) films grown on the lattice transformed Cu(001) substrates. The angular dependence indicated that the defect lines in the Cu(001) layers and the symmetry axes of magnetic defects in Fe(001) are along the ⟨100⟩ crystallographic directions of Fe(001). However, in these samples no significant enhancement of the in-plane four fold anisotropy was found. We recently observed this type of two magnon scattering in Heusler alloy NiMnSb(001) semi-metal films grown on InP(001) by the Molenkamp group [5.173]. In plan view TEM studies two sets of rectangular defects were observed along the ⟨100⟩ and ⟨110⟩ directions [5.174], and consequently the two magnon scattering showed a strong two magnon scattering along all principle crystallographic axes.

The above examples have shown clearly that the two magnon scattering mechanism is useful concept for the description of extrinsic damping in ultrathin films which have lattice defects. The in-plane symmetry of two magnon scaterring is given by the symmetry of lattice defects. Our extensive experience in FMR measurements has shown that *the angular dependent FMR linewidth (even with perfectly Lorenzian FMR peaks) is most likely caused by extrinsic damping which originates in a network of lattice defects.* In fact, the interpretation of the anisotropic FMR linewidth by the Stoner enhancement factor in spin pumping in Au/30Fe/40Au/4Pd/[Pd/Fe]$_5$/16Fe(001) and Au/30Fe/40Au/9Pd/16Fe(001) structures [5.25, 134, 140] was incorrect. In both structures the anisotropic part of the FMR linewidth was caused by the two magnon scattering described above.

5.7.2 Dry Magnetic Friction and Large Length Scale Inhomogeneities

Baltensberger and Helman have shown [5.175] that the presence of $\Delta H(0)$ can arise from a hysteretic rf loss which results from the more general concept of a dry friction [5.176]. The rf hysteretic loss can be created by magnetic anisotropies having irregular directions. During collective motion of the spin ensemble, a part of the magnetic system can depart from the path of uniform motion and may spiral down towards to a local potential minimum from which it is subsequently dragged out by exchange coupling with the surrounding magnetic moments. This represents an energy loss to the uniform motion even when the uniform motion is arbitrarily slow. The resulting hysteresis leads to zero frequency loss. Numerical simulations have shown that the onset of dry friction requires a large anisotropy energy which is an appreciable fraction of the exchange energy [5.175]. This is perhaps satisfied for rare-earth magnetic ions. It is not likely that dry friction plays an important role in soft magnetic 3d transition metals.

Acknowledgement. The author would like to thank to his colleagues Dr. R. Urban, Mr. G. Woltersdorf, Professor M. Scheinfein, and Professor J.F. Cochran for stimulating discussions and invaluable help in preparation of this chapter. I would like to particularly thank to Professor A.S. Arrott. My extensive scientific collaboration and association with Prof. Arrott has helped me greatly to shape my understanding of physics and magnetism in particular. I also owe a debt of gratitude to all those with whom I have held extensive discussions and provided a valuable help during preparation of this manuscript: K. Baberschke, G. Bauer, L. Berger, J.A.C. Bland, A. Brataas, C. Chappert, A. Fert, V. Kamberský, P. Kabos, J. Kirschner, J. Koetzler, J. Lynn, R. McMichael, C. Patton, R. Schäfer, T.J. Silva, E. Šimanek, M. Stiles, D. Schmool, Y. Tserkovniak, and A. Yelon. Without these discussions and help the chapter would have been less complete. I would like to thank Professor M. Scheinfein, Dr. M. Stiles, Professor Zangwill and Dr. R. McMichael for allowing me to include in this chapter their recent Figures and calculations which beautifully demonstrate the physics discussed. The field of magnetic relaxations is a very diversified field covering at least five decades. It is not possible for a single chapter to describe all the complexities of this field. I apologize to those who contributed to this field and whose results are not explicitly included in this chapter. The omission is not intentional but merely reflects my inability to include all studies in this chapter which has to be constrained to a reasonable length.

I would like to thank especially the Natural Science Engineering Research Council (NSERC) of Canada, and the Canadian Institute for Advanced Research (CIAR) for continued research funding which makes my work possible. I would like to express also my thanks to the Alexander von Humboldt Foundation and Professor J. Kirschner, Max Planck Institute in Halle, for providing me with a generous support during my summer research semesters in Germany where this manuscript was partly prepared.

References

5.1. J. MacDonald, Proc. Phys. Soc. A **64**, 968 (1951).
5.2. B. Heinrich, *Magnetic Ultrathin Film Structures II,* ed. B. Heinrich and J.A.C. Bland, Springer Verlag, 1994.
5.3. B. Heinrich and J. F. Cochran, Adv. Phys. **42**, 523 (1993).
5.4. B. Heinrich, J. Cochran, and M. Kowalewski, *Frontiers in Magnetism of Reduced Dimension Systems, edited by P. Wigen and V. Baryachtiar and N. Lesnik*, chapter V, pages 161–210, NATO-ASI, 1996.
5.5. R. Arias and D. L. Mills, Phys. Rev. B **60**, 7395 (1999).
5.6. D. Goerlitz and J. Koetzler, Eur. Phys. J. B **5**, 37 (1998).
5.7. C. Kittel, Phys.Rev. **71**, 270 (1947).
5.8. M. Sparks, *Ferromagnetic Relaxation Theory*, Mc-Graw-Hill, New York, 1966.
5.9. V. Safonov, J. Appl. Phys. **91**, 8653 (2002).
5.10. H. Mori, Prog.Theor.Phys. **33**, 423 (1965).
5.11. H. Mori and K. Kawasaki, Prog.Theor.Phys. **27**, 529 (1962).
5.12. B. Heinrich and A. Arrott, A.I.P. Conf. Proc. **34**, 119 (1976).
5.13. B. Heinrich, A. Arrott, and S. Hanham, J. Appl. Phys. **50**, 2134 (1979).
5.14. J. Koetzler, D. Goerlitz, M. Hartl, and C. Marx, IEEE Transactions on Magnetism **30**, 828 (1994).
5.15. B. Heinrich, A. Arrott, and S. Hanham, J. Appl. Phys. **50**, 2134 (1979).
5.16. A. Arrott and B. Heinrich, J. Mag. Mag. Mat. **31-34**, 1084 (1983).
5.17. B. Heinrich and A. Arrott, J. Appl. Phys. **55**, 2455 (1984).

5.18. D. Berkov and N. Gorn, J. Phys.: Cond. Matter **14**, L281 (2002).
5.19. J. L. Garcia-Palacios and F. Lazaro, Phys. Rev. B **58**, 477 (1998).
5.20. N. Smith, J. Appl. Phys. **92**, 3877 (2002).
5.21. A. Gurevitch, *Ferrites at rf frequencies*, chapter 1, pages 1–103, Springer, 1960.
5.22. B. Heinrich and J. A. C. Bland, *Ultrathin Magnetic Structures II.*, section 3.1, pages 195–222, Spinger-Verlag, 1994.
5.23. R. Urban, G. Woltersdorf, and B. Heinrich, Phys. Rev. Lett. **87**, 217204 (2001).
5.24. B. Heinrich, R. Urban, and G. Woltersdorf, J. Appl. Phys. **91**, 7523 (2002).
5.25. B. Heinrich, R. Urban, and G. Woltersdorf, IEEE Transactions on Magnetics **38**, 2496 (2002).
5.26. B. Heinrich, D. Fraitová, and V. Kamberský, Phys. Stat. Sol. **23**, 501 (1967).
5.27. Z. Frait and D. Fraitová, J. Mag. Mag. Mater. **15-18**, 1081 (1980).
5.28. Z. Frait and D. Fraitová, *Frontiers in Magnetism of Reduced Dimension Systems, edited by P. Wigen and V. Baryachtiar and N. Lesnik*, chapter IV, pages 121–151, NATO-ASI, 1996.
5.29. J. Cochran, B. Heinrich, and A. Arrott, Phys. Rev. B **34**, 7788 (1986).
5.30. N. Chan, V. Kamberský, and D. Fraitová, J. Mag. Mag. Mat. **214**, 93 (2000).
5.31. Z. Frait and H. MacFaden, Phys. Rev. **139**, 1173 (1965).
5.32. D. Fraitová, Czech. J. Phys. **B24**, 827 (1974).
5.33. B. Heinrich, J. F. Cochran, and R. Baartman, Can. J. Phys. **55**, 806 (1977).
5.34. W. Ament and G. Rado, Phys. Rev. **97**, 1558 (1955).
5.35. H. Suhl, IEEE Tran. on Magnetics **34**, 1834 (1998).
5.36. B. Heinrich, J. F. Cochran, and K. Myrtle, J. Appl. Phys. **53**, 2092 (1982).
5.37. T. Kobayashi, R. Barker, and A. Yelon, Phys. Rev. B **7**, 3286 (1973).
5.38. S. Vonsovskii and J. Izyumov, Fiz. Metallov i Metallovedenie **10**, 321 (1960).
5.39. R. J. Elliot, Phys. Rev. **96**, 266 (1954).
5.40. T. Valet and A. Fert, Phys. Rev. B **48**, 7099 (1993).
5.41. S. Dubois et al., Phys. Rev. B **60**, 477 (1999).
5.42. D. Zubarev, Sov. Phys. Usp. **3**, 320 (1960).
5.43. C. Kriesman and H. Callen, Phys. Rev. **94**, 837 (1954).
5.44. A. Elezzabi, M. Freeman, and M. Johnson, Phys. Rev. Lett. **77**, 3220 (1996).
5.45. D. Pines and C. Slichter, Phys. Rev. **100**, 1014 (1955).
5.46. S. Ingvarsson et al., Phys. Rev. B **66**, 214416 (2002).
5.47. V. Kamberský and C. E. Patton, Phys. Rev. B **11**, 2668 (1975).
5.48. V. Kamberský, Czech. J. Phys. **B26**, 1366 (1976).
5.49. K. Baberschke, Appl. Phys. A **62**, 417 (1996).
5.50. V. Kamberský, Can. J. Phys. **48**, 2906 (1970).
5.51. S. Vonsovskii, *Ferromagnetic resonance*, chapter E.A. Turov, Chap.V, GIMFL, Moscow, 1961.
5.52. S. M. Bhagat and P. Lubitz, Phys. Rev. B **10**, 179 (1974).
5.53. V. Korenman and R. Prange, Phys. Rev. B **6**, 2769 (1972).
5.54. J. Ziman, *Principles of the theory of solids*, second edition Chapter 8.7, page 607, Cambridge University Press, 1972.
5.55. B. Heinrich, D. J. Meredith, and J. F. Cochran, J. Appl. Phys. **50**, 7726 (1979).
5.56. G. Dewar, B. Heinrich, and J. Cochran, Can. J. Phys. **55**, 821 (1977).
5.57. V. Kamberský, private communication .
5.58. V. Kamberský, Czech. J. Phys. **34**, 1111 (1984).
5.59. V. Kamberský, J. Cochran, and J. Rudd, J. Mag. Mag. Mat. **104-107**, 2089 (1992).
5.60. J. Kuneš and V. Kamberský, Phys. Rev. B **65**, 212411 (2002).

5.61. T. Silva, C. Lee, T. Crawford, and C. Rogers, J. Appl. Phys. **85**, 7849 (1999).
5.62. S. Russek, R. McMichael, M. Donahue, and S. Kaka, *Topics in Applied Physics:Spin Dynamics in Confined Magnetic Structures II, ed. B. Hillebrands and K. Ounadjela*, chapter High Speed Switching and Rotational Dynamics in Small Magnetic Thin Film Devices, pages 93–154, Springer Verlag, 2003.
5.63. W. Bailey, P. Kabos, and S. Russek, IEEE Transactions on Magnetics **37**, 1749 (2001).
5.64. S. Russek et al., J. Appl. Phys. **91**, 8659 (2002).
5.65. P. Hammel et al., *Frontiers in Magnetism of Reduced Dimension Systems, edited by P. Wigen and V. Baryachtiar and N. Lesnik*, chapter X, pages 441–462, NATO-ASI, 1996.
5.66. D. Rugar, C. Yannoni, and J. Sidles, Nature **360**, 563 (1992).
5.67. P. Hammel et al., Proceedings of the IEEE **91**, 789 (2003).
5.68. J. Moreland, M. Loehndorf, P. Kabos, and R. McMichael, Rev. Scien. Instr. **71**, 3099 (2000).
5.69. M. Loehndorf, J. Moreland, and P. Kabos, App. Phys. Lett. **76**, 1176 (2000).
5.70. M. Loehndorf, J. Moreland, P. Kabos, and N. Rizzo, J. Appl. Phys. **87**, 5995 (2000).
5.71. J. Nibarger, R. Lopusnik, and T. Silva, Appl. Phys. Lett. **82**, 2112 (2003).
5.72. www.picoprobe.com .
5.73. S. Russek, S. Kaka, and M. Donahue, J. Appl. Phys. **87**, 7070 (2000).
5.74. M. Bailleul, *Propagation et confinement d'ondes de spin dans les microstructures magnétiques*, Service de Physique L'État Condensé-CEA Seclay, 2002.
5.75. B.-C. Choi and M. Freeman, *Ultrathin Film Magnetic Structures III, ed. B. Heinrich and J.F.C. Bland*, Springer Verlag, 2003.
5.76. A. Arrott, *Ultrathin Film Magnetic Structures IV, ed. J.F.C. Bland and B. Heinrich*, Springer Verlag, 2003.
5.77. H. Quach, A. Friedman, C. Wu, and A. Yelon, Phys. Rev. B **17**, 312 (1976).
5.78. J. Anderson, S. Bhagat, and F. Cheng, Phys. Stat. Sol. **45**, 357 (1971).
5.79. B. Heinrich and V. Mescheryakov, Sov. Phys. JETP Letters **9,11**, 378 (1969).
5.80. B. Heinrich and V. Mescheryakov, Sov. Phys. JETP **32,2**, 232 (1971).
5.81. J. Cochran, B. Heinrich, and G. Dewar, Can. J. Phys. **55**, 787 (1977).
5.82. J. Cochran, K. Myrtle, and B. Heinrich, J. Appl. Phys. **53**, 2261 (1982).
5.83. B. Heinrich, J. M. Rudd, K. Urquhart, K. Myrtle, and J. F. Cochran, J. Appl. Phys. **55**, 1814 (1984).
5.84. K. Myrtle, B. Heinrich, and J. Cochran, J. Appl. Phys. **52**, 2250 (1981).
5.85. J. Rudd, J. Cochran, K. Urquhart, K. Myrtle, and B. Heinrich, J. Appl. Phys. **63**, 3811 (1988).
5.86. J. Cochran, J. Rudd, W. Muir, G. Trayling, and B. Heinrich, J. Appl. Phys. **70**, 6545 (1991).
5.87. D. K. Cheng, *Field and Wave Electromagnetics*, second edition Chapter 9, pages 433–434, Addison-Wesley Publishing Company, 1982.
5.88. M. Britel, D. Menard, L. Melo, P. Ciureanu, and A. Yelon, Appl. Phys. Lett. **77**, 2737 (2000).
5.89. D. Menard and A. Yelon, J. Appl. Phys. **88**, 379 (2000).
5.90. E. Frey and F. Schwabel, Adv. Phys. **43**, 577 (1994).
5.91. P. Boeni, R. Roessli, D. Goerlitz, and J. Koetzler, Phys. Rev. B, **65**, 144434–1 (2002)
5.92. F. Mezei, Phys. Rev. Lett. **49**, 1096 (1982).
5.93. J. Koetzler, Phys. Rev. Lett. **51**, 833 (1983).
5.94. J. Lynn, Phys. Rev. B **11**, 2624 (1975).
5.95. H. Mook and J. Lynn, Phys. Rev. Lett. **57**, 150 (1986).

5.96. J. Cook, J. Lynn, and H. Davis, Phys. Rev. B **21**, 4118 (1980).
5.97. R. Vollmer, M. Etzkorn, P. A. Kumar, H. Ibach, and J. Kirschner, Phys. Rev. Lett. **91**, 147201 (2003).
5.98. C. Patton and F. Humphrey, J. Appl. Phys. **37**, 4269 (1966).
5.99. A. Hubert and R. Schäfer, *Magnetic domains. The analysis of magnetic microstructures*, chapter Domain theory, pages 215–355, Spinger, 2000.
5.100. K. D. Leaver and S. Vojdani, J. Appl. Phys. **3**, 729 (1970).
5.101. E. Fuchs, Z. Angew. Phys. **14**, 203 (1962).
5.102. T. Suzuki, C. Wolts, and C. Patton, J. Appl. Phys. **39**, 1983 (1968).
5.103. D. Rodbell, Phys. Rev. Lett. **13**, 471 (1964).
5.104. A. LaBonte, J. Appl. Phys. **40**, 2450 (1969).
5.105. A. Hubert, Phys. Stat. Solidi **32**, 519 (1969).
5.106. W. Rave and A. Hubert, J. Mag. Mag. Mat. **184**, 179 (1998).
5.107. A. Y. Dobin and R. Victora, Phys. Rev. Lett. **90**, 167203 (2003).
5.108. M. Sparks, *Ferromagnetic Relaxation Theory*, Mc-Graw-Hill, New York, 1966, Chapters 8.2-8.3.
5.109. H. Suhl, Phys. Rev. Lett. **6**, 174 (1961).
5.110. J. A. Katine, F. J. Albert, R. A. Buhrman, E. B. Myers, and D. C. Ralph, Phys. Rev. Lett. **84**, 3149 (2000).
5.111. M. Tsoi et al., Phys. Rev. Lett. **80**, 4281 (1998).
5.112. J. C. Slonczewski, J. Magn. Magn. Mater. **159**, 1 (1996).
5.113. J. C. Slonczewski, J. Magn. Magn. Mater. **195**, 261 (1999).
5.114. A. Fert et al., J. Mag. Mag. Mat.,272–276, 1706 (2004).
5.115. L. Berger, Phys. Rev. B **54**, 9353 (1996).
5.116. L. Berger, J. Appl. Phys. **89**, 5521 (2001).
5.117. A.G.Aronov, JETP Lett. **24**, 32 (1976).
5.118. M. Stiles and A. Zangwill, Phys. Rev. B **66**, 014407 (2002).
5.119. E. Šimanek, Phys. Rev. B **63**, 224412 (2001).
5.120. M. Stiles and A. Zangwill, J. Appl. Phys. **91**, 6812 (2002).
5.121. A. Brataas, Y. Nazarov, and G. Bauer, Europ. Phys. J. **B22**, 99 (2001).
5.122. C. Heide, P. Zilberman, and R. Elliot, Phys. Rev. B **63**, 064424 (2001).
5.123. F. Albert, J. Katine, R. Buhrman, and D. Raph, Appl. Phys. Lett. **77**, 3809 (2000).
5.124. M. Tsoi et al., Nature **406**, 46 (2000).
5.125. M. Puffal, W. Rippard, and T. Silva, Appl. Phys. Lett. **83**, 323 (2003).
5.126. S. Kiselev et al., Nature **425**, 380 (2003).
5.127. W. Rippard, M. Puffal, S. Kaka, S. Russek, and T. Silva, Phys. Rev. Lett. **92**, 027201 (2004).
5.128. J. Shi, *Ultrathin Film Magnetic Structures IV, ed. J.F.C. Bland and B. Heinrich*, Springer Verlag.
5.129. M. Scheinfein, llgmicro.home.mindspring.com.
5.130. E. Myers et al., Phys. Rev. Lett. **89**, 196801 (2002).
5.131. F. Albert, N. Empley, E. Myers, D. Ralph, and R. Buhrman, Phys. Rev. Lett. **89**, 226802 (2002).
5.132. A. Enders et al., J. Appl. Phys. **89**, 7110 (2001).
5.133. L. Berger, J. Appl. Phys. **90**, 4632 (2001).
5.134. B. Heinrich, G. Woltersdorf, R. Urban, and E. Šimanek, J. Mag. Mag. Mat. **258**, 376 (2003).
5.135. Y. Tserkovnyak, A. Brataas, and G. Bauer, Phys. Rev. Lett. **88**, 117601 (2002).
5.136. P. Brouwer, Phys. Rev. **B58**, R10135 (1998).

5.137. A. Brataas, Y. Tserkovnyak, G. Bauer, and B. Halperin, Phys. Rev. B **66**, id060404 (2002).
5.138. K. Xia, P. Kelly, G. Bauer, A. Brataas, and I. Turek, Phys.Rev. B **65**, 220401 (R) (2002).
5.139. B. Heinrich et al., Phys. Rev. Lett. **90**, 187601 (2003).
5.140. B. Heinrich, G. Woltersdorf, R. Urban, and E. Šimanek, J. Appl. Phys. **93**, 7545 (2003).
5.141. Y. Yafet, Phys.Rev. **B 36**, 3948 (1987).
5.142. S. Doniach and E. Sondheimer, *Green's Functions for Solid State Physics*, MW.A. Benjamin, Inc., 1974.
5.143. E. Šimanek and B. Heinrich, Phys. Rev. B **67**, 144418 (2003).
5.144. S. Mizukami, Y. Ando, and T. Miyazaki, J. Mag. Mag. Mat. **226**, 1640 (2001).
5.145. R. Urban, B. Heinrich, and G. Woltersdorf, J. Appl. Phys. **93**, 8280 (2003).
5.146. E. Schlömann, Phys. Chem. Sol. **6**, 257 (1958).
5.147. R. LeCraw, E. G. Spencer, and C. S. Porter, Phys. Rev. **110**, 1311 (1958).
5.148. S. Geschwind and A. M. Clogston, Phys. Rev. **108**, 49 (1957).
5.149. M. J. Hurben, D. R. Franklin, and C. E. Patton, J. Appl. Phys. **81**, 7458 (1997).
5.150. E. Schloemann, Phys. Rev. **182**, 632 (1969).
5.151. C. E. Patton, C. H. Wilts, and F. B. Humphrey, J. Appl. Phys. **38**, 1358 (1967).
5.152. D. Mattis, *Theory of Magnetism I. Statics and dynamics*, W.A. Benjamin, Inc, 1988.
5.153. B. Heinrich, J. F. Cochran, and R. Hasegawa, J. Appl. Phys. **57**, 3690 (1985).
5.154. R. McMichael, D. Twisselmann, and A. Kunz, Phys. Rev. Lett. **90**, 227601 (2003).
5.155. R. D. McMichael, M. D. Stiles, P. J. Chen, and W. F. Egelhoff, Jr., J. Appl. Phys. **83**, 7037 (1998).
5.156. R. McMichael, private communication .
5.157. C. E. Patton and C. H. Wilts, J. Appl. Phys. **38**, 3537 (1967).
5.158. C. E. Patton, Z. Frait, and C. H. Wilts, J. Appl. Phys. **46**, 5002 (1975).
5.159. G. Counil et al. J. Appl. Phys. **95**, 05646 (2004) private communication .
5.160. G. Woltersdorf, B. Heinrich, J. Woltersdorf, and R. Scholtz, J. Appl. Phys. (in press).
5.161. R. Arias and D. L. Mills, J. Appl. Phys. **87**, 5455 (2000).
5.162. J. Lindner et al., Phys. Rev. B **68**, 060102(R) (2003).
5.163. M. J. Hurben and C. E. Patton, J. Appl. Phys. **83**, 4344 (1998).
5.164. G. Woltersdorf and B. Heinrich, Phys. Rev. B **69**, 1844xx (2004).
5.165. B. Heinrich et al., Phys. Rev. Lett. **59**, 1756 (1987).
5.166. F. Schreiber, J. Pflaum, Z. Frait, T. Mühge, and J. Pelzl, Solid State Comm. **93**, 965 (1995).
5.167. R. Urban et al., Phys. Rev. B **65**, 020402R (2002).
5.168. D. Twisselmann and R. McMichael, J. Appl. Phys. **93**, 6903 (2003).
5.169. B. Heinrich et al., Phys. Rev. **38**, 1287 (1988).
5.170. Z. Celinski and B. Heinrich, J. Appl. Phys. **70**, 5935 (1991).
5.171. W. Wulfhekel, F. Zavaliche, F. Porrati, H. Oepen, and J. Kirschner, Europhysics Letters **49**, 651 (2000).
5.172. B. Heinrich et al., Mat. Res. Soc. Symposium Proceedings **313**, 485 (1993).
5.173. B. Heinrich et al., J. Appl. Phys. (in press).
5.174. A. Koveshnikov et al., J. Appl. Phys. , submitted for publication (2003).
5.175. W. Baltensperger and J. Helman, J. Appl. Phys. **73**, 6516 (1993).
5.176. C. Kittel and J. Galt, *Solid State Physics*, chapter Vol. 3, page 437, Academic, New York edited by F. Seitz and D. Turnbull, 1956.

6

Nonequilibrium Spin Dynamics in Laterally Defined Magnetic Structures

B.C. Choi and M.R. Freeman

Understanding the magnetism in low-dimensional magnetic materials, in which a variety of novel phenomena involving both the static and the dynamic aspects occur, has become a major challenge in fundamental physics. From the technological point of view, material engineering, including artificially fabricating of magnetic materials on a nanometer scale, has made significant progress in recent years, and the continued development of high-performance magnetic information technologies requires detailed understanding of the magnetization dynamics in lithographically structured magnetic elements. The investigation of such small magnetic structures relies increasingly on magnetic imaging techniques, since the relevant properties can vary over length scales from micrometers to nanometers. This chapter is intended to provide a detailed description of an experimental method for imaging nonequilibrium magnetic phenomena in the picosecond temporal regime and with sub-micrometer spatial resolution. The method employs a stroboscopic scanning Kerr microscope, and is capable of measuring simultaneously all three components of the magnetization vector. Therefore, this experimental approach allows direct insight into the spatiotemporal evolution of magnetization dynamic processes. A few experimental examples are presented, illustrating dynamic micromagnetic processes during magnetization switching and spontaneous magnetic domain pattern formation in small magnetic elements.

6.1 Introduction

The study of confined magnetic structures, where both the thickness and lateral dimensions may be on the nanometer scale, has become fertile ground for novel magnetic phenomena in low dimensions [6.1–4]. The goal in this research field is to understand how magnetic properties change as the dimensions of magnetic systems approach the ultimate limits of miniaturization. Recently attention has focused on the dynamic behavior of such small magnetic elements [6.5–8]. This is because magnetization dynamics in thin continuous films on short time scales is different in many aspects from the static case [6.3, 9]. Moreover, magnetization dynamics in small patterned elements differs substantially from that in continuous films, as experienced in the magnetostatics of edges which modify the equilibrium states of the element in terms of the magnetic moment distribution [6.10, 11] (the coupling between the geometric shape of the element and its magnetic properties as represented by 'shape anisotropy' or 'configurational anisotropy' [6.12]). From a practical point of view, understanding magnetization dynamics on nano- and pico-second time scales in small elements with dimensions in the micrometer size regime and below has become crucial, owing to the increasing demands for higher speeds and densities from conventional data storage technologies, and for newer approaches such as magnetic random access memories (MRAM) [6.13, 14] or exploiting phenomena such as giant magnetoresistance [6.15]. The broader potential of novel magnetic phenomena for technological applications is apparent from considering that most of the information we deal with is stored magnetically at some stage.

Motivated by these accumulated interests, dynamic behavior in micro- and nano-sized magnets is being actively studied by a number of groups [6.16–20]. In order to elucidate magnetization dynamics in small elements, it has long been recognized that direct observations of magnetization processes with simultaneous spatial and temporal resolutions are most desirable. Imaging of micromagnetic configurations, however, has been carried out mostly by magnetic force microscopy (MFM) [6.21], Lorentz transmission electron microscopy [6.22] and ballistic electron magnetic microscopy [6.23, 24] in addition to magneto-optic microscopy [6.25]. These techniques provide good spatial resolution, but are generally focused on static magnetic imaging. For the study of dynamic phenomena, it has been demonstrated that very high spatial and temporal resolution can be achieved by employing stroboscopic scanning Kerr microscope with pulse excitation [6.26–28]. This technique has been proved to be a powerful tool for dynamic micromagnetic imaging in small structures, and will be described in this chapter.

The first high bandwidth spatially resolved optical experiments date back to the sixties, when J.F. Dillon and coworkers "saw" ferromagnetic resonance in a sample of 45 μm thick $CrBr_3$ using microwave optical technique [6.29]. In the late 1960s Kryder et al. presented first spatially and temporally resolved dynamic magnetization configuration in magnetic thin films with "A Nanosecond Kerr Magneto-Optic Camera" [6.30]. This Kerr photo-apparatus enabled to capture magnetization reversal process in $Ni_{83}Fe_{17}$ films with a 10 μm spatial resolution and a 10 ns temporal resolution. This spatiotemporal resolution was an enormous achievement even by today's standard. The most recent achievements along this specific avenue of attack were reviewed in [6.31].

The time-resolved imaging technique, however, has not been extensively used at higher speeds due to the complex instrumental requirements. Revival of interest in fast imaging techniques began to take place in the mid-1980s. Kasiraj et al. reported magnetic domain imaging with a scanning Kerr effect microscope [6.32], in which a digital imaging technique was used with a spatial resolution of less than 0.5 μm. This allowed the observation of the nucleation and growth of magnetic domains in the pole tip region of a recording head. In this implementation, a time resolution of 50 ns was achieved. Subsequently, an ultrafast magneto-optical technique was developed [6.33] which exploited advances that had been made in the field of electro-optic sampling. In this work ps-scale magnetic field pulses were launched by photoconductive switches, and applied to samples which were subsequently probed by time-delayed optical pulses, yielding temporal information in a conventional stroboscopic manner. This work extended the so-called "pump-probe" technique to a broad range of magnetic materials, providing an experimental tool for the study of ps time-resolved magnetization reversal dynamics. Since then, a number of spatiotemporally resolved experiments to directly measure the dynamical evolution of magnetization configuration have been reported, in particular in microstructured elements [6.5–8, 28, 34–36].

6.2 Experimental Methods

A simple way to probe the magnetization at magnetic surfaces or magnetic thin films is by means of interactions of light with magnetic medium. When a linearly polarized light is reflected from a magnetic surface, the incident light is transformed into elliptically polarized light. Thus, the final state of polarization can be characterized by both a rotation of the major axis θ and an ellipticity δ defined as the ratio between minor and major axis. Both θ and δ are proportional to the magnetization of the material. This effect is known as the magneto-optical effect [6.37], and has provided an important tool to visualize a detailed picture of the magnetization configuration of a broad range of magnetic materials. S.D. Bader and J.L. Erskine have covered details of the magneto-optical effect in Chap. 4 of this series [6.3].

Magneto-optical effects can be further used for exploring time-resolved studies of the ultrafast magnetic phenomena by employing short light pulses as a light source. This section outlines the experimental details of time-resolved scanning Kerr microscopy for the study of nonequilibrium dynamics in magnetic materials on picosecond time scales. Such experiments involve many techniques, such as generation of fast magnetic pulses, precise control of time intervals between the pump and probe beam, and modulation of magnetic signals by means of high frequency methods.

6.2.1 Pump-and-Probe Methods

Time-resolved experiments are to directly measure the dynamic evolution of systems, away from or towards equilibrium, in response to sudden perturbations. One of the ways to observe such a fast process, which is out of the limitations of our five senses, is to use a very short flashing light to freeze action and make photos. H. Edgerton perfected the stroboscopic photograph in the 1920s and 30s, allowing direct observation of quickly moving objects in a series of as-if frozen images. The related approach to time-resolved magnetic measurements was first seen in the inspired studies of dilute magnetic semiconductors by Awschalom et al. [6.38].

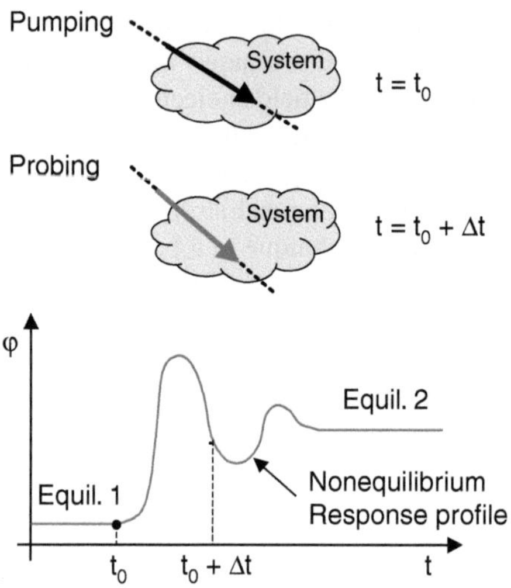

Fig. 6.1. Schematic illustration of the "pump-and-probe" technique. At time $t = t_0$, the system is excited out of its equilibrium with the ump pulse. After a short time interval Δt, the system is probed with the probe pulse. The time delay between the probe and pump beam is then changed to give a new Δt temporal position and the signal built-up again. This procedure is repeated until the entire nonequilibrium profile is measured

The method is schematically illustrated in Fig. 6.1. The experimental implementation of time-resolved methods relies, for example, on ultrashort laser pulses. One part of the pulses is used to excite a nonequilibrium state in the system, φ, at the time $t = t_0$ ("pumping"), while the delayed part of light pulses is used to detect the corresponding change in the system at $t = t_0 + \Delta t$ ("probing"). The time interval Δt can be created, for example, just by making the optical path of the probe beam longer than that of the pump beam. After setting a time point t, the perturbation of the system, $\varphi(t)$, is detected. The probe beam path is then changed to give a new Δt temporal position, and the probe beam detects the corresponding change in the system again. This procedure is repeated until the entire nonequilibrium response profile of the system is measured. The nonequilibrium state $\varphi(t)$, for example, corresponds to the magnetization switching process excited by magnetic pulses in the case of magnetization reversal dynamics. In such studies, a synchronously triggered transient magnetic pulse is propagated past the sample under study, perturbing the magnetization system, and the subsequent evolution of the magnetization configuration is monitored through its interaction with a time-delayed probe beam. The experimental details of the time-resolved magneto-optical Kerr effect technique are described in the next section.

6.2.2 Experimental Setup

Figure 6.2 illustrates the schematic diagram for the entire system, including the optical and electronic layouts. The experimental arrangements include an ultrafast solid-state laser plus associated optics, a piezo-driven flexure stage for scanning the sample, and electronics controlling the time-delay of probe beam and magnetic pulse generation.

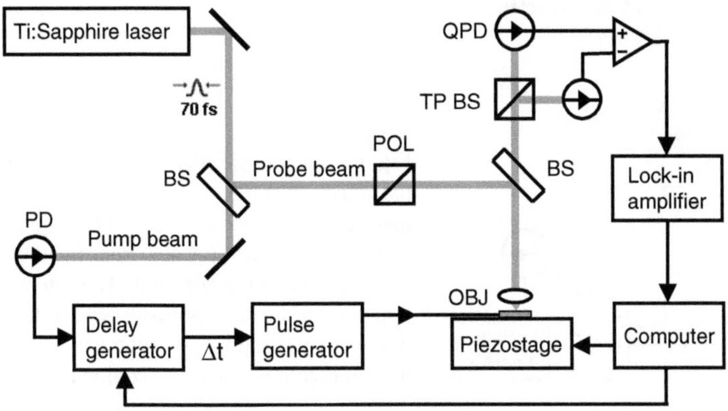

Fig. 6.2. A block diagram of the layout of the time-resolved scanning Kerr microscope (TR-SKM). A linear polarized laser beam is split into a pump and a probe beam with a 50/50 beam splitter (BS). The probe beam is focused on to the sample using a microscope objective (OBJ), and then reflected from the sample. The change of the polarization of the reflected light is analyzed using Thomson polarizing beam splitter (TP BS), combined with the signal detection using quadrant photodiodes (QPD). A small polarization rotation of the reflected light induced by the magnetization in the sample is turned into intensity shift by the Thomson polarizing beam splitter (TP BS), in which the intensity is measured. The pump beam is directed for triggering magnetic pulses. After electrical pulses are generated by a photodiode (PD), the output signals are synchronized by a computer controlled variable delay generator (VDG). Magnetic pulses are generated by an electronic pulser, which provides fast electrical pulses of 45–50 V, 0.25–0.5 ns rise times, and pulse widths of 10 ns. These pulses are delivered to micro stripe lines, which create magnetic field pulses. (cf. inset in Fig. 6.5 showing such stripe lines). Samples are mounted on a computer controlled piezo-driven flexure stage, which enables a raster scanning of the sample surface to build a spatial image

Optical Setup and Signal Detection

The pulsed light source usually used is a mode-locked Ti:Sapphire laser, which provides 70 fs long pulses of near-infrared light ($\lambda = 800$ nm) with a repetition rate of 82 MHz. During measurements, laser beam is split into two beams (i.e., a pump and a probe beam) with a 50/50 beam splitter (BS), which lets through equal amounts of *s*- and *p*-polarized light. The probe beam passes through a linear polarizer (POL) and is deflected toward the sample and is focused on to the sample using an infinity-corrected microscope objective (OBJ), while the pump beam is directed for triggering magnetic pulses (as will be described more in detail below). The optical power of the probe beam is reduced before being brought to a sharp focus on a sample in order to avoid permanent damage to the sample surface. Typically, an average power of 30 μW is focused onto the sample through a microscope objective. The spatial resolution (d) is determined by the numerical aperture (N.A.) of the objective lens and wavelength (λ) of the laser beam, given by the diffraction limited Rayleigh criterion, $d = 0.82\lambda/\text{N.A.}$ In our experimental setup, a spatial resolution down to 0.9 μm is yielded using the 0.75 N.A. microscope objective and near-infrared light

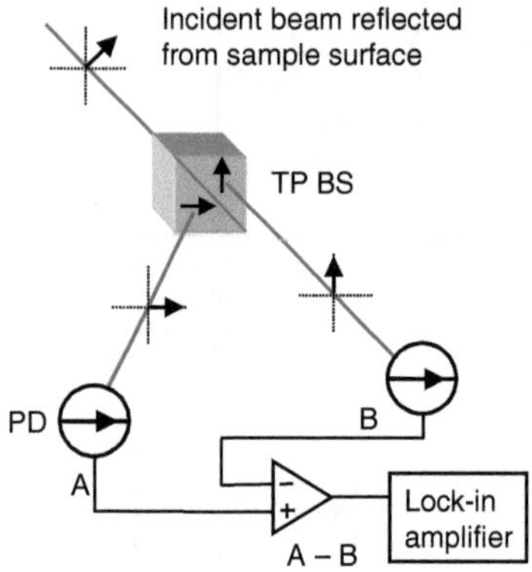

Fig. 6.3. Schematic illustration describing differential detection of polarization rotations induced by the magneto-optical Kerr effect. Both outputs of a polarizing beam splitter (at 45° from incident polarization) are used in subtraction. The intensity in each arm rises or falls, respectively with rotations of the plane of polarization. The subtracted signal (A-B) is only non-zero when a polarization rotation occurs

source. A detailed discussion of resolution in scanning optical microscopy has been presented by Mansuripur [6.39].

After the probe beam is reflected from the sample, magnetization measurements are accomplished through the polarization analysis of the reflected light in an optical bridge. A particular detection method using quadrant photodiodes (QPD) has been recently developed to allow for simultaneous detection of all three magnetization components (i.e., vector magnetometry) [6.28]. This approach is adopted from static Kerr imaging [6.40, 41], and works equally well in time-resolved measurements. The principle of the differential detection method is schematically depicted in Fig. 6.3. The probe beam reflected from the sample surface is split into two orthogonal polarization states by the Thomson polarizing beam splitter (TP BS), which is set at 45° to the incident polarization plane. Consequently, equal intensities are sent to the photodiodes. If there is no polarization rotation in the incident beam, each portion of the split beam will be of equal intensity and differential subtraction of the outputs coming from the two photodiodes will result in a zero signal. A small polarization rotation induced by the magnetization in the sample, however, will be turned into intensity shift by the Thomson analyzer, in which the intensity in one PD increases while decreasing in the other. This differential detection technique, where each PD signal is used as a reference to the other, has the advantage of common mode rejection of laser noise while doubling the signal [6.42].

An important issue in the signal detection is the signal-to-noise ratio, which has always been a problem in Kerr imaging due to the weak magneto-optic Kerr effect. In static Kerr measurements, wide-field imaging with digital enhancement is generally used, in which the image with magnetization pattern is digitally subtracted from an image of magnetization saturation [6.43]. This technique removes edge defect and optical system polarization artifacts. In our experimental setup, signal enhancement is achieved at each pixel of a raster scan using synchronous modulation of the magnetic excitation and lock-in signal detection technique. The modulated magnetic excitation is accomplished by modulating the trigger electronics, which generates magnetic pulses, at a low frequency (1∼4 kHz). The modulation on the magnetization in the sample itself isolates the signal from other artifacts, such as depolarizing effects, in the system.

Another important issue relating to the noise is the question whether the entire dynamics measured is perfectly repeatable, since the stroboscopic scanning technique by its nature captures only the repetitive part of the process being imaged. Nonrepetitive instabilities, such as thermal fluctuation of the individual spins, will lead to the averaging the temporal response. In addition to varying scan rate, number of averages, etc., the most sensitive test for underlying stochastic behavior so far comes from spectrum analysis of the noise on the magneto-optic signal. An example of noise imaging has been reported in which details of random magnetic switching are exposed in stroboscopically averaged time-resolved experiments [6.44].

Synchronization and Magnetic Field Pulse Generation

In stroboscopic method the temporal excitation ("pumping") of the system must be repetitive and triggered synchronously with the probe pulses. This is because interactions of many probe pulses and repetitive excitation events are averaged and represented as a single event. In our case, such synchronization is achieved as a portion of laser pulse itself triggers magnetic field pulses, which are used to excite the sample out of equilibrium state, while another portion of pulse probes (Fig. 6.2).

In the experimental setup, schematically described in Fig. 6.2, pump pulses are directed to a fast photodiode (e.g. ThorLabs DET210), which creates and sends clock signals to the variable electronic delay generator (Stanford Research Systems DG535). At this stage, the repetition rate of the pulsed beam is reduced from 82 MHz to 0.8 MHz via pulse picking of the mode-locked laser pulse train. This is required since the maximum trigger rate of the typical delay generator electronics is limited to 1 MHz. On the other hand, the delay electronics creates the propagation delay of the order of 100 ns. Therefore, an additional time delay between the pump and probe beam is required to achieve temporal synchronization. This can be achieved by delaying clock signals by an equivalent amount by propagation, for example, through a length of coaxial cable, until current pulse is actually synchronized with the laser pulse immediately following the one it was triggered by. After setting a time delay Δt, the delay generator sends the clock pulses on to the pulse generator.

The electronic delay method is very convenient, particularly when delay ranges of 10 ns or more are needed, but adds undesirable electronic jitter of about 50 ps rms. This trigger jitter is the main limiter of temporal resolution in this case. Alternatively, an optical delay line can be used for the synchronization of the probe beam and magnetic pulse, in which the travel path of probe beam with respect to pump beam is computer-controlled using mirrors mounted on a slider. Figure 6.4 shows the full experimental set-up, in which details of the alignment of optical components and optical delay line are illustrated by a photo realistic rendering. In this set-up, the optical configuration remains identical except for that electrical pulses generating magnetic field pulses are made by a fast photodiode (ThorLab Model # SV2-FC), which is launched by pump pulse itself. Details are provided in the captions of Fig. 6.2 and Fig. 6.4. This technique is inherently jitter-free and is generally beneficial to measurements for

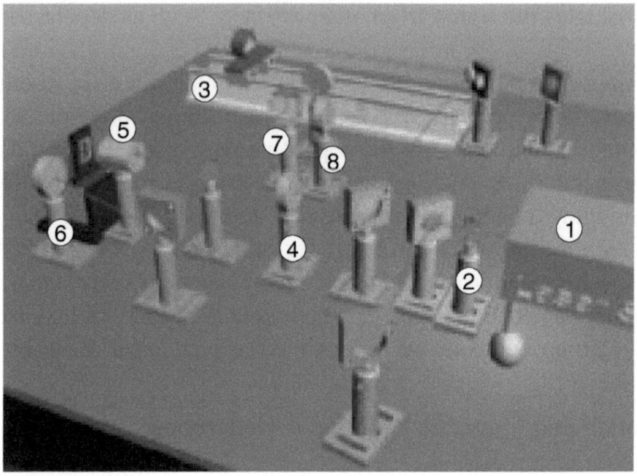

Fig. 6.4. Photo realistic rendering of experimental setup, in which an optical delay line is used for the synchronization of the probe beam and magnetic pulse. The travel path of probe beam with respect to pump beam is computer-controlled using mirrors mounted on a slider (3). Magnetic field pulses are generated by a fast photodiode (6), which is launched by pump pulse itself. Other components include pulsed laser (1), 50/50 beam splitter (2), linear polarizer (4), infinity-corrected microscope objective (5), Thomson polarizing beam splitter (7), and quadrant photodiodes (8)

faster (low ps regime) processes. However, in practice the time delay range usually spans only a few nanoseconds due to the limited length of delay line.

The magnetic pulse generation has been usually made using a photoconductive switch [6.45], in which the switch is closed by photons above the bandgap energy, freeing up carriers between two biased, metallized regions in a semiconductor substrate. This method generates fast and jitter free pulses, but it is generally known that the electric pulse shape is hard to control. In most cases in our experiments, the generation of magnetic pulses relies on the current driver, which is based on the avalanche transistor pulser (Model 2000D Pulse Generator, Picosecond Pulse Labs®) using the technique of discharging a transmission line. Pulses from this source have 0.25–0.5ns rise times, 1.0–1.5ns fall time, and pulse widths of 10 ns with the amplitude of 45–50V. The current pulses are synchronously triggered by Ti-sapphire fs laser pulses ($\lambda = 800$ nm, 0.8 MHz repetition rate) and are delivered to micro stripe lines, which create magnetic field pulses. The inset in Fig. 6.5 shows an image for such stripe lines, on or near which magnetic elements are placed. To excite the sample with an out-of-plane magnetic field pulse, samples are situated between lines and for an in-plane pulse, on top of a line. Stripe lines are fabricated using lithographic technique, and have the width of 20 μm and thickness of 300 nm in this case. The stripe lines create magnetic pulses as high as 24 kA/m. The temporal shape of field pulses can be measured by a commercially available 2 GHz inductive probe (Tektronix CT-6) or by measuring Faraday rotation in a garnet indicator film [6.20]. A garnet film allows optical measurement of the current waveforms in a very high

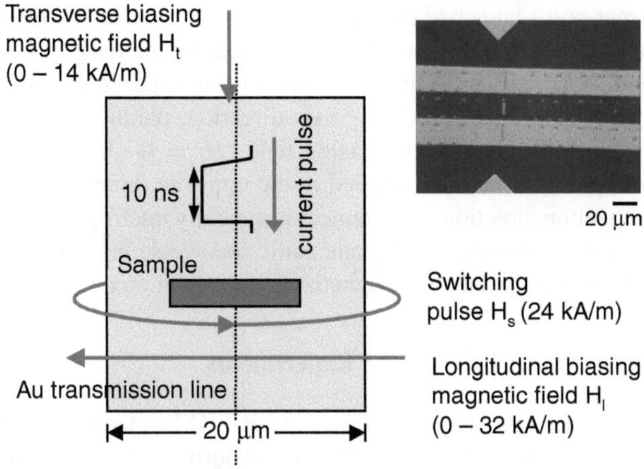

Fig. 6.5. Schematic measurement configuration of a 180° dynamic magnetization reversal experiment for microstructure excitation. H_s, H_l, and H_t indicate the switching field, longitudinal (easy-axis) biasing field, and transverse (hard-axis) biasing field, respectively. In the inset an image using an optical microscope is given, showing magnetic elements on top or near of the gold transmission lines

bandwidth (over 50 GHz), in addition to providing an absolute time reference for the time-resolved magnetic measurements [6.28].

Sample Preparation and Magnetic Field Configuration

The samples investigated in our experiments are polycrystalline permalloy ($Ni_{80}Fe_{20}$) thin film elements with various dimensions and shapes and typical thicknesses of 15 nm. The permalloy is a soft magnetic alloy used, for example, throughout information storage industry. The $Ni_{80}Fe_{20}$ film was deposited using DC magnetron sputtering onto 300 nm thick Au on sapphire substrate, at a growth rate of 0.1 μm/s in a high-vacuum system with a base pressure of 5×10^{-8} Torr. Magnetic microstructures were fabricated by electron beam lithography and lift-off techniques. The right hand side of Fig. 6.5 shows an optical micrograph of the magnetic structures and transmission lines. The patterned elements are made on (or close to) a transmission line that carries a fast current pulse. In most cases the element shapes are nicely defined, demonstrating an excellent lithographic resolution, which is currently about currently 50 nm. Atomic Force Microscope (AFM) and Scanning Electron Microscope (SEM) are also employed to inspect the completed structures. In most cases the lithographic structures reveal a very smooth surface with minor burrs at the sample edge [6.46]. During deposition an external magnetic field with the strength of 12 kA/m was applied in the plane of the substrate in order to induce the uniaxial magnetic anisotropy.

The geometric configuration of biasing and switching magnetic fields used is schematically illustrated on the left hand side of Fig. 6.5, where the stripe line carries a current pulse launched by a pulse generator in order to create an in-plane switching field pulse (H_s) of, typically, 24 kA/m along the long axis of the sample. In experiments, 180° magnetic field configuration is used, in which the sample is first magnetically saturated in the easy axis direction, parallel to the long sides of the elements, by an in-plane static biasing field ($H_l = 0$–32 kA/m). An in-plane switching field pulse, H_s, is then applied in the opposite direction to H_l in order to flip the magnetization direction. The element is optically interrogated while switching is taking place. Additionally, an in-plane static transverse biasing field, H_t, can be applied in order to manipulate the magnetization reversal process.

6.2.3 Operation Modes in TR-SKM Experiments

Two operation modes are usually employed in TR-SKM experiments. One of these is temporal-resolving mode, where one obtains a majority of the information very efficiently by measuring the dynamic response of the 'local' magnetization of a sample as a function of time. In this mode the probe beam is focused on a particular place (usually at the center) of the sample surface, and then the time delay Δt is changed. This mode is suitable for quick local characterization of the magnetic dynamics, but ignores spatial information. Temporal resolution is ultimately limited by the laser pulse width, but practically limited by the trigger jitter from the delay electronics, as described above. The Kerr signal is detected after each time step, building up the time-dependent profile for selected magnetization components. One example is given in Fig. 6.6, in which time traces for the magnetic responses, i.e., M_x, M_y, and M_z measured for three orthogonal in-plane magnetization components, are shown. The data have been measured at the center of a 15 nm thick square ($10 \times 10\,\mu m^2$) $Ni_{80}Fe_{20}$ element, and have been offset for clarity. This example clearly demonstrates that the magnetization is thrown out of its initial saturation state and into magnetization oscillation about the effective magnetic field upon application of the switching pulse. The magnetization oscillations result from magnetic precession about the effective switching field axis, which occurs generally when the magnetization vector experiences the torque due to the abruptly changed applied magnetic field [6.16]. A closer inspection of the dataset reveals that the oscillations are not in phase, supporting the precessional oscillation of the magnetization vector about a new equilibrium direction (cf. ferromagnetic resonance in Chap. 3, volume II in this series [6.3]). In the present case the magnetization vector undergoes a large angle motion, as schematically shown in the inset of Fig. 6.6. Experiments of this kind provide guidelines for understanding the local vectorial response from nonequilibrium magnetic systems.

More detailed information on the magnetization dynamics can be obtained from the spatiotemporal-resolving operation mode. After the time-dependent profile of the magnetization is measured, the sample surface can be scanned at a particular fixed time delay in order to obtain two-dimensional images mapping the magnetization configuration. This is required, since the dynamic response of the most magnetic systems studied can be spatially nonuniform [6.5, 6]. Figure 6.7 displays an example of such nonuniformity, in which spatial-time scanned domain images are captured during magnetization reversal in a thin rectangular ($10 \times 2\,\mu m^2$) permalloy film

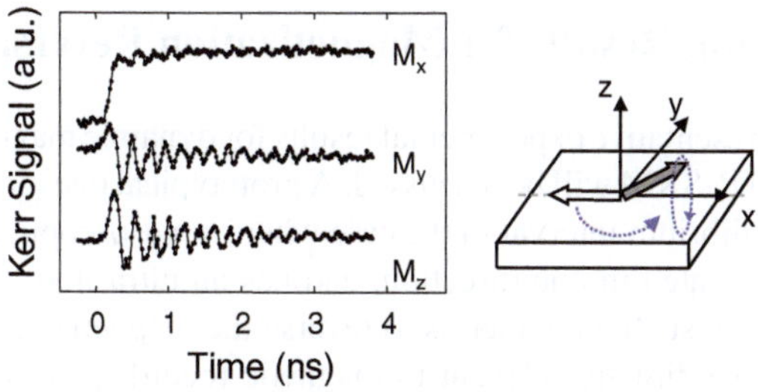

Fig. 6.6. Example of time-resolved magnetization components from a permalloy ($Ni_{80}Fe_{20}$) rectangular element. Data are from the magnetization response at the center of the sample versus time, and show phase shifted oscillations in x, y, and z magnetization components. These phase shifts indicate precessional motion. Inset shows a schematic cartoon of idealized precessional switching induced by the application of an abrupt magnetic pulse

element. The images are on a color scale to render the change in magnetization components with red corresponding to the maximum reversal. The evolution of spatial profiles with increasing time, associated with the magnetic domain nucleation, reveals that the magnetic response occurs highly local at different times spatially across the sample. (Details about this experiment will be discussed in Sect. 6.3.1.) During this measurement mode, the sample is placed on computer-controlled piezo-driven flexure stage providing scanning motion at a typical scan rate of 8 pixels/s (which typically corresponds to about 0.3 μm/s). A time record of few hundred separate sample measurements of the average magnetization was collected for each pixel in the image. Critical in this operation mode is the stability of the piezostage over long time intervals, since the quality of this stage also contributes to the effective spatial resolution of the system. The ThorLabs flexure stage used for the data-acquisition presented in Sect. 6.3 is relatively stable, but can drift at rates on the order of 0.1 μm/hour in response to the ambient temperature fluctuations in our laboratory. Auto-centering and auto-focusing have been implemented in the control software to compensate for this difficulty.

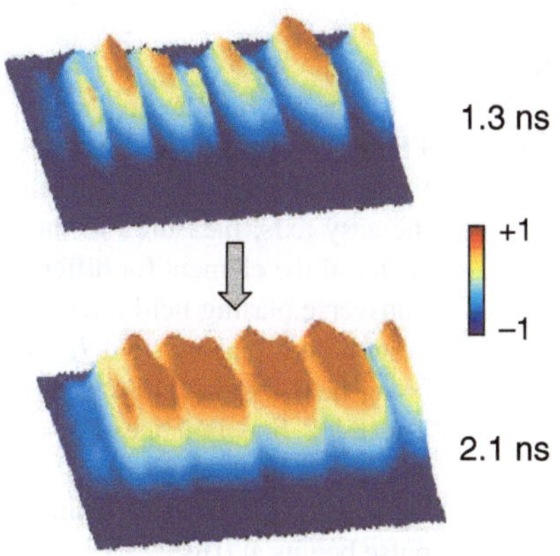

Fig. 6.7. Spatial-time scanned data captured during magnetization reversal. Spatial profiles demonstrate that the magnetic response occurs highly local at different times spatially across the sample. Movies showing complete sequences of reversal processes are found on http://nanoscale.phys.ualberta.ca

6.3 Experimental Results for Magnetization Reversal Dynamics

In this section, representative experimental results for dynamic magnetization reversal obtained through TR-SKM will be discussed. A prototypical question is what happens during the nonequilibrium interval in the case where a mesoscopic magnetic element is magnetically saturated in one direction, and has an ultrashort rise time magnetic field pulse applied in such a manner as to reverse the magnetization direction. Such experiments are also directly relevant to magnetic recording, where information is written on a magnetic medium by reversing the magnetization using a recording head. With bits now written and read in less than 1 ns, better understanding of the temporal dynamics in small magnetic elements becomes more crucial for the development of high-performance information storage systems.

6.3.1 Picosecond Time-Resolved Magnetization Reversal Dynamics

Figure 6.8 shows the local magnetic response of a 15 nm thick $Ni_{80}Fe_{20}$ rectangular element with a dimension of $10 \times 2\,\mu m^2$. Measurements were made with the 0.9 μm focus spot positioned at the center of the structure. The data represents the time traces of easy axis magnetization components M_x, compared for different transverse biasing fields H_t, while H_l is kept at 4.8 kA/m. Note the biasing and switching field configuration as depicted in Fig. 6.5. Applying no transverse field, i.e., $H_t = 0\,kA/m$ as indicated by the thick line, a definite delay in the magnetic response after the beginning of the pulse is observed, and the subsequent dynamics are relatively slow with the magnetization fully reversed after 3.5 ns. When a transverse biasing field H_t is applied perpendicular to the easy axis, a striking change in the magnetic response is observed. The first point to note is that the magnetic response becomes much faster with respect to the case without applying H_t. In such cases the magnetization switches within 1 ns after the pulse is given, and a transverse biasing field strength as low as 1.4 kA/m is found to be sufficient to cause such an abrupt switching. The qualitative explanation for this effect is that when a transverse field H_t is applied, the equilibrium position of the magnetization vector $\boldsymbol{M}$ is away from the initial easy axis, hence the

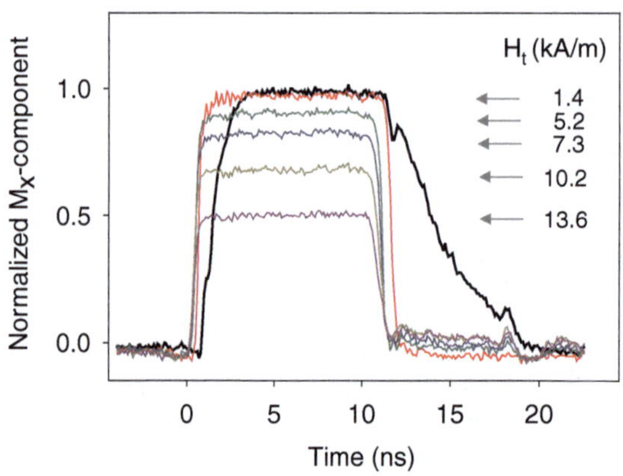

Fig. 6.8. Time-resolved magnetization components (M_x) along the easy axis, measured in the center of the element for different transverse biasing field (H_t). The longitudinal biasing field H_l is being held fixed at 4.8 kA/m. The thick line indicates the M_x component measured at $H_t = 0\,kA/m$. The switching pulse begins at 0 ns

reversing field exerts torque on the magnetization vector immediately. In addition, the magnetization is away from the minimum anisotropy energy state along the easy axis, so the effective coercive field is lower than for $H_t = 0\,\mathrm{kA/m}$. Consequently, lower longitudinal Zeeman energy or smaller switching field strength is required to overcome the energy barrier. We note that this is well known from quasistatic studies and modeling [6.25], and our study is extending to the fast dynamic regime.

Further insight into the temporal evolution of the magnetization reversal is obtained by direct time-domain imaging. Figure 6.9 shows a sequence of time-resolved images representing the easy axis magnetization components (M_x) for $H_l = 4.8\,\mathrm{kA/m}$ at selected time points, demarcated in nanoseconds relative to the initial application of the switching pulse. For $H_t = 0\,\mathrm{kA/m}$ (the left column), the reversal is mainly governed by a domain nucleation process. In the very beginning (0.5 ns) a stripelike instability is observed inside the sample, from nucleation occurring in the same regions. The main dynamical reversal, however, is first initiated from the demagnetized edges (0.9 ns), is followed by expansion of the nucleated

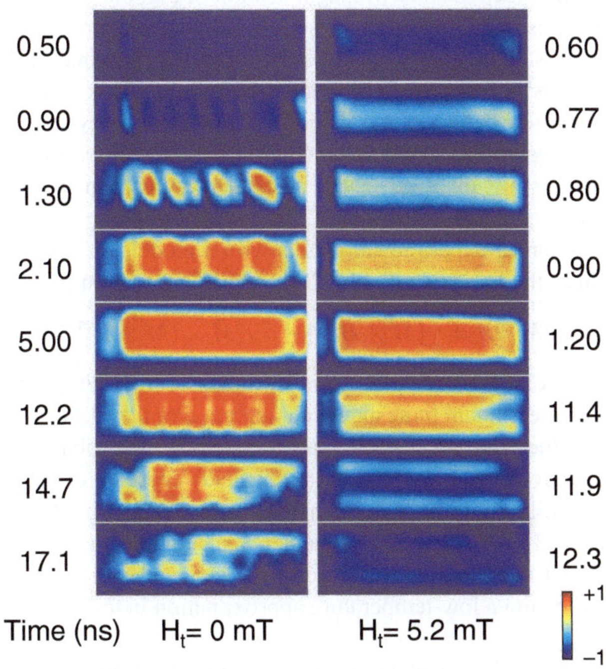

Fig. 6.9. Spatial magnetization profiles of the M_x component as a function of time after the magnetic pulse was applied. Each panel corresponds to 12×4 mm field of view, and contains the entire 10 × 2 μm sample. The numbers by the frames indicate the time in nanoseconds at which the measurement was made, relative to the initial application of the switching pulse. The comparison of time domain images clearly reveals a sensitive transition between two distinct magnetization reversal mechanisms, i.e., domain nucleation and domain wall motion, depending on the effective biasing field configuration

domains (1.3 and 2.1 ns), and finally leads to a uniform distribution of fully reversed magnetization, excluding the left and right edge regions (5 ns). These edge regions correspond to free magnetic poles related to the demagnetized areas in a ferromagnet of finite size. On the back reversal, the stripe instability is also pronounced (12.2 ns). From the spatiotemporally resolved magnetic images, it becomes clear how the magnetization switching in small elements occurs: The finite domain nucleation limits switching time to about 3.5 ns.

This nucleation-dominant reversal process can be drastically manipulated through an application of an additional transverse biasing field. This is demonstrated by time domain images, shown in the right column of Fig. 6.9. Applying a transverse biasing field H_t =5.2 kA/m, the 180° domains at the short edges are formed (0.6 ns), but there appears no stripelike distribution inside. The edge domains expand quickly in the easy axis direction to form a long, narrow domain parallel to the easy axis (0.77 ns). In the next stage, this elongated domain expands by parallel shifts in the hard direction towards the long edge (0.8 and 0.9 ns) until saturation is reached (1.20 ns). This type of reversal, which is characteristic of domain wall motion, is considerably faster as revealed in the time dependence of magnetization in Fig. 6.8.

The differences in the time domain sequences demonstrate that the formation of nuclei inside the sample can be easily avoided by the presence of H_t and that the nucleation process is replaced by domain wall motion. This result clearly reveals a sensitive transition between two distinct magnetization reversal mechanisms in a very "same" magnetic element, in which switching occurs over longer times when the stripe domains are involved in the reversal process than if pure domain wall motion occurs.

For comparison of experimental results with a model, we resort to a numerical approximation of the Landau-Liftshitz-Gilbert (LLG) equation [6.47]

$$\mathrm{d}\boldsymbol{M}/Dt = \gamma[\boldsymbol{M} \times \boldsymbol{H}^{\mathrm{eff}}] - \alpha\gamma[\boldsymbol{M} \times (\boldsymbol{M} \times \boldsymbol{H}^{\mathrm{eff}})] , \tag{6.1}$$

where γ is the gyromagnetic ratio and α is the dimensionless damping parameter. The simulation is a time domain finite element integration of the LLG equation, under the assumption that the effective field $\boldsymbol{H}^{\mathrm{eff}}$ is produced by the exchange, magnetostatic, crystalline anisotropy, and Zeeman (external field) energies of the magnetization. The sample is broken into a two dimensional array of 512 by 128 blocks making cells $19.5 \times 15.6 \times 15\,\mathrm{nm}^3$ each. Micromagnetic calculations have been carried out at 0 K. We note that the Curie temperature of the element is much higher than room temperature, allowing a low-temperature approximation in the statistical mechanical sense [6.29].

The temporal evolution of the easy axis magnetization component simulated for the biasing field conditions of $H_t = 5.2\,\mathrm{kA/m}$ and $H_l = 4.8\,\mathrm{kA/m}$ is compared to the experiment in Fig. 6.10. The simulated time trace is averaged over an area with the diameter of 0.9 μm at the center in the element, reflecting the experimental condition. The simulation agrees remarkably well with the experiment even in the absolute time scale, as represented by expanded views for the rising and falling ends of the switching shown in Fig. 6.10 (a) and (b). From the comparison, the experimental traces are found to be slightly lagging behind,

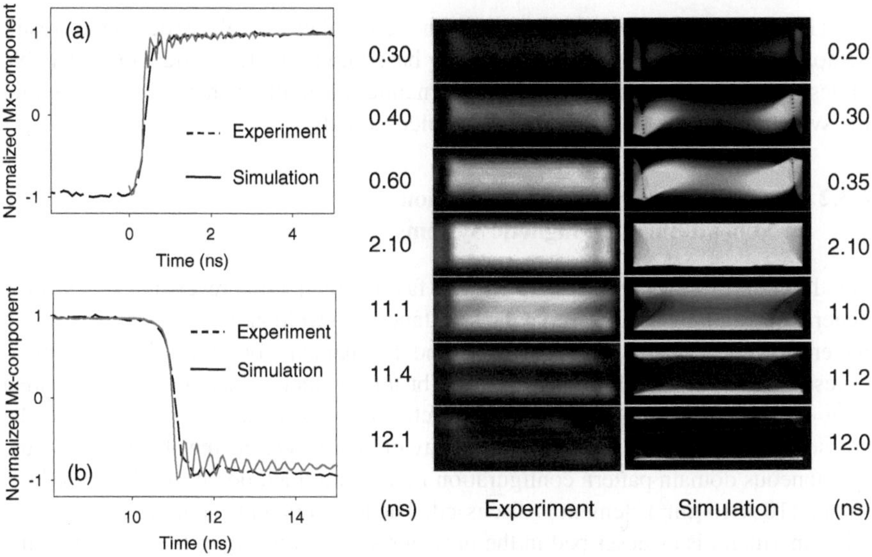

Fig. 6.10. Comparison of the temporal evolution ((**a**) and (**b**)) and spatial magnetization profiles of the easy axis magnetization component simulated for biasing field conditions of $H_t = 5.2\,\mathrm{kA/m}$ and $H_l = 4.8\,\mathrm{kA/m}$ to the experimental results. The good agreement between simulation and experiment demonstrates that the micromagnetic modeling based upon the LLG equation is capable of describing the magnetization reversal dynamics in small magnetic elements. The small oscillations seen in the rising (**a**) and falling (**b**) ends are due to magnetization precession [6.5]

with the initial and final slopes steeper in the simulation. Noticeable in the simulation are small oscillations found both in the rising and falling ends. These are on account of the precessional motion of the magnetization vector as discussed previously, and are less pronounced in the experiment due to the experimental temporal resolution limited by the ~50 ps RMS jitter of the electronic delay generator.

The right column of Fig. 6.10 compares the experimentally obtained time domain images to those obtained by the simulation. Frames show a sequence of time-resolved images representing the easy axis magnetization components measured and simulated for $H_t = 5.2\,\mathrm{kA/m}$ and $H_l = 4.8\,\mathrm{kA/m}$ at selected time points.

The images are on a linear gray scale to render the change in magnetization components with black corresponding to no change and white corresponding to the saturation level or maximum possible reversal. In remarkable agreement with the experiment, simulation reveals that short edges of the sample initiate reversal ($t = 0.2\,\mathrm{ns}$) and then propagate toward the center until the saturation reaches ($t = 0.3-2.1\,\mathrm{ns}$). Quite similar to the experimental case, the long edges are found to remain pinned along the initial magnetization direction ($t = 11.1-11.2\,\mathrm{ns}$) and are reversed last ($t = 12.0\,\mathrm{ns}$).

The overall agreement of the micromagnetic simulations to the experimental result demonstrates the micromagnetic modeling based upon the LLG equation is capable of describing the magnetization reversal dynamics in small magnetic elements within the switching time scale of few hundreds picoseconds.

6.3.2 Dynamic Domain Pattern Formation in Nonequilibrium Magnetic Systems

Another interesting aspect associated with fast magnetization reversal process is the general question of what governs the spontaneous development of dynamic domain patterns. This concerns the time scale and mechanism for removal of the initial excess Zeeman energy from the nonequilibrium magnetic system. In order to shed light on the complex spatiotemporal structures which arise, we can map out the crossover from quasi-static to dynamic behavior. In this section, the observation of the spontaneous domain pattern configuration in a 15 nm thick polycrystalline $Ni_{80}Fe_{20}$ square ($10 \times 10\,\mu m^2$) element is discussed. The magnetic field configuration used in this experiment is as described in the previous section (cf. Fig. 6.5), except that the rise time of the switching pulse is varied between 0.24 and 8.6 ns.

Figure 6.11 shows the spatio-temporal evolution of the magnetization component in response to short magnetic pulses with the rise time of 240 ps. The magnetic domain images are measured along the magnetic easy axis at selected time points after applying a switching pulse. The contrast in the images reflects the local degree of magnetization reversal, with dark red areas corresponding to fully reversed regions. The domain configurations show a complex spatial appearance, which for purposes of compact description we characterize as a labyrinth pattern. A particularly well-developed labyrinth pattern appears for the frame measured at 400 ps. We note that the qualitatively same behavior is observed in other samples. Emergence of the labyrinth formation is seen already at the very beginning of the reversal process. Near the initial state (100 ps), the magnetization switch has started at the element ends and is accompanied by the nucleation of branch-like fine structures visible in the interior regions. Labyrinth domain patterns evolve out of these fine nucleation sites, and once a pattern forms, it is quasi-stable and the completion of reversal is mainly governed by a gradual expansion of the reversed domains.

It is remarkable that this level of detail is observed in a stroboscopic (i.e., repetitively averaged) measurement. It remains to be understood how the nonequilibrium dynamics, presumably acting in concert with sample imperfections or pinning sites, select out a pattern that largely recurs from reversal to reversal. Note that the spatial contrast one observes in this experiment represents only the lower limit of what one might find in a "single shot" temporal observation, or with finer spatial resolution in the stroboscopically-averaged case.

The evolution of the domain configuration has been measured as a function of the rise time of the magnetic switching pulse, as shown in Fig. 6.12. The upper panel shows the temporal evolution of the magnetization at the center of the same $10 \times 10\,\mu m^2$ element, excited by magnetic pulses with different rise time (i.e. t_{rise} = 8.6 ns, 5.1 ns, 2.8 ns, and 0.24 ns). All traces show a change of magnetization from

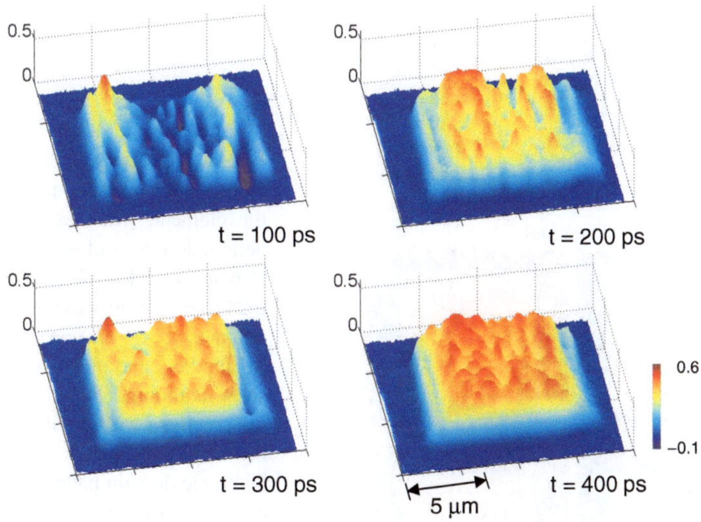

Fig. 6.11. Spatio-temporally resolved domain images representing the easy axis magnetization component at selected time points after the onset of a magnetic switching field of 0.24 ns rise time (0–90%). The 3D topography and color map both render the magnitude of the magnetic signal from unchanged (*blue*) to fully reversed (*red*). A complex labyrinth-like domain pattern evolves during the reversal, unlike anything observed at slower switching speeds

one saturated state to another, and magnetic responses show additional delay with increasing rise time, with oscillations following the primary switch.

The bottom panel of Fig. 6.12 shows characteristic domain patterns imaged at time points at which the reversal reaches about 50 % of the full M_x-intensity change. The situations for slower switching are very different from the fast case ($t_{\mathrm{rise}} = 0.24$ ns), for example, for $t_{\mathrm{rise}} = 5.1$ ns the labyrinth domain pattern is no longer observed and instead the domain configuration has a more regular character in which stripe domains predominate. Thus, accelerating the switching process is required to cause the more intricate structure to form, as seen in the domain pattern evolution from $t_{\mathrm{rise}} = 8.6$ to 0.24 ns. This observation clearly illustrates that the speed at which the magnetization reversal is driven plays an essential role in determining the domain patterns.

The appearance of labyrinth-like patterns in these in-plane $Ni_{80}Fe_{20}$ thin film elements is also notable because the occurrence of this kind of domain structure is a characteristic feature of the domain configurations of ferromagnetic films for which the magnetization is normal to the plane of the surface [6.48, 49]. Labyrinth pattern formation contrasts greatly to the quasi-static cases for in-plane magnetized thin film elements [6.8]. For example, Gomez et al. have shown that the remanent and switching domain configurations in $Ni_{80}Fe_{20}$ thin film elements display a variety of domain patterns [6.21], none of which resemble the dynamic patterns we observe.

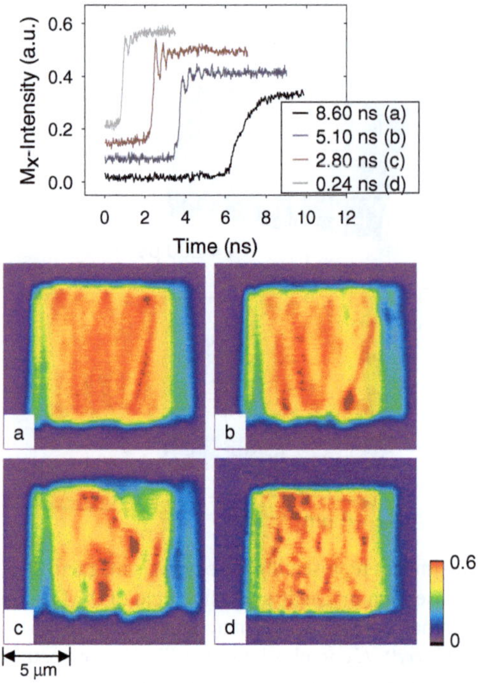

Fig. 6.12. (*Top panel*) Time traces of the magnetization at the center of the element, when excited by magnetic field pulses of different rise time (8.6 ns (**a**), 5.1 ns (**b**), 2.8 ns (**c**), and 0.24 ns (**d**)). Traces show the change of magnetization from one saturated state to another, and magnetic responses show additional delay with increasing rise time, with oscillations following the primary switch. (*Bottom panel*) Dynamic domain patterns revealed as a function of the magnetic pulse rise time. The spontaneous domain configuration is sensitively dependent on the switching speed. A transition from stripe- to labyrinth-like domains occurs with increasing speed

The sensitive dependence of the spatial pattern on switching speed can be discussed in analogy with other pattern forming systems. The Landau-Lifshitz-Gilbert phenomenology [6.50], which has been very successful in describing essentially all micromagnetic cases to which it has been applied in detail, is closely related to the complex Landau-Ginzburg equation, a known generator of spatiotemporal complexity [6.51]. Finite-element simulations to model the present experiment also show dramatic changes in the domain patterns as the switching speed of the field increases towards $(0.24\,\text{ns})^{-1}$ [6.52]. The spatial structures in the model for the fast rise time regime do not closely resemble those of the experiment, however. The characteristic pattern length scale changes as the system is driven further from equilibrium, reminiscent of other nonequilibrium pattern forming systems [6.53]. Ultimately the emergence of patterns in the magnetic system derives from the competition between the strong, short-range exchange interaction favoring parallel spin alignment and the weak, long-range dipolar interaction preferring antiparallel alignment. The additional energy gives rise to additional line length of domain wall within the sample in the nonequilibrium state.

Another remarkable phenomenon arising in the stroboscopic observation of these patterns is demonstrated in Fig. 6.13, where time domain images captured during magnetization oscillation are presented. Some of the fine spatial structure has vanished before even the first cycle of precession is complete, but an alternating change of domain pattern contrast is observed clearly during the magnetization precession, and without significant changes of the pattern shape. This becomes immediately apparent

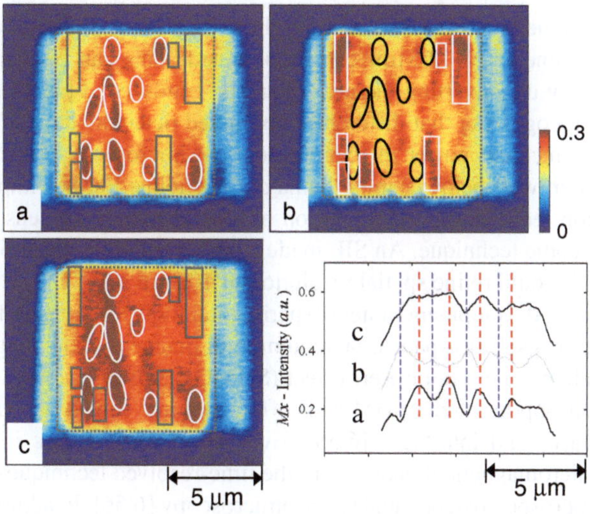

Fig. 6.13. Comparing spatio-temporally resolved domain images measured at the peaks (**a** and **c**) and dip (**b**) during magnetization oscillation. A striking feature is the alternation of contrast during the magnetization precession without the spatial pattern changing significantly. Through a cycle of oscillations, the already fully reversed regions at the oscillation peak (**a**) simply turn into not-fully-reversed regions at the dip (**b**). In the next oscillation peak (**c**), one observes the same pattern with progressing reversal. (*inset*) Horizontal cross-sectional views of the domain images

by comparing domain images measured at the peak and dip of the magnetization oscillations (compare to the time traces in the inset). During oscillations, the already fully reversed regions, measured at the oscillation peak (a), simply turn into not-fully-reversed regions at the dip (b). In the next oscillation peak (c), one observes a same domain pattern with progressing reversal. The alternating magnetic contrast during precession implies that the selected domain pattern is maintained while the remaining excess Zeeman energy introduced by the applied switching field is damped from the precessional magnetization motion. The micromagnetic simulations share this qualitative characteristic of strong temporal oscillations at the highest switching speeds.

6.4 Conclusion and Outlook

In summary, the current state of TR-SKM technique is reviewed, showing its significance for metrology in fundamental studies and industrial applications today. The unique combination of temporal and spatial resolution puts it foremost in elucidating the ultrafast magnetization reversal mechanism.

Future prospects include the improvement of the spatial resolution. In order to obtain a higher spatial resolution, an oil immersion objective can be employed. With a 1.3 N.A. oil immersion objective and a light source ($\lambda = 800$ nm), for example, the Airy disk diameter of 0.65 µm can be obtained. This can be again improved by a factor two by upconverting the near-infrared femtosecond pulses into the blue ($\lambda = 400$ nm), recognizing that differently optimized detectors have to be used and that the average incident power in scanning optical microscopy has to be reduced in order to remain well below sample damage thresholds. Further improvement of spatial resolution requires the incorporation of a solid immersion lens (SIL) [6.54] into the stroboscopic technique. An SIL made from a very high index materials such as GaP ($n = 3.4$) could refine spatial resolution to 100 nm.

The continuing development of new experimental tools for imaging fast magnetization dynamics is itself fast-paced. Spatiotemporal magneto-optic imaging with the second-harmonic magneto-optic Kerr effect (SHMOKE) is an exciting complement to linear magneto-optics, since SHMOKE offers extreme sensitivity to the magnetization at surfaces and interfaces [6.55]. Even stronger prospects in the field of experimental micromagnetic dynamics are the time-resolved techniques based on x-ray magnetic microspectroscopy and spectromicroscopy [6.56]. In addition to spatial resolution already in the low-10 s of nm range, these techniques provide elemental selectivity obtained as the x-ray energy is tuned to the adsorption edge of the desired element, very useful for resolving individual layers within multilayer structures composed of different materials [6.57, 58].

Acknowledgement. We would like to thank M. Belov and G. Ballentine for sample preparation and micromagnetic modeling, and D. Fortin for technical help.

References

6.1. B. Heinrich. *Canadian Journal of Physics*, 78:161, 2000.
6.2. P. Grünberg. *Physics Today*, May:31, 2001.
6.3. B. Heinrich and J.A.C. Bland, editors. volume II. Springer-Verlag, Berlin Heidelberg, 1994.
6.4. J.A.C. Bland and B. Heinrich, editors. volume I. Springer-Verlag, Berlin Heidelberg, 1994.
6.5. B.C. Choi, M. Belov, W.K. Hiebert, G.E. Ballentine, and M. R. Freeman. *Phys. Rev. Lett.*, 86:728, 2001.
6.6. Y. Acremann, C.H. Back, M. Buess, O. Portmann, A. Vaterlaus, D. Pescia, and H. Melchior. *Science*, 290:492, 2000.
6.7. R.H. Koch, J.G. Deak, D.W. Abraham, P.L. Trouilloud, R.A. Altman, Yu Lu, W.J. Gallagher, R.E. Scheuerlein, K.P. Roche, and S. S. P. Parkin. *Phys. Rev. Lett.*, 81:4512, 1998.
6.8. W.K. Hiebert, A. Stankiewicz, and M. R. Freeman. *Phys. Rev. Lett.*, 79:1134, 1997.
6.9. S.W. Yuan and H. N. Bertram. *J. Appl. Phys.*, 73:5992, 1993.
6.10. A.F. Popkov, L.L. Savchenko, N.V. Vorotnikova, S. Tehrania, and J. Shi. *Appl. Phys. Lett.*, 77:277, 2000.
6.11. K.J. Kirk, J.N. Chapman, and C. D. W. Wilkinson. *Appl. Phys. Lett.*, 71:539, 1997.
6.12. R.P. Cowburn, A.O. Adeyeye, and M. E. Welland. *Phys. Rev. Lett.*, 81:5414, 1998.

6.13. J. M. Daughton. *Thin Solid Films*, 216:162, 1992.
6.14. W.J. Gallagher, S.S.P. Parkin, Yu Lu, X.P. Bian, A. Marley, K.P. Roche, R.A. Altman, S.A. Rishton, C. Jahnes, T.M. Shaw, and G. Xiao. *J. Appl. Phys.*, 81:3741, 1997.
6.15. M.N. Baibich, J.M. Broto, A. Fert, F. Nguyen Van Dau, F. Petroff, P. Etienne, G. Creuzet, A. Friederich, and J. Chazelas. *Phys. Rev. Lett.*, 61:2472, 1988.
6.16. C.H. Back, D. Weller, J. Heidmann, D. Mauri, D. Guarisco, E.L. Garwin, , and H. C. Siegmann. *Phys. Rev. Lett.*, 81:3251, 1998.
6.17. C. Stamm, F. Marty, A. Vaterlaus, V. Weich, S. Egger, U. Maier, U. Ramsperger, H. Fuhrmann, and D. Pescia. *Science*, 282:449, 1998.
6.18. M. Hehn, K. Ounadjela, J.-P. Bucher, F. Rousseaux, D. Decanini, B. Bartenlian, and C. Chappert. *Science*, 272:1782, 1996.
6.19. R.P. Cowburn, D.K. Koltsov, A.O. Adeyeye, M.E. Welland, and D. M. Tricker. *Phys. Rev. Lett.*, 83:1042, 1999.
6.20. A.Y. Elezzabi, M.R. Freeman, and M. Johnson. *Phys. Rev. Lett.*, 77:3220, 1996.
6.21. R.D. Gomez, T.V. Luu amd A.O. Pak, K.J. Kirk, and J. N. Chapman. *J. Appl. Phys.*, 85:6163, 1999.
6.22. J.N. Chapman, P.R. Aitchison, K.J. Kirk, S. McVitie, J.C.S. Kools, and M. F. Gillies. *J. Appl. Phys.*, 83:5321, 1998.
6.23. W.H. Rippard and R. A. Burhman. *Appl. Phys. Lett.*, 75:1001, 1999.
6.24. W.H. Rippard, A.C. Perrella, P. Chalsani, F.J. Albert, J.A. Katine, and R. A. Buhrman. *Appl. Phys. Lett.*, 77:1357, 2000.
6.25. A. Hubert and R. Schäfer. *Magnetic Domains*. Springer-Verlag, Berlin, 1998.
6.26. T.M. Crawford, T.J. Silva, C.W. Teplin, and C. T. Rogers. *Appl. Phys. Lett.*, 74:3386, 1999.
6.27. G. Ju, A.V. Nurmikko, R.F.C. Farrow, R.F. Marks, M.J. Carey, and B. A. Gurney. *Phys. Rev. Lett.*, 82:3705, 1999.
6.28. G.E. Ballentine, W.K. Hiebert, A. Stankiewicz, and M. R. Freeman. *J. Appl. Phys.*, 87:6830, 2000.
6.29. Jr.J.F. Dillon, H. Kamimura, and J. P. Remeika. *J. Appl. Phys.*, 34:1240, 1963.
6.30. M.H. Kryder and F. B. Humphrey. *Rev. Sci. Instrum.*, 40:829, 1969.
6.31. M.H. Kryder, P.V. Koeppe, and F.H. Liu. *IEEE Trans. Magn.*, 26:2995, 1990.
6.32. P. Kasiraj, D. Horne, and J. Best. *IEEE Trans. Magn.*, MAG-23:2161, 1987.
6.33. M.R. Freeman, R.R. Ruf, and R.J. Gambino. *IEEE Trans. Magn.*, 27:4840, 1991.
6.34. A. Stankiewicz, W.K. Hiebert, G.E. Ballentine, K.W. Marsh, and M. R. Freeman. *IEEE Trans. Magn.*, 34:1003, 1998.
6.35. M.R. Freeman, W.K. Hiebert, and A. Stankiewicz. *J. Appl. Phys.*, 83:6217, 1998.
6.36. R.J. Hicken and J. Wu. *J. Appl. Phys.*, 85:4580, 1999.
6.37. S. D. Bader. *J. Magn. Magn. Mater.*, 100:440, 1991.
6.38. D.D. Awschalom, J.-M. Halbout, S. von Molnar, T. Siegrist, and F. Holtzberg. *Phys. Rev. Lett.*, 55:1128, 1985.
6.39. M. Mansuripur. *Classical Optics and its Applications*. Cambridge University Press, Cambridge, 2002.
6.40. W.W. Clegg, N.A.E. Heyes, E.W. Hill, and C. D. Wright. *J. Magn. Magn. Mat.*, 95:49, 1991.
6.41. T.J. Silva and A. B. Kos. *J. Appl. Phys.*, 81:5015, 1997.
6.42. P. Kasiraj, R. Shelby, J. Best, and D. Horne. *IEEE Trans. Magn.*, 22:837, 1986.
6.43. P.L. Trouilloud, B. Petek, and B. E. Argyle. *IEEE Trans. Magn.*, 30:4494, 1994.
6.44. M.R. Freeman, R.H. Hunt, and G. M. Steeves. *Appl. Phys. Lett.*, 77:717, 2000.
6.45. D. H. Auston. *Appl. Phys. Lett.*, 26:101, 1975.

6.46. J. Lohau, S. Friedrichowski, and G. Dumpich. *J. Vac. Sci. Technol. B*, 16:1150, 1998.
6.47. M. Mansuripur. *J. Appl. Phys.*, 63:5809, 1988.
6.48. P. Fischer, T. Eimüller, G. Schütz, G. Denbeaux, A. Pearson, L. Johnson, D. Attwood, S. Tsunashima, M. Kumazawa, N. Takagi, M. Köhler, and G. Bayreuther. *Rev. Sci. Instrum.*, 72:2322, 2001.
6.49. R. Allenspach, M. Stampanoni, and A. Bischof. *Phys. Rev. Lett.*, 65:3344, 1990.
6.50. J.G. Zhu and H. N. Bertram. *J. Appl. Phys.*, 63:3248, 1988.
6.51. G. Nicolis. *Introduction to Nonlinear Science*. Cambridge University Press, Cambridge, 1995.
6.52. B.C. Choi, G.E. Ballentine, M. Belov, and M. R. Freeman. in preparation.
6.53. P. Ball. *The self-made tapestry: Pattern formation in nature*. Oxford University Press, Oxford, 1999.
6.54. J.A.H. Stotz and M. R. Freeman. *Rev. Sci. Instrum.*, 68:4468, 1997.
6.55. J. Reif, J.C. Zink, C.M. Schneider, and J. Kirschner. *Phys. Rev. Lett.*, 67:2878, 1991.
6.56. G. Schütz, W. Wagner, W. Wilhelm, P. Kienle, R. Zeller, R. Frahm, and G. Materlik. *Phys. Rev. Lett.*, 58:737, 1987.
6.57. C.T. Chen, Y.U. Idzerda, H.-J. Lin, G. Meigs, A. Chaiken, G.A. Prinz, and G. H. Ho. *Phys. Rev. B*, 48:642, 1993.
6.58. F. Nolting, A. Scholl, J. Stöhr, J.W. Seo, J. Fompeyrine, H. Siegwart, J.-P. Locquet, S. Anders, J. Lüning, E.E. Fullerton, M.F. Toney, M.R. Scheinfein, and H. A. Padmore. *Nature*, 405:767, 2000.

7

Polarised Neutron Reflection Studies of Thin Magnetic Films

J. A. C. Bland and C. A. F. Vaz

7.1 Introduction

The magnetic moment is the most fundamental quantity in magnetism and yet the most experimentally challenging quantity to measure in the case of thin films. Magnetic order corresponds to the microscopic ordering of the electron angular momentum, be it the localised atomic orbital and spin moment and/or the spin moment of itinerant band electrons. Such ordered states correspond to states of lower energy compared with higher energy thermally excited states. The magnitude of the magnetic moment depends, on the one hand, on the magnitude of the angular moment (in particular for localised systems) and on the other on the spin imbalance between majority and minority electron bands. For the $3d$ transition metals the orbital moment is quenched by the strong crystal field and the magnetic moment is mostly due to the electron spin. For the $4f$ series, the orbital moment is not quenched, and the main contribution to the magnetic moment comes from the atomic angular moment. Additionally, the mechanisms responsible for the magnetic order are different in these two situations.

Due to changes in electronic structure with reduced dimensions, the magnetic moment ariadoes not necessarily scale with the volume of the system. In fact, it is expected that as the physical dimensions of the system are reduced, the break in symmetry induces a localisation of the wavefunction of those atoms close to the interface. While the effect of this symmetry break on the magnetic anisotropy is very pronounced, the effect on the magnetic moment is in general smaller and more difficult to observe. Also, metastable crystal structures can be stabilised in thin epitaxial films or small particles, which often exhibit magnetic moments which are different from those corresponding to the bulk equilibrium phase. Furthermore, it is now increasingly recognized that the interface moment is crucial in determining the behaviour of magnetic devices, e.g., magnetic tunnel junctions and in particular 'buried interfaces' are critical to the successful performance of such devices. While the magnetic moment of materials have been measured routinely in the bulk, the equivalent studies for systems of reduced dimensions are more challenging and require the use of more sensitive techniques.

Among the magnetometry techniques employed for the measurement of the magnetic moment in thin films, polarised neutron reflection (PNR), x-ray magnetic circular dichroism (XMCD) and SQUID (superconducting quantum interference device) magnetometry are often employed due to their sensitivity and their specific capabilities. XMCD is an element specific technique (it measures the magnetic-dependence of light absorption at the absorption edges) and allows the separate determination of both the spin and orbital components of the magnetic moment. However, XMCD does not independently yield an absolute determination of the magnetic moment as it requires knowledge of the exact number of holes, n_h, which may vary with film thickness [7.1], and also relies on an atomic model for its selection rules. The power of XMCD lies in its chemical selectivity and high sensitivity and its relative insensitivity to structural quality [7.2, 3]. PNR on the other hand is a layer selective, self-calibrating technique that relies on the reflectivity response of a magnetic step-like potential (a consequence of the magnetic dipole interaction between the neutron spin and the magnetisation of the medium), giving the magnetisation profile of the film and also the structural parameters of the film (layer thickness and interface roughness). The surface sensitivity of this method is a consequence of the phenomenon of total reflection; this results from the fact that the real part of the refractive index is slightly smaller than unity for neutrons (and x-rays) [7.4–6]. However, PNR is based on the 'optical' interference of cold neutrons reflected at the different interfaces and so is critically dependent on the sample structural quality. Nevertheless, PNR is unique in allowing a quantitative determination of the sample moment to be obtained in the monolayer range. This is the combined result of significant developments in the PNR technique over the last 10 years ranging from (i) improved sample flatness and reduced interface roughness, (ii) introduction of step wedge samples (same growth conditions) which allow high precision comparative measurements, (iii) vector magnetometry, (iv) polarisation analysis and (v) also an increase in the number of neutron reflectometers available. SQUID magnetometry, although very accessible and simple to use, has the disadvantage of providing the average magnetic moment of the whole structure, which in many cases of interest involves a thick substrate with a very large dia- or paramagnetic contribution superimposed on the small magnetic signal from the film. However, for some systems, *in situ* SQUID techniques have been successfully employed to measure the absolute magnetic moment of ultrathin magnetic films with very high accuracy [7.7].

Furthermore, the measurement of the magnetic moment in thin films has been stimulated by the results of many ab initio numerical studies. These studies allow for a detailed understanding of the microscopic origin of the distinct properties of ultrathin film systems and often provide clues for which systems may exhibit the largest deviations from bulk properties. This may shed additional light on the understanding of the electronic mechanisms underlying the magnetism of such systems, as well as a check on the numerical calculations and structural plus chemical contamination. Although these numerical studies usually involve approximations that are difficult to implement in practice (e.g., flat surfaces/interfaces, defect free films) and are calculated for the ground state of the system, ab initio calculations constitute an invaluable guide to a correct understanding of the experimental results.

In this article a review of the basic theory of PNR and some of the experimental aspects of this technique are presented in Sect. 7.2, followed by an overview of PNR results for thin magnetic films (Sect. 7.3.1), spin-valve systems (Sect. 7.3.2) and multilayers (Sect. 7.3.3). PNR results on superconductors, rare-earth systems as well as studies relating to diffuse scattering and the phase problem are excluded from this review. Shorter introductions to PNR can be found in [7.5, 8–15].

7.2 Theoretical Basis

Polarised neutron reflection (PNR) magnetometry is based on the magnetic interaction between the neutron magnetic moment and the medium magnetisation, allowing therefore the measurement of the magnetic moments per atom while giving simultaneously structural information such as film thickness and interface roughness. The geometry of the measurement limits this technique to smooth and flat samples, such as thin films and multilayers supported by a thick substrate. For a typical sample area of 1 cm^2, the magnetic moment from a 1 ML of Fe is $\approx 1.4 \times 10^{-8}$ emu. This is well within the reach of the PNR sensitivity, which can be compared with the sensitivity of commercial SQUID (superconducting interference device) magnetometers, of the order of 10^{-7} emu. However, for ultrathin films supported by a thick substrate, the dia/paramagnetic contribution from the substrate and overlayers usually overwhelm the signal from the magnetic layer; and an exact knowledge of the sample area and the film thickness is required in order to obtain the value of the magnetisation of the film. If two or more magnetic layers are present, it can only provide the average value of the magnetisation. MOKE (magneto-optic Kerr effect) magnetometry, on the other hand, can easily reach such sensitivities, but unlike the other two techniques, cannot provide a quantitative measure of the magnetisation and is limited to surface layers. In contrast, PNR is a self-calibrating technique, providing independently values of both the magnetic moment *and* film thickness; it is layer selective and can probe buried layers in a multilayer system. In the simplest arrangement, the magnetisation of the sample is saturated along the direction of the neutron spin, and the reflectivity for the two neutron spin states is measured. Due to the sensitivity of the neutron reflection amplitude to the surface roughness, an estimate of the interface roughness of the film structure can be obtained (within the range 0–50 Å); in addition, the fringe spectra produced by multiple interference of the neutron beam within the sample layers enables the thickness of each layer to be determined with high accuracy (better than 1–10% for layer thicknesses in the range 100–0.1 nm, respectively). Furthermore, the theory of neutron reflection is relatively simple allowing quite straightforward data modelling. Contrary to x-ray scattering, the nuclear scattering length varies considerably from element to element with the consequence that neutron reflectometry offers a larger layer contrast than the equivalent x-ray reflectometry technique [7.16, 17]. For these reasons, polarised neutron reflection is a technique particularly suited when both structural (thickness and interface roughness) and magnetic (saturation magnetic moments) information is needed in magnetic films and multilayers (magnetic moments can be obtained with an accuracy better than 5–8% for films in the few monolayers range, which can be improved by combination with other techniques, such as SQUID and/or x-ray reflectometry). In addition, for non-collinear magnetisation configurations, spin-flip processes occur changing the spin up and down reflected intensities and allowing this technique to provide vectorial information even in the absence of polarisation analysis of the reflected beam [7.18–21].

7.2.1 Theory: Basics of Polarised Neutron Reflection

The neutron has several properties that makes it unique when used for probing the condensed state of matter. It is almost devoid of any electrical properties, having no electric charge, no electric dipole moment and no electric polarisability (at least for the range of energies where the internal structure of the neutron can be disregarded). It has mass and a magnetic moment. Its interaction with matter is dominated by the interaction with the nucleus of the atoms (due to nuclear forces) [7.22]. A neutron plane wave, when scattered by a single fixed nucleus, appears at a distance r from the scatterer with a relative amplitude $f(\theta, k)/r$ in the direction given by θ. The point-like pseudo-potential for a nucleus in a position R can be described by a delta function of the form [7.23–25],

$$V(\boldsymbol{r}) = 2\pi\hbar^2\mu^{-1}a\delta(\boldsymbol{r} - \boldsymbol{R}),$$

where μ is the reduced mass of the system and a is the scattering length. Although this does not correspond to the real neutron-nuclei interaction, it is a way of expressing the effective nuclear interaction as a small perturbation to the Hamiltonian of the non-interacting system (a rigorous treatment of this problem is given in [7.26]). The solution in the Born approximation gives, for slow neutrons, $f(\theta) = -a$, independently of the angle θ, as the scattering occurs only for small values of the angular momentum, $l = 0$ (this results from the small range of the strong interaction with respect to the neutron wavelength [7.25]). The relation between the total scattering cross section σ and the scattering length is given by $\sigma = 4\pi a^2$ [7.27].

In the case of a solid and for neutrons with wavelengths large in comparison with the distance between scatterers, the neutrons 'feel' the mean potential

$$V = 2\pi\hbar^2 m_{\rm n}^{-1} Nb \tag{7.1}$$

where N is the number of scatterers per unit volume and b the effective scattering length ($m_{\rm n} = 1838.6 m_{\rm e}$ is the neutron mass). Following [7.22], the notation for the scattering length was changed from a for the free atom to b, for the bound atom, with $b = (1 + m_{\rm n}/A)a$, where A is the atomic mass of the atom. Interference between incoming and scattered neutron waves and between the scattered waves from a set of nuclei is possible if the scattering by each individual nucleus is elastic and coherent. Fluctuations of the individual phase shift of the scattered waves produce an incoherent component of scattering (which can be caused by resonant scattering with different spin states and by scattering on an ensemble of various isotopes of a chemical element or of various chemical elements); consequently, the scattering length a is separated into a coherent and an incoherent part, $a_c = \langle a \rangle$ and $a_i = (\langle a^2 \rangle - \langle a \rangle^2)^{1/2}$ respectively, where the brackets represent averaging over the individual scattering lengths. The incoherent scattering length gives rise to isotropic neutron scattering, and does not produce interference fringes since the phase information is lost; in studying the elastic scattering of neutrons, only the coherent scattering length has to be considered in (7.1) [7.28, 29].

In the case of magnetic materials, the magnetic dipole moment of the neutron interacts with the effective magnetic induction of the medium, $V_{mag} = -\boldsymbol{\mu}_n \cdot \boldsymbol{B}$. Unlike the nuclear scattering, the magnetic scattering is not isotropic, due to the axial symmetry of the dipole field. In fact, the scattering amplitude due to this interaction in the Born approximation is given by [7.26, 30]

$$b_m \equiv -f(\boldsymbol{Q}) = \frac{m_n}{2\pi\hbar^2} \int V_{mag}\, e^{-i\boldsymbol{Q}\cdot\boldsymbol{r}'}\, dv' \qquad (7.2)$$

where $\boldsymbol{Q} = \boldsymbol{k} - \boldsymbol{k}'$ is the momentum transfer, or:

$$b_m = \frac{m_n}{2\pi\hbar^2} [-\boldsymbol{\mu}_n \cdot \boldsymbol{B}(\boldsymbol{Q})] \qquad (7.3)$$

where $\boldsymbol{B}(\boldsymbol{Q})$ is the Fourier transform of $\boldsymbol{B}(\boldsymbol{r})$. Further, it can be shown that [7.26, p. 157]

$$\boldsymbol{B}(\boldsymbol{Q}) = \hat{\boldsymbol{Q}} \times (\boldsymbol{M} \times \hat{\boldsymbol{Q}}) = \boldsymbol{M} - \hat{\boldsymbol{Q}}(\boldsymbol{M} \cdot \hat{\boldsymbol{Q}}) = \boldsymbol{M}_\perp(\boldsymbol{Q}) \qquad (7.4)$$

where $\boldsymbol{M}_\perp(\boldsymbol{Q})$ is the component of $\boldsymbol{M}$ perpendicular to $\boldsymbol{Q}$. We see therefore that the magnetic interaction is only effective in scattering neutrons when there is a component of the magnetisation perpendicular to the momentum transfer. This can also be seen by recalling that the divergenceless of $\boldsymbol{B}$ can be written in the momentum space as $\boldsymbol{Q} \cdot \boldsymbol{B}(\boldsymbol{Q}) = 0$ [7.27, 29, 31, 32].

Therefore, as with MOKE [7.33, 34], we can distinguish three scattering geometries according to the orientation of the magnetisation with respect to the momentum transfer direction and to the quantisation axis (Fig. 7.1):

1. the polar geometry, where there is no magnetic interaction and therefore the scattering is due only to the nuclear potential of the magnetic material;
2. the transverse geometry, where the magnetisation is perpendicular to the momentum transfer wavevector and parallel to the quantisation axis and for which geometry the Hamiltonian commutes with the spin operator, and
3. the longitudinal geometry, where the magnetisation is perpendicular to the neutron spin ($\boldsymbol{\mu}_n \cdot \boldsymbol{M} = 0$) and the Hamiltonian of the system does not commute with the spin operator (spin-flipping scattering processes occur, thus making this situation different from the polar geometry) [7.35].

It is also worth mentioning that the neutron mass interaction with the gravitational field of the Earth can be comparable in magnitude with the nuclear and magnetic interactions for very slow neutrons; in fact a neutron at a height z from a reference plane has a potential energy given by $m_n g z$, or 100 neV/m [7.29] (values for the nuclear and magnetic interactions for selected materials are given in Table 7.1).

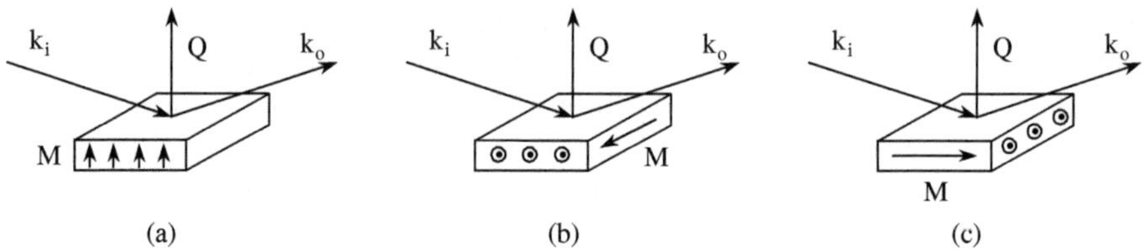

Fig. 7.1. Scattering geometries for polarised neutron reflection: **(a)** polar geometry, **(b)** transverse geometry and **(c)** longitudinal geometry

Table 7.1. Values for the nuclear and magnetic potential of neutron elastic scattering (selected materials). Relevant parameters are also shown, where $\langle b \rangle$ is the coherent nuclear scattering length (from [7.36]), ϱ the atomic density, N the atomic density (at room temperature), $I_s = 4\pi M_s$ the saturation magnetisation, V_N and V_{mag} the nuclear and magnetic terms of the nuclear potential, respectively, and q_c the critical wavevector for total reflection

Element	$\langle b \rangle$ (fm)	ϱ (g/cm^3)	N (10^{28} at./m^3)	I_s (kG)	V_N (neV)	V_{mag} (neV)	q_c +/− (10^{-3} Å^{-1})
Si	4.149	2.329	4.97	—	53.77	—	5.09
Cr	3.635	7.194	8.30	—	78.68	—	6.16
Mn	−3.73	7.43	8.144	—	−79.22	—	—
Fe	9.54	7.873	8.46	21.580	210.47	129.8	12.81/6.24
Co	2.50	8.800	8.99	17.970	55.81	107.9	8.89
Ni	10.3	8.907	9.10	6.556	244.42	39.43	11.70/9.95
$Ni_{80}Fe_{20}$	10.148	8.700	8.972	9.183	237.43	55.2	11.88/9.38
Cu	7.718	8.933	8.46	—	170.27	—	9.06
Ge	8.193	5.323	4.42	—	94.43	—	6.75
Pd	5.91	11.995	6.785	—	104.57	—	7.10
Ag	5.992	10.500	5.84	—	91.25	—	6.64
Pt	9.63	21.450	6.62	—	166.25	—	8.96
Au	7.63	19.281	5.87	—	116.80	—	7.51
GaAs	6.934	5.316	4.43	—	79.92	—	6.21
InAs	5.223	5.66	3.593	—	48.94	—	4.86

7.2.1.1 Kinematical Theory of Neutron Reflection (Elastic Scattering)

The theory of neutron reflection can be studied within the framework of neutron optics [7.37, 38]. Its basic assumption is that for long wavelength neutrons the interaction potential can be considered to vary slowly from point to point in space. This is not strictly true with the nuclear interaction of neutrons with the nuclei, as mentioned above. However, the wavelength of slow neutrons is typically four orders of magnitude larger than the range of the (attractive) nuclear interaction and therefore the scattered waves are spherical and isotropic. These waves, emanating from each nuclear site combine coherently and add up to a single resultant wave characterised by the average interaction potential [7.37]. (A more refined justification, which relies on the validity of the Fermi potential to represent the nuclear potential, is given in [7.25], chapter 2.) Thus, neutrons can be seen as interacting with a uniform nuclear and magnetic potential when scattering with matter. In quantum mechanics this corresponds to the simple and soluble case of a potential barrier, which we consider next [7.19, 30, 39–41].

The Schrödinger equation for the wavefunction of a neutron in a medium is given by

$$\left[-\frac{\hbar^2}{2m_\mathrm{n}}\nabla^2 + V(\boldsymbol{r})\right]\psi(\boldsymbol{r}) = E\psi(\boldsymbol{r}) \tag{7.5}$$

where m_n is the neutron mass and $V(\boldsymbol{r})$ the potential energy. Let us consider an interface between two different media and translational symmetry in the plane xy of the interface, which defines the zero of the z-direction, see Fig. 7.2 (the y-direction is defined as the axis of quantisation). Then the neutron wavefunction $\psi(\boldsymbol{r})$ can be written in terms of the parallel (to the interface) component of $\boldsymbol{k}$:

$$\psi(\boldsymbol{r}) = \psi(z)\,\mathrm{e}^{\mathrm{i}\boldsymbol{k}_\parallel\cdot\boldsymbol{r}} \tag{7.6}$$

by defining $\boldsymbol{k}_\parallel \equiv k_\parallel \boldsymbol{e}_x$, and including $k_\perp$ in $\psi(z)$. The Schrödinger equation becomes:

$$\left\{\frac{\mathrm{d}^2}{\mathrm{d}z^2} + \left[\frac{2m_\mathrm{n}}{\hbar^2}(E-V) - k_\parallel^2\right]\right\}\psi(z) = 0 \tag{7.7}$$

which can be written as[1]

$$\left(\frac{\mathrm{d}^2}{\mathrm{d}z^2} + q^2\right)\psi(z) = 0\,, \quad q \equiv \sqrt{\frac{2m_\mathrm{n}}{\hbar^2}(E-V) - k_\parallel^2}\,, \tag{7.8}$$

where q is the component of the wavevector parallel to the z direction. Equation (7.8) corresponds to the "equation of motion" of the neutron along the z-direction, with the general solution

$$\psi(z) = A\,\mathrm{e}^{\mathrm{i}qz} + B\,\mathrm{e}^{-\mathrm{i}qz}\,, \tag{7.9}$$

where the coefficients A and B are determined by imposing the boundary conditions for ψ and continuity of $\mathrm{d}\psi/\mathrm{d}z$. We have then (using the subscripts 1 and 2 for regions $z < 0$ and $z > 0$), with $A_1 = 1$ and $B_2 = 0$:

$$\begin{cases} 1 + B_1 = B_2 \\ q_1(1 - B_1) = q_2 A_2 \end{cases} \Rightarrow \begin{cases} B_1 = (q_1 - q_2)/(q_1 + q_2) \\ A_2 = (2q_1)/(q_1 + q_2) \end{cases}. \tag{7.10}$$

[1] For ease of notation, q corresponds to the wavevector component perpendicular to the interface (i.e., $q = k_\perp$). Some authors use $\boldsymbol{q}$ to represent the momentum transfer, which here is denoted by $\boldsymbol{Q}$.

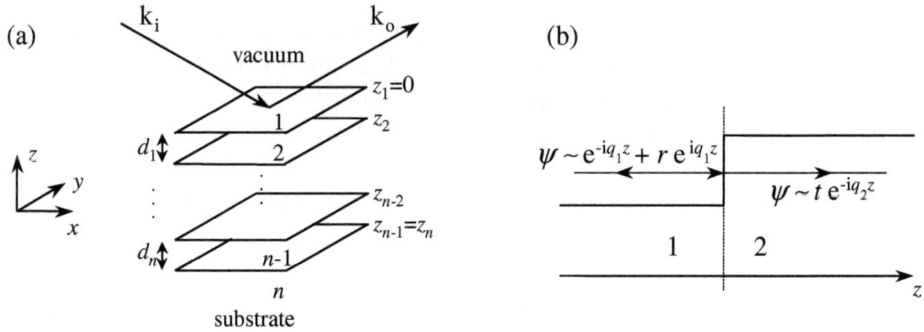

Fig. 7.2. **(a)** Neutron reflection geometry and **(b)** the correspondent step potential for the interface between the media 1 and 2. The momentum transfer $\boldsymbol{Q} = \boldsymbol{k}_\mathrm{i} - \boldsymbol{k}_o$ is parallel to the z-direction and is equal to $2k_\perp = 2k_\mathrm{i}\sin\theta$, where θ is the angle between $\boldsymbol{k}_\mathrm{i}$ and the plane of the interface

The wavefunction is of the form:

$$\psi = \begin{cases} \mathrm{e}^{\mathrm{i}q_1 z} + r\,\mathrm{e}^{-\mathrm{i}q_1 z}\,, & z < 0 \\ t\,\mathrm{e}^{\mathrm{i}q_2 z}\,, & z > 0 \end{cases} \tag{7.11}$$

where $r \equiv B_1$ is the amplitude of the reflected wave and $t \equiv A_2$ is the amplitude of the transmitted wave. The transmission and reflection coefficients are defined as [7.42] $T = J_\mathrm{t}/J_\mathrm{i}$ and $R = J_\mathrm{r}/J_\mathrm{i}$, respectively, where J_i, J_r, J_t are the incident, reflected and transmitted flux density, which is given by [7.43]:

$$\boldsymbol{J}(\boldsymbol{r}, t) = \mathrm{Re}\left[\psi^* \frac{\hbar}{im}\nabla\psi\right] . \tag{7.12}$$

We have, from (7.11),

$$J_\mathrm{i} = \frac{\hbar^2}{m_\mathrm{n}}|q_1|\,, \quad J_\mathrm{r} = |r|^2\frac{\hbar^2}{m_\mathrm{n}}|q_1|, \quad J_\mathrm{t} = |t|^2\frac{\hbar^2}{m_\mathrm{n}}|q_2|\,, \tag{7.13}$$

and therefore,

$$R = |r|^2 = \left|\frac{q_1 - q_2}{q_1 + q_2}\right|^2; \quad T = \frac{|q_2|}{|q_1|}|t|^2 = \left|\frac{2q_2}{q_1 + q_2}\right|^2 . \tag{7.14}$$

This is the equivalent of Fresnel's law in optics.

Considering the case when $V_1 = 0$ and making $q_1^2 = 2m_\mathrm{n}E/\hbar^2 - k_\parallel^2$, we have for the reflection coefficient (where $v_2 = 2m_\mathrm{n}V_2/\hbar^2$):

$$R = \left|\frac{q_1 - \sqrt{q_1^2 - v_2}}{q_1 + \sqrt{q_1^2 - v_2}}\right|^2 . \tag{7.15}$$

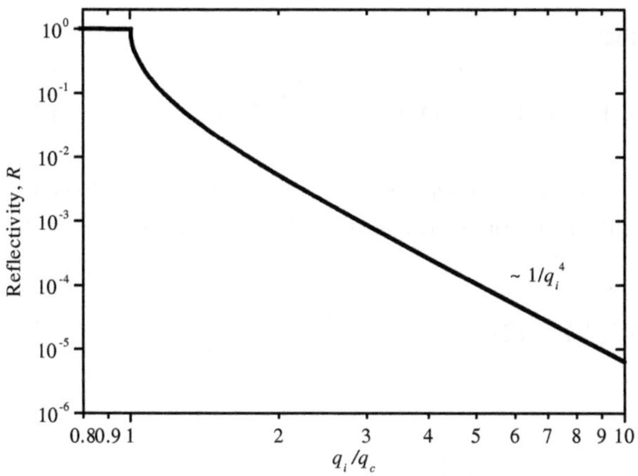

Fig. 7.3. Variation of the reflectivity R with the incident neutron wavelength. For $q_\mathrm{i} \gg q_\mathrm{c}$, $R \sim q_\mathrm{i}^{-4}$

As is easily seen, for $q_1^2 - v_2 < 0$, $R = 1$, which defines the critical wavevector q_c for total reflection (Fig. 7.3). The lengthscale defined by this critical wavevector, $\lambda_e = 2\pi/q_\mathrm{c}$ (of the order of 50–100 nm for the materials listed in Table 7.1) is the penetration depth, the distance probed by the incident neutron wave: if the layer is thinner than this value, the probability of transmission is non-zero and total reflection no longer occurs. The phenomenon of total reflection of neutrons was first observed by Fermi and co-workers [7.44, 45].

The problem can be easily generalised for a multilayer system, corresponding to the problem of successive barrier potentials. Denoting the vacuum and the substrate with the subscripts 1 and n respectively (see Fig. 7.1), the generalisation of (7.11) for the n-layer case is

$$\psi_\alpha(z) = a_\alpha \,\mathrm{e}^{\mathrm{i}q_\alpha z} + b_\alpha \,\mathrm{e}^{-\mathrm{i}q_\alpha z} \quad (z_\alpha < z < z_{\alpha+1}\,, \quad 1 < \alpha < n)\,, \tag{7.16}$$

where $a_1 = 1$, $b_\mathrm{r} = 1$, $b_n = 0$ and $a_n = t$, q_α is given by (7.8) with $V = V_\alpha$ corresponding to the nuclear potential for layer α ($q_\alpha = \sqrt{q_1^2 - v_\alpha}$). Continuity of the wavefunction and of its derivative at each interface imposes conditions on the a_α and b_α that are given by the set of equations, in matrix notation:

$$\begin{bmatrix} \mathrm{e}^{\mathrm{i}q_\alpha z_\alpha} & \mathrm{e}^{-\mathrm{i}q_\alpha z_\alpha} \\ q_\alpha \,\mathrm{e}^{\mathrm{i}q_\alpha z_\alpha} & -q_\alpha \,\mathrm{e}^{-\mathrm{i}q_\alpha z_\alpha} \end{bmatrix} \begin{bmatrix} a_\alpha \\ b_\alpha \end{bmatrix} = \begin{bmatrix} \mathrm{e}^{\mathrm{i}q_{\alpha+1} z_\alpha} & \mathrm{e}^{-\mathrm{i}q_{\alpha+1} z_\alpha} \\ q_{\alpha+1} \,\mathrm{e}^{\mathrm{i}q_{\alpha+1} z_\alpha} & -q_{\alpha+1} \,\mathrm{e}^{-\mathrm{i}q_{\alpha+1} z_\alpha} \end{bmatrix} \begin{bmatrix} a_{\alpha+1} \\ b_{\alpha+1} \end{bmatrix} \tag{7.17}$$

The 2×2 matrix in the right hand side of (7.17) can be written as a product of two matrices with the form $\mathrm{K}(\alpha)\mathrm{D}(\alpha)$, where

$$\mathrm{D}(\alpha) = \begin{bmatrix} \mathrm{e}^{\mathrm{i}q_\alpha z_\alpha} & \mathrm{e}^{-\mathrm{i}q_\alpha z_\alpha} \\ q_\alpha \,\mathrm{e}^{\mathrm{i}q_\alpha z_\alpha} & -q_\alpha \,\mathrm{e}^{-\mathrm{i}q_\alpha z_\alpha} \end{bmatrix}, \tag{7.18}$$

$$\mathrm{K}(\alpha) = \begin{bmatrix} (\mathrm{e}^{-\mathrm{i}q_\alpha d_\alpha} + \mathrm{e}^{\mathrm{i}q_\alpha d_\alpha})/2 & (\mathrm{e}^{-\mathrm{i}q_\alpha d_\alpha} - \mathrm{e}^{\mathrm{i}q_\alpha d_\alpha})/2q_\alpha \\ q_\alpha(\mathrm{e}^{-\mathrm{i}q_\alpha d_\alpha} - \mathrm{e}^{\mathrm{i}q_\alpha d_\alpha})/2 & (\mathrm{e}^{-\mathrm{i}q_\alpha d_\alpha} + \mathrm{e}^{\mathrm{i}q_\alpha d_\alpha})/2 \end{bmatrix}, \tag{7.19}$$

with $d_\alpha = z_{\alpha+1} - z_\alpha$ such that

$$\begin{bmatrix} a_\alpha \\ b_\alpha \end{bmatrix} = \mathrm{D}^{-1}(\alpha)\mathrm{K}(\alpha+1)\mathrm{D}(\alpha+1)\begin{bmatrix} a_{\alpha+1} \\ b_{\alpha+1} \end{bmatrix} \tag{7.20}$$

with the proviso that $z_n \equiv z_{n-1}$, in which case $\mathrm{K}(n) = \mathbf{1}$. The relation between the coefficients of the wavelength in region 1 (vacuum) and the substrate, region n, is then:

$$\begin{bmatrix} a_1 \\ b_1 \end{bmatrix} = \mathrm{D}^{-1}(1)\mathrm{K}(2)\mathrm{D}(2)\mathrm{D}^{-1}(2)\ldots\mathrm{K}(n)\mathrm{D}(n)\begin{bmatrix} a_n \\ b_n \end{bmatrix} = \mathrm{M}\begin{bmatrix} a_n \\ b_n \end{bmatrix}, \tag{7.21}$$

with

$$\mathrm{M} = \mathrm{D}^{-1}(1)\prod_{i=2}^{n-1}\mathrm{K}(i)\mathrm{D}^{-1}(n) \tag{7.22}$$

since $\mathrm{K}(n) \equiv \mathbf{1}$ as mentioned above; $\mathrm{K}(i)$ is the characteristic matrix of the medium i and M is called the transfer matrix. In terms of the transfer matrix elements, and recalling that $b_n = 0$, we have

$$\begin{cases} a_1 = \mathrm{M}_{11}a_n \\ b_1 = \mathrm{M}_{21}a_n \end{cases}. \tag{7.23}$$

The reflection and transmission coefficients are obtained in a similar way as before, taking into account that the incident, reflected and transmitted waves are given by:

$$\psi_\mathrm{i} = a_1\,\mathrm{e}^{\mathrm{i}q_1(z-z_1)} \qquad (z < z_1) \tag{7.24}$$

$$\psi_\mathrm{r} = b_1\,\mathrm{e}^{-\mathrm{i}q_1(z-z_1)} \qquad (z < z_1) \tag{7.25}$$

$$\psi_\mathrm{t} = a_n\,\mathrm{e}^{\mathrm{i}q_n z} \qquad (z > z_n)\,. \tag{7.26}$$

Therefore,

$$T = \frac{J_\mathrm{t}}{J_\mathrm{i}} = \frac{|a_n|^2\,q_n^2}{|a_1|^2\,q_1^2} = \frac{1}{|M_{11}|^2}\left|\frac{q_n}{q_1}\right|, \tag{7.27}$$

$$R = \frac{J_\mathrm{r}}{J_\mathrm{i}} = \frac{|b_1|^2}{|a_1|^2} = \frac{|\mathrm{M}_{21}|^2}{|\mathrm{M}_{11}|^2}\,. \tag{7.28}$$

In general, the potential energy in the α-th region V_α in (7.8) is given as a sum of a nuclear and a magnetic term,

$$V_\alpha = \frac{2\pi\hbar^2}{m_\mathrm{n}}N_\alpha b_\alpha - 4\pi\boldsymbol{\mu}_\mathrm{n}\cdot\boldsymbol{M}_{\perp\alpha} \tag{7.29}$$

where N_α, b_α, $\boldsymbol{\mu}_\mathrm{n}$ and $\boldsymbol{M}_{\perp\alpha}$ are, respectively, the atomic density, the nuclear scattering length, the neutron magnetic moment and the α-layer component of the magnetisation

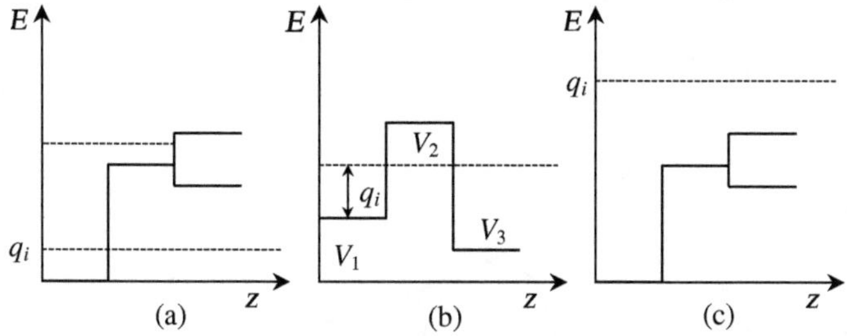

Fig. 7.4. Different regimes for neutron reflection: **(a)** total reflection; **(b)** quantum tunnelling and **(c)** partial reflection. (The two levels in the medium 1 and 3 represent those situations where the nuclear potential is either bigger or smaller than the nuclear potential in medium 2)

normal to the momentum transfer. Here we are interested in the case where the neutrons are spin polarised along the y-direction, with the magnetisation of the sample either parallel or anti-parallel to the direction of the neutron spin (this situation is often implemented in practice, although not the most general case). With these conditions, spin-flip scattering does not occur, since the Hamiltonian (7.5) is diagonal in the neutron spin basis, and we can then consider separately the case when the spin is parallel to the magnetisation (spin up $\uparrow$, neutron moment anti-parallel to $\boldsymbol{M}_{\perp\alpha}$), and the opposite case when the spin is anti-parallel to the magnetisation, spin down $\downarrow$ (neutron moment parallel to $\boldsymbol{M}_{\perp\alpha}$). The potential V_α assumes the values, respectively for the spin up and down neutron states:

$$V_\alpha^\uparrow = \frac{2\pi\hbar^2}{m_\mathrm{n}} N_\alpha b_\alpha + 4\pi|\mu_\mathrm{n}|M_\alpha \tag{7.30}$$

$$V_\alpha^\downarrow = \frac{2\pi\hbar^2}{m_\mathrm{n}} N_\alpha b_\alpha - 4\pi|\mu_\mathrm{n}|M_\alpha \tag{7.31}$$

for the α region. Table 7.1 shows the values for the nuclear and magnetic potential for some elements, as well as other parameters that enter in its calculation.

Consider, in particular, the case $n = 3$. Inspection of the energy barrier diagram, Fig. 7.4, shows that three different situations may occur, regarding the relative height of each barrier and the energy of the incident neutrons. When q_i is smaller than the critical wavevector we have total reflection and when the potential of the intermediate medium exceeds the other two, quantum tunnelling may occur for a certain range of the incident neutron energies. Finally, when the neutron energy is high enough, transmission occurs. The expressions for the reflectivity in this case are given by

$$R = 1, \quad \text{for } V_1 < V_2 \text{ or } V_3 \tag{7.32}$$

$$R = \frac{(1 - q_3/q_1)^2\cos^2\theta_2 + (q_3/q_2 - q_2/q_1)^2\sin^2\theta_2}{(1 + q_3/q_1)^2\cos^2\theta_2 + (q_3/q_2 + q_2/q_1)^2\sin^2\theta_2}, \quad \text{for } V_1 > V_2, V_3 \tag{7.33}$$

$$R = \frac{(1 - q_3/q_1)^2\cosh^2\theta_2 + (q_3/q_2 - q_2/q_1)^2\sinh^2\theta_2}{(1 + q_3/q_1)^2\cosh^2\theta_2 + (q_3/q_2 + q_2/q_1)^2\sinh^2\theta_2}, \quad \text{for } V_3 < V_1 < V_2, \tag{7.34}$$

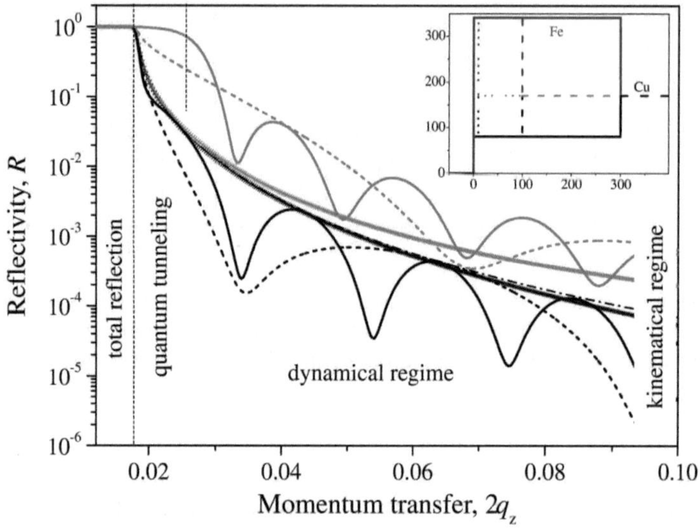

Fig. 7.5. Calculated reflectivity curves for a Fe layer on a Cu substrate, for spin up and down neutron states and three values of the Fe layer thickness (10, 100 and 300 Å). For $q_c^{Cu} < q_1 < q_c^{Fe\uparrow}$ quantum tunnelling occurs, while for $q_1 < q_c^{Cu}$ there is total reflection of the neutron beam. The reflectivity curve for the Cu substrate is also shown (*dot-dash curve*). The effect of increasing the layer thickness results in the appearance of an increasing number of interference fringes for the same q range. For the 300 Å Fe↑ case, the critical wavevector for total reflection gets closer to that of Fe, q_c^{Fe} (quantum tunnelling becomes less significant across this thicker Fe layer). For this simple case, the period of the oscillations is directly related to the thickness of the Fe layer, as is clear from the figure. The inset shows the energy potential profile across the structure for the different film thicknesses and for the two neutron spin states

where V_i are the potentials for each region, $q_i = \sqrt{q_1^2 - v_i}$ and $\theta_2 = d_2 q_2$. For a magnetic layer deposited onto a non-magnetic substrate, with $V_1 = 0$ (vacuum), only V_2 is spin dependent, as in (7.30)–(7.31). From (7.32)–(7.34) we can infer that the critical wavevector is given by $q_1^2 = V_3$, but when quantum tunnelling occurs, the transition from total reflection can be smooth (see Fig. 7.5). The sines and cosines in expression (7.33) show that the reflectivity spectrum consists of oscillations which vary, in the general case, in a complex fashion with the incident wavevector but with a frequency given by the layer thickness. It corresponds to the interference between the scattered neutrons at the various interfaces and will appear as fringes in the measured reflectivity, as first observed by Hayter et al. [7.46]. In fact, for $q_1 \gg q_2, q_3$, (7.33) reduces to

$$R \approx \frac{1}{16q_1^4}[v_3^2 + 4v_2(v_2 - v_3)\sin^2 d_2 q_1] \tag{7.35}$$

(in particular, for $v_2 = v_3$ the fringes disappear, as expected). This expressions shows that the oscillating term gives the thickness of the layer while the amplitude of the oscillations depend on the scattering potential contrast between the two media. The experimental problem is the rapid $1/q_1^4$ decay of the reflectivity. To measure the thickness of a thin layer requires that the maximum value of q_1 is large enough for there to be a measurable interference effect, preferably enough for a complete fringe to be measured, i.e., $q_{1\max} \sim 2\pi/d_2$. For a thick layer, the angular resolution must be less than the fringe separation, i.e., $\Delta q_1 < 2\pi/d_2$. The amplitude of the fringes must also be sufficient for them to stand out against the rest of the signal; this is best achieved when the nuclear potential contrast is as large as possible (this can be enhanced by using isotopic substitution) [7.47]. Expression (7.35) also shows that the reflectivity profile may be obtained by different combinations of the ordering of the same nuclear potentials (e.g., for $-v_2$ and $-v_3$); this is known as the phase problem [7.48] and is a result of the loss of phase information in the measured reflectivities. (This is only a limitation when no details of the sample structure are known, but interest in this problem remains [7.48–50].) Examples of reflectivity spectra are presented in Fig. 7.5 for Fe films of different layer thickness.

For polarised neutron reflection studies, it is customary to define the spin-asymmetry as

$$SA = \frac{R_\uparrow - R_\downarrow}{R_\uparrow + R_\downarrow} \tag{7.36}$$

where $R_{\uparrow(\downarrow)}$ are the spin up (down) reflectivities; it gives a direct measure of the difference in the reflectivity due to the spin dependent magnetic interaction. (It also cancels out spurious effects, such as instrumental and normalisation factors and is also less sensitive to the interface roughness, although it still needs to be corrected for the degree of polarisation of the neutron beam and spin-flipper efficiency [7.51, 52].)

7.2.1.2 The Diffraction Limit

If the incident kinetic energy associated with the perpendicular wavevector is large compared with the potential barriers of each layer in a multilayer system then we approach the conditions for kinematical diffraction to occur [7.19, 40]. We can expect that in this limit multiple reflections become less important and we may consider the approximation where only one wave is reflected at each successive interface. Then, for a given layer α,

$$q_\alpha = q_1 - v_\alpha/(2q_1) + o(v/q_1^2) \tag{7.37}$$

where q_1 and v_α were defined in (7.15) and at the α, $\alpha + 1$ interface the reflection coefficient r is (cf. (7.14))

$$r_{\alpha,\alpha+1} = \frac{q_\alpha - q_{\alpha+1}}{q_\alpha + q_{\alpha+1}} = \frac{1}{4q_1^2}(v_{\alpha+1} - v_\alpha) + o(v/q_1^2) \tag{7.38}$$

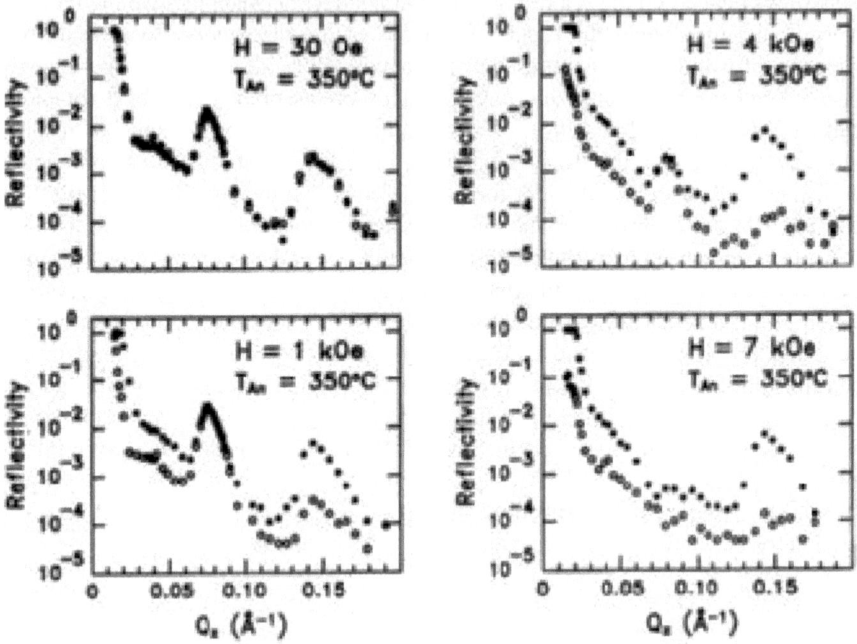

Fig. 7.6. Spin dependent reflectivities of a (30 Å Fe/10 Å Cr)$_{30}$ multilayer annealed at 350 °C, at different fields. Full and open circles correspond to spin up and down reflectivities. Note the disappearance of the antiferromagnetic diffraction peak as the antiferromagnetic order is suppressed by the applied field. (After [7.53])

and, in first order of approximation (assuming single reflection at each interface), we have for a multilayer:

$$r = \frac{1}{4q_1^2} \sum_{\alpha=1}^{n-1} e^{2iq_1 d_\alpha} (v_{\alpha+1} - v_\alpha) , \tag{7.39}$$

where the phase factor is introduced by the travelling of the wave across the layer. This expression is formally identical to the expression for the scattering amplitude for x-ray or electron diffraction in the first Born approximation (kinematical theory).

For a superlattice where the magnetisation of all layers is aligned along the quantisation axis, the nuclear potential is represented by a periodic rectangular wave, and the above expression gives rise to diffraction peaks with a basic period defined by the period of the rectangular wave (the first peak occurring at $q_1 = \pi/T$, where T is the period of the potential). For the antiparallel state, a new periodicity appears in the potential, which gives rise to new half-order diffraction peaks, the first at roughly half the wavevector value where the first-order diffraction peak appears. This is the principle behind the use of PNR to probe the magnetic configuration in magnetic multilayers, and is exemplified in Fig. 7.6 for the case of a superlattice exhibiting an antiferromagnetic alignment at remanence, and which is forced to a ferromagnetic state by the applied magnetic field. As the field amplitude is increased, the half order diffraction peak disappears while the ferromagnetic peak (which coincides with the structural peak) remains unchanged.

7.2.1.3 General Case: Spin-flip Analysis

So far we have considered the case where the Hamiltonian of the system commutes with the spin operator (in which case the $|\uparrow\rangle$ and $|\downarrow\rangle$ states are eigenstates of the Hamiltonian). In the general case where the magnetisation of the medium is not collinear with the neutron spin the wave function is, in general, a linear combination of these eigenstates. This is to say that the probability of spin-flip processes is non-zero, unlike the case when the quantisation axis coincides with the direction of the neutron spin. The Hamiltonian in the general case is:

$$H = \frac{-\mathrm{i}\hbar}{2m_\mathrm{n}}\nabla^2 + V_\mathrm{n} - \mu_\mathrm{n}\boldsymbol{\sigma}\cdot\boldsymbol{M}_\parallel \tag{7.40}$$

where $\boldsymbol{\sigma}$ has components given by the Pauli spin matrices, V_n is the nuclear potential and $\mathbf{M}_\parallel$ is the component of the magnetisation not collinear with the momentum transfer (in-plane component of the magnetisation).[2] The σ_x component of the spin operator (which can be expressed in terms of spin creation and annihilation operators) makes the probability of spin flipping non-zero. The wavefunction space is given by the direct product of the spin and momentum space. The calculations follow along a similar line as that for the simpler case of the Hamiltonian diagonal in the spin operator, by taking account of the rotation in the quantisation axis, and changing the transfer and the characteristic matrices to properly describe this more general situation. These calculations are detailed in [7.19, 41, 59, 60]. Here, we consider instead the simple case of a single magnetic layer [7.35]; in this case, the incident spin wave function can be written in terms of the components parallel and antiparallel with respect to the quantisation axis defined by the direction of the magnetisation (at an angle ϕ relative to the direction of the *up* neutron spin):

$$|\uparrow_\mathrm{i}\rangle = \cos(\phi/2)|\uparrow\rangle - \sin(\phi/2)|\downarrow\rangle \tag{7.41}$$

$$|\downarrow_\mathrm{i}\rangle = \sin(\phi/2)|\uparrow\rangle + \cos(\phi/2)|\downarrow\rangle\,. \tag{7.42}$$

The two components parallel and antiparallel to this axis then interact independently with the magnetisation in the system and hence the intensities are given by the weighted sums of the spin conserving reflectivities:

$$R^\uparrow(q) = \cos^2(\phi/2)|r^+|^2 + \sin^2(\phi/2)|r^-|^2 \tag{7.43}$$

$$R^\downarrow(q) = \sin^2(\phi/2)|r^+|^2 + \cos^2(\phi/2)|r^-|^2\,. \tag{7.44}$$

[2] It has been noted, however, that the Zeeman splitting induced by the component perpendicular to the film surface produces a deflection of the spin-flipped reflected neutron beam, which is dependent on the magnitude of the applied magnetic field and on the neutron wavelength [7.54–58]. The scattering angle of the non-spin-flipped beam is not affected.

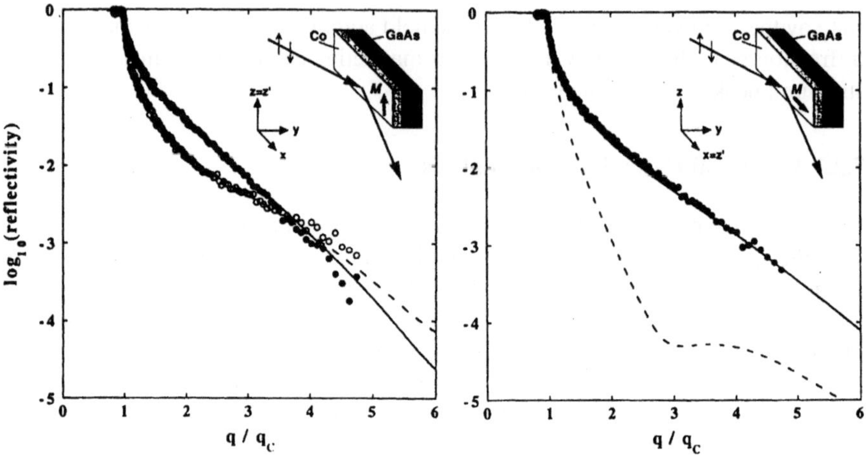

Fig. 7.7. *Left:* measured and calculated intensities as a function of the reduced wavevector from a 80 Å Co/GaAs(001) film for the conventional ('transverse') geometry. Open and full circles correspond respectively to spin up and down reflectivities and the lines are the calculated reflectivities. *Right:* Measured and calculated reflectivities for the same film, with the magnetisation aligned perpendicular to the incident neutron spin (circles and full line, respectively). The dashed line correspond to the reflected intensity expected from the nuclear potential only. (After [7.35])

In conventional PNR experiments ($\phi = 0$) only one incident spin polarisation contributes to the reflected intensity. In the other special geometry, when the spin polarisation is perpendicular to the sample magnetisation ($\phi = \pi/2$, 'longitudinal geometry') the measured intensity (without spin polarisation analysis) is identical to that obtained for an unpolarised incident neutron beam in the conventional geometry:

$$R^{\uparrow}(q) = R^{\downarrow}(q) = \frac{1}{2}(|r^+|^2 + |r^-|^2) = \frac{1}{2}(R^{\uparrow}_{\mathrm{conv}} + R^{\downarrow}_{\mathrm{conv}}) = R_{\mathrm{unpol}} \, , \qquad (7.45)$$

but is different from the situation where the magnetic potential is absent (which would be the case if the magnetisation is aligned along the momentum transfer direction, perpendicular to the film plane). This mechanism, entirely quantum mechanical in nature, has been observed experimentally, as shown in Fig. 7.7.

7.2.1.4 Interface Roughness

We assumed so far that the interfaces of the system have a perfect 2D symmetry; this approximation is not a good one in practice, as the interface between any two systems is very seldom perfect. Defects, roughness, interdiffusion or grading [7.61] introduce a break in the 2D symmetry of the interface that can lead to diffuse scattering of the incoming beam. In optics, for multiple beam interference to occur in a multilayer stack the surfaces must be optically smooth such that the variation in the phase is small, $\Delta(2q_\alpha t_\alpha) \ll 2\pi$. This condition is met if the interface (thickness) fluctuations Δt_α satisfy

$$\Delta t_\alpha \ll \lambda/(2\theta) \tag{7.46}$$

at large wavevectors for which $q \gg q_{c\alpha}$ (θ is the angle of incidence, measured from the film surface). For light, this implies that the surfaces be smooth on a length-scale corresponding to a fraction of a wavelength. For neutrons the situation is different since θ_c (and therefore θ) is small and we require that the fluctuations are small on the scale of the perpendicular wavevector, i.e., $\Delta t_\alpha \ll 0.02\,\mu$m for cold neutrons ($\lambda \sim 10\,$Å). This is fortunate since PNR studies of real ultrathin metal films would be impossible if the refractive index was as large as in the optical case.

Another important consideration concerns the neutron wave coherence lengths, both transversal and longitudinal to the sample plane, which determine respectively the thickness over which multiple interference can occur across the multilayer stack and the length over which any two points can be considered as illuminated by a single plane wave. The uncertainty in the perpendicular wavevector in terms of the wavelength and angular spread of the incident beam, $\Delta\lambda$ and $\Delta\theta$, assuming that they are uncorrelated, is given by ($k^\perp = k\sin\theta$):

$$(\Delta k^\perp)^2 = k^2 \sin^2\theta[(\Delta k/k)^2 + (\cos\theta\Delta\theta/\sin\theta)^2] \tag{7.47}$$

and for $\theta \ll 1$,

$$(\Delta k^\perp)^2 = (k^\perp)^2[(\Delta k/k)^2 + (\Delta\theta/\theta)^2]\,. \tag{7.48}$$

We can write $\Delta k^\perp l_c^t = 2\pi$, where l_c^t is the transverse coherence length. With $\Delta k/k$, $\Delta\theta/\theta \sim 0.1$ and $k\theta \sim 0.01\,$Å^{-1}, we have $l_c^t \sim 0.5\,\mu$m. For the longitudinal coherence length we have similarly ($k^\parallel = k\cos\theta$):

$$(\Delta k^\parallel)^2 = (\Delta k\cos\theta)^2 + (k\Delta\theta\sin\theta)^2 \approx k^2[(\Delta k/k)^2 + \theta^2(\Delta\theta/\theta)^2] \tag{7.49}$$

for $\theta \ll 1$, and with $\Delta k^\parallel l_c^\parallel = 2\pi$, $l_c^\parallel \sim l_c^t/\theta$ ($\theta \ll 1$) or $l_c^\parallel \sim 100\,\mu$m using the previous values and $\theta = 5\times10^{-3}$ rad.

It is difficult to account rigourously for the effect of interface roughness in the reflectivity; in general, a heuristic approach is followed by modifying the reflected and transmitted coefficients at the interface to account for the decrease in the intensity due to these defects. Several expressions have been suggested [7.62–65] and workable expressions can be obtained in the limit of either large or very small correlation lengths[3] [7.67].

---

[3] Compared with the *extinction length*, $L_e = V/\lambda b$, which is the length after which a measurable phase shift of the radiation is achieved (λ in the neutron wavelength, V the nuclear potential and b the scattering length); for thermal neutrons it is of the order of $L_e \approx 10^5\lambda$ [7.66].

For the case of a slowly varying interface height (assuming a gaussian distribution), the wave function is still locally given by (7.11), and the reflection coefficient is given by the average over the surface [7.67]:

$$r^{\text{rough}} = \left\langle r \frac{\mathrm{e}^{\mathrm{i}q_1 z(x)}}{\mathrm{e}^{-\mathrm{i}q_1 z(x)}} \right\rangle_x = r \langle \mathrm{e}^{2\mathrm{i}q_1 z(x)} \rangle_x \approx r\,\mathrm{e}^{-2q_1^2 \langle z^2 \rangle} , \tag{7.50}$$

where $\langle z^2 \rangle = \sigma^2$ is the roughness; this is the Debye-Waller factor [7.64, 68], and results from the fact that the correlation length is large enough so that the incident and reflected beams still have a precise phase relationship. For the opposite case, when the correlation length is much smaller than the extinction length, there are no short-scale correlations between the beams but only an overall perturbation of the wavefunction, which can be written as the combination of upwards and downwards propagating plane waves, whose amplitude depends on the roughness:

$$\psi(x, z) = a_{\alpha,\text{eff}}\,\mathrm{e}^{\mathrm{i}q_\alpha z} + b_{\alpha,\text{eff}}\,\mathrm{e}^{-\mathrm{i}q_\alpha z} \tag{7.51}$$

where the $a_{\alpha,\text{eff}}$ and $b_{\alpha,\text{eff}}$ are the (unknown) effective amplitudes for the rough interface. The reflection coefficient is defined as

$$r^{\text{rough}} = b_{1,\text{eff}}/a_{1,\text{eff}} . \tag{7.52}$$

Assuming that the phase relationships between the waves across the interface are only valid on average, we obtain, from the continuity condition at the interface:

$$\begin{cases} a_{1,\text{eff}}\,\mathrm{e}^{\mathrm{i}q_1 z} + b_{1,\text{eff}}\,\mathrm{e}^{-\mathrm{i}q_1 z} = a_{2,\text{eff}}\,\mathrm{e}^{\mathrm{i}q_2 z} \\ q_1 a_{1,\text{eff}}\,\mathrm{e}^{\mathrm{i}q_1 z} - q_1 b_{1,\text{eff}}\,\mathrm{e}^{-\mathrm{i}q_1 z} = q_2 a_{2,\text{eff}}\,\mathrm{e}^{\mathrm{i}q_2 z} \end{cases} \tag{7.53}$$

or

$$\begin{cases} 2q_1 a_{1,\text{eff}} = (q_1 + q_2) a_{2,\text{eff}} \langle \mathrm{e}^{\mathrm{i}(q_2 - q_1) z} \rangle_x \\ 2q_1 b_{1,\text{eff}} = (q_1 - q_2) a_{2,\text{eff}} \langle \mathrm{e}^{\mathrm{i}(q_2 + q_1) z} \rangle_x \end{cases} \tag{7.54}$$

and therefore,

$$r^{\text{rough}} = r^{\text{flat}}\,\mathrm{e}^{-2q_1 q_2 \langle z^2 \rangle} \tag{7.55}$$

which corresponds to the Croce-Névot factor [7.62, 64]. The roughness for each interface can be accounted for by multiplying the corresponding transfer matrix with this Croce-Névot factor. This is the factor implemented in the fitting program used in our data analysis. The Croce-Névot result is valid in PNR when the gaussian roughness amplitude is small as defined above, uncorrelated (i.e., the fluctuations are random with respect to each other) in-plane and the average interface fluctuation is zero. This does not often correspond to the experimental situation for single crystals metal substrates, although for structures supported by optically flat semiconductor substrates the Croce-Névot expression is appropriate. A more complete discussion of the effects of roughness on the reflectivity spectra can be found in [7.4, 40, 41, 65, 69]. For small roughness amplitudes, the effect on the reflected intensity can be accounted

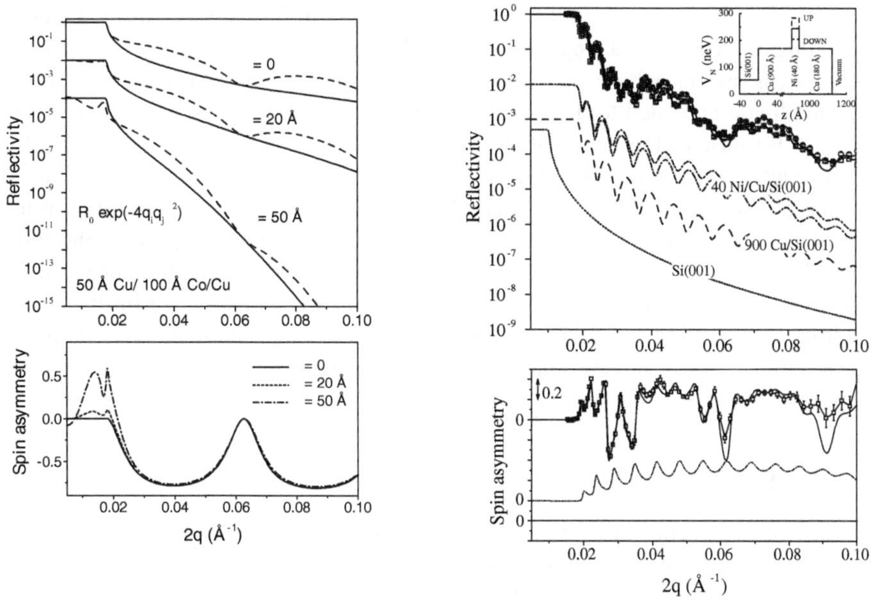

Fig. 7.8. *Left:* dependence of PNR spectra on the interface roughness. Although the effect of roughness is very marked in the reflectivities, the spin asymmetry is not very much affected, except for low q values. *Right:* example of a PNR spectrum for a magnetic structure with nominal composition of the form Cu(180 Å)/Ni(40 Å)/Cu(900 Å)/Si(001) (full lines correspond to the fit to the data)

by the Croce-Névot model and the effect on the values of the magnetic moment is small. For larger roughness values, diffuse scattering processes cannot be neglected and this has the effect of reducing the spin asymmetry in practice. This is because the diffuse cross section in the specular direction has a weaker dependence than the specular reflectivity; besides, it dominates the specular reflectivity for sufficiently large wavevector values [7.40, 69]. One way of avoiding the complications introduced by diffuse scattering is by using optically flat substrates.

Simply speaking, roughness has the effect of decreasing the reflected intensity (see Fig. 7.8, left panel) without affecting the critical wavevector value for total reflection and for uncorrelated roughnesses not to affect the spin asymmetry (this is the case for roughness amplitudes smaller than 10 Å rms). The sensitivity of the reflected spectrum to the film roughness makes neutron reflection measurements a powerful tool for the structural characterisation of thin films and multilayers [7.70].

Figure 7.8 (right panel) shows the evolution of the PNR spectrum at different stages of the sample structure, and the comparison with experimental data (in fact, the data was fitted first, and from these parameters the other curves were obtained). For these measurements the magnetisation of the Ni film was either saturated parallel and anti-parallel to the neutron spin, giving the spin-down and spin-up reflectivities, respectively. The sample has the nominal composition Cu(180 Å)/Ni(40 Å)/Cu(900

Å)/Si(001); for the Si substrate, a simple q^{-4} decay in intensity is expected, corresponding to the case of a single interface. When the thick Cu buffer layer is added, oscillations appear, whose period correspond to the Cu layer thickness and where the amplitude is related to the relative height of the nuclear potential (according to (7.35)). Because of the large value of the Cu thickness, the critical wavevector for total reflection has been shifted from that corresponding to Si ($2q_c = 0.010$ Å^{-1})) to that of Cu ($2q_c = 0.018$ Å^{-1}). The Ni magnetic layer induces a splitting in the spin down and up reflectivities, also shown in the spin-asymmetry plot. To the short period oscillations (from the Cu underlayer), a larger period oscillation due to the Ni thickness can be identified. Finally, the addition of a Cu capping layer introduces an additional oscillation period, modulating the oscillations from the Ni layer. This has the advantage of introducing a large variation in the reflected intensities, allowing for a more accurate fit to the data.

7.2.2 Experimental Setup

Several neutron reflectometers have been equipped with polarised beams, that allow for polarised neutron reflection studies. Table 7.2 lists (perhaps not extensively) several neutron reflectometers where PNR studies were conducted (see also Felcher [7.71]). Two techniques are usually used for the monochromatisation of the neutron beam, namely by crystal diffraction and by time-of-flight (TOF) methods, which define two basic geometries of measurements: for the TOF methods the sample is set at a fixed angle and neutrons with the right wavevector are selected by the TOF choppers; in the other case, only a monochromatic neutron beam is available, and the neutron (perpendicular) momentum transfer amplitude is changed by varying the angle of incidence.

Here, we describe briefly the CRISP neutron reflectometer at the Rutherford-Appleton Laboratory (U.K.), where a pulsed spallation neutron source is installed [7.88, 89]. A schematic of the setup is presented in Fig. 7.9 [7.16, 70]. The neutron beam wavelength is determined by a time-of-flight (TOF) method, and ranges from 0.5 to 6.5 Å. The neutron beam passes first through a ^{3}He detector (D_1) that monitors the incident beam intensity, and is collimated by the set of slits S_1. Neutrons of very long wavelength that pass through the choppers of the TOF system are deflected by a Ni mirror (inclined at $\sim 1.3°$) and the transmitted neutrons are polarised by reflection at a Fe-Co-V:TiN$_x$ supermirror (inclination angle of $\sim 0.4°$) [7.90]. A two coil non-adiabatic Drabkin spin flipper is used to reverse the spin direction of the polarised neutron beam, which is then guided to the sample, after being collimated by S_2. Detector D_2 monitors the neutron flux before arriving at the sample. The reflected beam is then collimated and detected by another ^{3}He detector, D_3. Slits S_1 and S_2 define the wavelength resolution of the measurements [7.91].

In our studies, a Bitter type electromagnet is used to align the magnetisation of the sample in the direction parallel to the sample film and perpendicular to the neutron momentum transfer so that the neutron spin is either parallel or anti-parallel to the sample magnetisation. The sample is placed at an angle such that the momentum transfer [given by $Q = 2k_i \sin\theta = (4\pi/\lambda)\sin\theta$] includes the critical wavelength

Table 7.2. List of neutron reflectometers with polarised neutron beams. The type of reflectometer tells whether the beam is monochromatised by a crystal analyser (CA) or by a time-of-flight (TOF) method (in which case the neutron wavelength range is given)

Reflectometer	Type	Wavelength (Å)	Reference
CRISP, RAL, U.K.	TOF	0.5–6.5	[7.72]
ADAM, ILL, Grenoble, France	CA	4.4; 2.2	[7.73–75]
D17, ILL, Grenoble, France	CA	5	[7.75]
EVA, ILL, Grenoble, France	CA	5.5;2.75	[7.75–77]
PADA, LLB, Saclay, France	CA	4	[7.78]
AMOR, PSI, Switzerland	TOF	1.3–13	[7.79]
ROG, TU Delft, Netherlands	TOF	0.7–15	[7.80, 81]
V6, BENSC, Berlin, Germany	CA	4.66	[7.82]
V14, BENSC, Berlin, Germany	CA	4.7	
TOREMA-2, GKSS, Geethacht, Germany	CA	4.3	[7.83]
HADAS, FRJ-2, Jülich, Germany	CA	4.8	[7.84]
ASTERIX, LANSCE, Los Alamos, USA	TOF	1–12	[7.85]
NG1REFL, NIST, Gaithersburg, USA	CA	4.75	
NG7REFL, NIST, Gaithersburg, USA	CA	0.235, 0.407, 0.47, 0.55	
POSY I, IPNS, Argonne, USA	TOF	3–15	[7.18]
C5, NPMR, Chalk River, Canada	CA	2.37	
PORE2, KENS, Tsukuba, Japan	TOF	3–16	
REFLEX-P, JINR, Dubna, Russia	TOF	0.7–10	
SPN, IBR-2, Dubna, Russia	TOF	0.7–10	
DHRUVA (Trombay), India	CA	4.06	[7.86, 87]

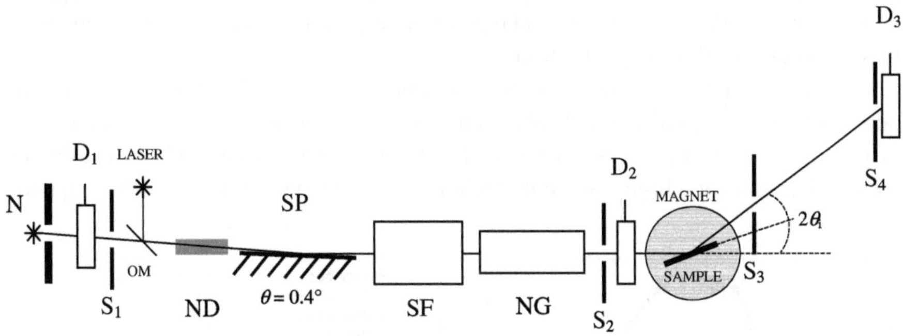

Fig. 7.9. Schematic diagram of the CRISP reflectometer (drawing not to scale). The neutron beam N flows from *left* to *right*; D_1 to D_3 are ^{3}He detectors, OM1 is an optical mirror (Si wafer), ND is a neutron deflector for long wavelength neutrons, SP is the supermirror, SF the spin flipper, NG a neutron guide and S_1 to S_4 are adjustable Cd slits. Detector D_2 and ND are moved away for sample alignment, which is performed with the aid of the laser beam. A magnetic field is present at the sample position applied in the direction normal to the plane of the page (the circular shadow corresponds to the pole pieces of the electromagnet)

value q_c (typically around 0.35°). The reflected beam intensity decays strongly away from the critical value (with q^{-4}) which means that long acquisition times are required in order to get good statistics for high q-values. One way of achieving this is to increase θ, shifting the momentum transfer to a higher region. An overlap between the previous q-range allows the data sets to be combined.

7.3 Polarised Neutron Reflection Magnetometry

In this section, an overview of selected experimental results on thin magnetic films, multilayers (in particular, spin-valve systems) and superlattices is presented, with an emphasis on the technical evolution of the neutron reflectometers and the corresponding improvement in the data quality. A comparison with results from other experimental techniques is made, wherever possible.

7.3.1 Ultrathin Magnetic Films

Most of research work involving PNR has either focused on the measurement of the magnetic moment of ultrathin films or on the study of the magnetic configuration of the magnetic layers in superlattices; these two cases are usually studied in two different scattering regimes, the first operating close to the extinction edge (total reflection), with wavelengths of the order of the total film thickness, and the latter case operating in the diffraction regime, using wavelengths that are much shorter, and that probe the long-range magnetic order defined by the successive magnetic layers. We shall deal with the second type of systems later and focus here on the PNR magnetometry of thin magnetic films.

The main incentive in probing the magnetic moment of thin films stems, on the one hand, from the prediction of enhanced magnetic moments in systems with broken symmetry (such as interfaces) suggested by ab initio calculations, and on the other by the requirement for a high sensitivity technique for the measurement of the magnetic

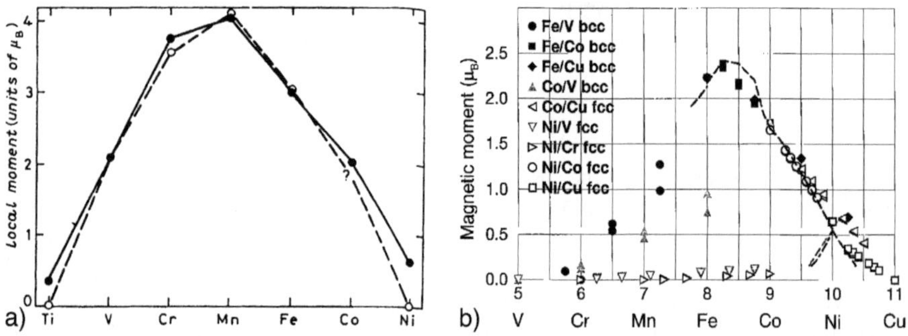

Fig. 7.10. Slater-Pauling curves for **(a)** monolayers of 3d transition metals on Ag(001) (after [7.92]) and **(b)** for the average interface moments for different magnetic interfaces (*dashed lines* correspond to experimental moments for FeV, FeCo, CoNi and NiCr alloys; after [7.93])

moments of ultrathin films, in the monolayer thickness regime. As mentioned before, PNR is a technique that neatly fulfils this requirement due to its sensitivity and with the added advantage of being a self-calibrated technique. Most of the pioneering work with PNR was done in the late 1980's, early 1990's, and was concentrated on magnetic transition metals, for which results from ab initio calculations were then made available for comparison (e.g., Fig. 7.10). These systems have since long attracted much interest due to their relative magnetic instability and their intermix of localised and itinerant magnetism.

7.3.1.1 Magnetic Moment of Thin Fe Films

Fe shows a remarkable variety of magnetic phases and magnetic moment configurations depending on ultrathin film thickness, lattice parameter, crystal structure and temperature [7.94–97].

The magnetic moment of ultrathin Fe films have been studied for a range of interface materials and for different crystallographic phases and orientations. These structures are often fabricated by atom deposition from the vapour phase onto the substrate in ultrahigh vacuum conditions by heteroepitaxy. This is possible for a range of materials that exhibit a close crystallographic match to the Fe lattice, such as the (100) surface of Ag, where the Fe layer grows in the bcc phase, with the (001) plane rotated by 45° relative to the Ag(001) (lattice mismatch of 0.8%). When deposited on Cu(001), the Fe adopts a distorted fcc phase for the first few monolayers. These films are single crystalline, as opposed to the polycrystalline films that result from evaporation onto an amorphous or polycrystalline surface. Epitaxial films have a much larger crystallite size and a preferential crystal orientation, although some degree of texture (deviation from the ideal crystallographic structure) is always present in macroscopic films. They exhibit therefore less structural defects than polycrystalline films and have an effective magnetic anisotropy which is usually averaged out (and therefore absent) in polycrystalline films. Needless to say, the magnetic properties depend sensitively on the preparation conditions, since contaminants can change the magnetic and/or structural properties of thin films, and can therefore mask the intrinsic properties of the system under study. In particular it is important to rule out chemical reaction or interdiffusion between the magnetic film and the substrate. In fact, one of the motivations behind the study of the Fe system on Ag and Cu substrates using PNR was the possibility of growing high quality, optically flat films (with interface roughness of ~15 Å and correlation lengths of the order of 200 Å for continuous Fe films [7.98, 99]) with limited interdiffusion at the interface [7.100–102] (although with quite complex structural changes for the case of Fe/Cu [7.100, 101, 103–109] and in both cases complete coverage of the substrate is achieved only for films thicker than ~5 ML, as deduced from the appearance of RHEED oscillations at this stage of growth [7.101, 102, 104, 108, 110]).

Table 7.3 gives a compilation of the values of the magnetic moment of Fe for different contact interfaces and also different crystallographic phases measured by polarised neutron reflection. The very early studies [7.111–115] consist chiefly of preliminary demonstrations of the capabilities of the PNR technique for measuring

Table 7.3. Values for the magnetic moment of Fe thin films for different interface materials (thickness in ML unless specified otherwise)

Sample	phase	μ^{Fe} (μ_B)	T (K)	Ref.
15 Å Au/12 Å Fe/55 Å Au/glass [a]	bcc, poly.	$1.4^{+1.2}_{-0.6}$	300	[7.112]
50 Å Cu/5 Fe/ Rh(001)	fcc	< 0.1	300	[7.111]
30 Å Au/8 Fe/Cu(001)	fcc	0.15	300	[7.111]
Cu/3 Fe/Cu(001)	fcc	1-1.5	4	[7.113]
150 Å Ag/1 Fe/Ag(001)	bcc	0 [b]	4	[7.115]
90 Å Ag/8 Fe/Ag(001)	bcc	1.0	4	[7.114, 115]
20 Au/7 Ag/5.5 Fe/Ag(001)	bcc	2.58 ± 0.09	4	[7.99, 116–119]
20 Au/7 Ag/10.9 Fe/Ag(001)	bcc	2.33 ± 0.05	4	"
52 Au/ 5.7 Fe/Ag(001)	bcc	2.5 ± 0.1	4	"
20 Au/9 Fe/Ag(001)	bcc	2.3 ± 0.2	4	"
20 Au/7 Cu/ 5.8 Fe/Ag(001)	bcc	2.48 ± 0.08	4	"
42 Au/8 Cu/ 5.7 Fe/Ag(001)	bcc	2.5 ± 0.1	4	"
20 Au/7 Pd/ 5.6 Fe/Ag(001)	bcc	2.66 ± 0.05	4	"
42 Au/7 Pd/ 5.7 Fe/Ag(001)	bcc	2.6 ± 0.2	4	"
24 Au/3 Ni/ 5 Fe/Ag(001)	bcc	2.6 ± 0.1 [c]	4	"
200 Å Au/4 Å Fe/MgO(001)	bct	0 (2.2 ± 0.2)	RT (40)	[7.132]
200 Å Au/6 Å Fe/MgO(001)	bct	< 0.5 (2.2 ± 0.2)	RT (40)	"
200 Å Au/8 Å Fe/MgO(001)	bct	2.0 (2.2 ± 0.2)	RT (40)	"
100 Å W/6 Å Fe/W(110) [d]	bcc	1.80 ± 0.05	RT	[7.129]
70 Å Au/Fe/140 Å Ni/SiO_2 [e]	fcc	~0.2	RT	[7.128]
70 Å Au/Fe/140 Å Ni/SiO_2 [f]	bcc	~2.1	RT	"
6 Å Fe(110)/V(110) [g]	bcc	0	80	[7.133, 134]
10 Å Fe/V(110)	bcc	1.3	80	"
19.5 Å Fe/V(110)	bcc	1.8	110, 300	"
30 Å Au/315 Å Cr /5.6 Å Fe/ V(110)	bcc	1.1	2	[7.135]
300 Å Cr/8.5 Å Fe / V(110)	bcc	1.5	2	[7.135]
300 Å Cr/12 Å Fe / V(110)	bcc	2.0	2	[7.135]

[a] $P_{growth} \sim 10^{-7}$ Torr.
[b] PMA.
[c] $\mu^{Ni} = 0.6 \pm 0.1\ \mu_B$.
[d] 100 Å W/6 Å Fe(110)/550 Å W/$Al_2O_3(11\bar{2}0)$; $\mu^{Fe} = 2.1 \pm 0.1\ \mu_B$ at 0 K, extrapolated from SQUID data.
[e] $t^{Fe} = 7, 13, 17, 24, 26$ Å; $\mu^{Ni} \approx 0.5\ \mu_B$.
[f] $t^{Fe} = 49, 53, 84, 87$ Å; $\mu^{Ni} \approx 0.5\ \mu_B$.
[g] *In situ* PNR.

the magnetic moments of ultrathin films. Due to limitations of the beam intensity and instrumental efficiency, the data statistics were often limited, and the data analysis was further complicated by roughness effects (namely diffuse scattering). For severe roughness (i.e., large amplitude, short parallel correlation length) the spin asymmetry is strongly perturbed and cannot be fitted by the Névot-Croce formalism, as in the early studies of Ag/Fe/Ag(001) [7.114, 115].

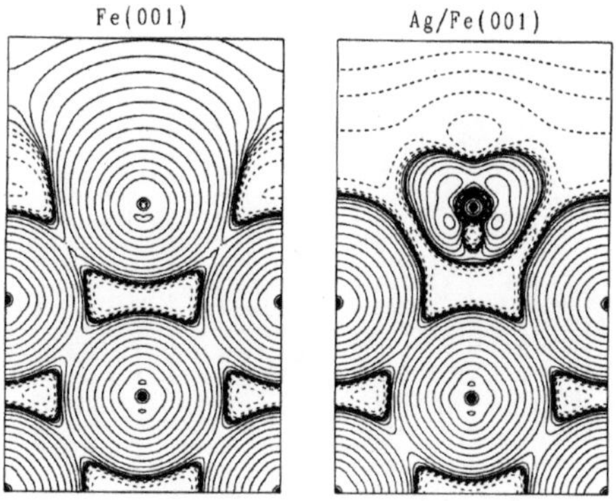

Fig. 7.11. Total spin densities for the Fe(001) and the Ag/Fe(001) surfaces. Solid (dotted) lines mark the zero and positive (negative) 2^n ($n = 0, 1, 2...$) contours in units of 10^{-4} electrons/a.u. While the spin density for the layers below the interface are identical for both surfaces, at the interface it is significantly modified by the presence of the Ag overlayer (after [7.120])

More reliable results were possible after some of these limitations were overcome, by improvements in both the instrument efficiency and sample quality. This was the case with the systematic study of the magnetic moment of bcc Fe/Ag(001) in contact with different materials, Ag, Au, Cu, Pd and Ni [7.99, 116–119]; in all cases the magnetic moment per atom of Fe is found to be significantly enhanced with respect to the bulk value, in agreement with ferromagnetic resonance (FMR) measurements performed on the same samples [7.99, 118, 119]. Figure 7.12 shows

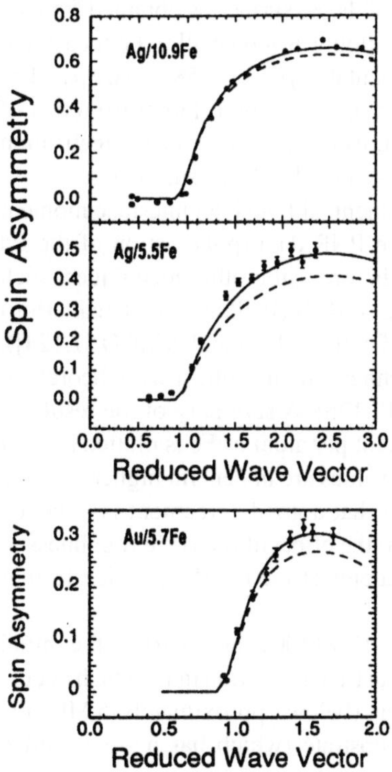

Fig. 7.12. Spin asymmetry obtained from reflectivity measurements at low temperature (4 K) for (*top panel*) 20 ML Au/7 ML Ag/10.9 ML Fe/Ag(001), (*middle panel*) 20 ML Au/7 ML Ag/5.5 ML Fe/Ag(001), and (*bottom panel*) 52 ML Au/ 5.7 ML Fe/Ag(001). The dashed and solid lines correspond to model fits assuming a bulk and enhanced moment for the Fe films, respectively. (After [7.99])

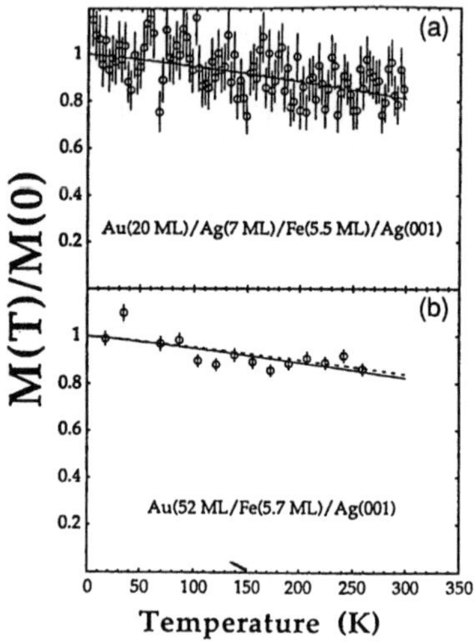

Fig. 7.13. Normalised temperature dependence of the saturation magnetisation determined from PNR for (a) the Ag/5.5 ML Fe, (b) Au/5.7 ML Fe samples (after [7.99])

the spin-asymmetry data for three of the samples studied, where the best fit to the data is compared with that expected from the bulk moment values. For the Ag/5.5 ML Fe/Ag(001) sandwich, Ohnishi et al. [7.120] predicted a moment of 0.08 μ_B for the interface Ag layer and a layer averaged moment per Fe atom of 2.4 μ_B (Fig. 7.11 shows the effect of the Ag overlayer on the spin density of the Fe surface as obtained from ab initio calculations [7.120]); while the Ag polarisation was too small to be measured, the value for Fe is close to that observed experimentally, $\mu^{Fe} = 2.58 \pm 0.09\,\mu_B$. The interface moment of Fe in contact with Au is not significantly different from that of the Ag/Fe/Ag structure. In Fig. 7.13 the temperature dependence of the magnetisation for these two samples is shown, as determined from the PNR measurements.

The Pd/Fe sample exhibits the largest net moment, although an interface moment for the Pd layer could not be excluded from the PNR data (polarisation of the Pd atoms at the Pd/Fe interface has been reported before using other techniques, such as VSM (vibrating sample magnetometry) [7.121], BLS (Brillouin light scattering) and FMR (ferromagnetic resonance) [7.122], SQUID [7.123] and XMCD [7.124]). For the Cu/Fe system, the moment of Fe is still enhanced, in contrast with theoretical predictions of a reduction in the moment of Fe [7.125]. A summary of the results is presented in Fig. 7.14, where the magnetic moment per atom of Fe is plotted against the Fe thickness. It is seen that the experimental values are in general higher than the predicted values for the 5–6 ML thick films, but that a good agreement is obtained for the $\sim$ 10 ML films [7.99]. This is attributed to the effect of interface roughness in reducing the effective local coordination of Fe atoms at the interface which further enhances the magnetic moment in ultrathin films.

For Fe grown on fcc substrates such as Cu(001), Rh(001) and Ni(001) the initial studies were complicated by the complex structural transformation in Fe which occurs with increasing thickness. For very thin Fe films (below approximately 5 ML) the Fe adopts the fct structure of the underlying substrate (which has a small lattice mismatch to the lattice parameter of the bulk fcc phase of Fe, as inferred from

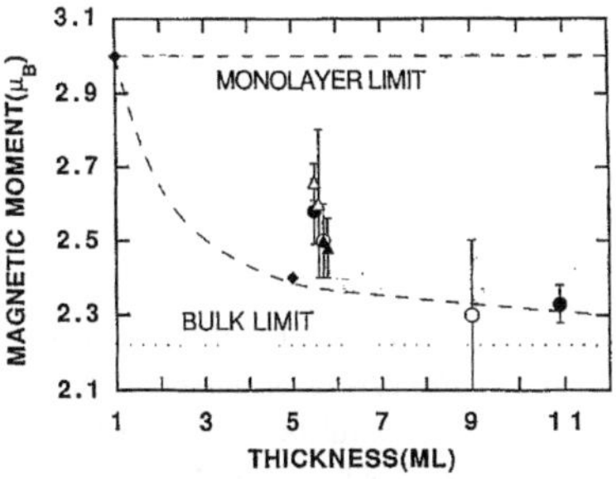

Fig. 7.14. Values for the layer averaged moment per bcc Fe atom deduced from PNR measurements reported in [7.99], compared with the predictions by [7.120] for the layer averaged moment of a 1 ML Fe/Ag(001) and 5 ML Fe/Ag shown as solid diamonds. The Fe/Ag data are shown as full circles, the Cu/Fe as solid triangles and the Pd/Fe as open triangles. The dashed line is a guide to the eye only (after [7.99])

the lattice constant value extrapolated from high temperatures, $a^{\mathrm{fcc\,Fe}} = 3.59$ Å [7.105, 126, 127]) and before reverting to the equilibrium bcc phase (above 11 ML), it is believed to pass through a non-magnetic, non-strained fcc phase [7.108, 127] (see however [7.109]. This behaviour was only later understood, and the early works already reported the strongly reduced magnetism of thin fcc Fe films [7.111, 113]. The study by Li et al. [7.128] on thin Fe films on Ni using PNR focused on the evolution of the Fe moment with increasing thickness. The polycrystalline 140 Å thick Ni films were deposited on SiO_2 in ultrahigh vacuum, and Fe films from 7 to 83 Å were subsequently deposited and capped with a thick Au layer. It is observed that the magnetic moment of Fe increases abruptly at ~ 32 Å from 0.2 μ_B to the bulk value (see Fig. 7.15). These authors used x-ray reflectivity measurements to

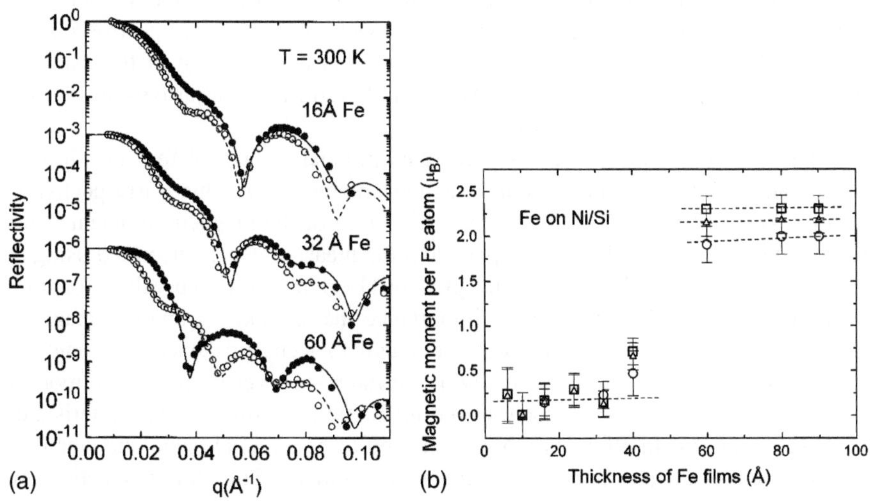

Fig. 7.15. **(a)** Spin polarised neutron reflectivities for Fe films sandwiched between a Au overlayer and the Ni substrate, for different values of the Fe film thickness. The full and open symbols correspond to spin up and down reflectivities respectively, and the lines are fit to the data. **(b)** Variation of the Fe magnetic moment with thickness. Circles correspond to PNR data while triangles and squares are from SQUID magnetometry, at 300 K and 5 K, respectively. (After [7.128])

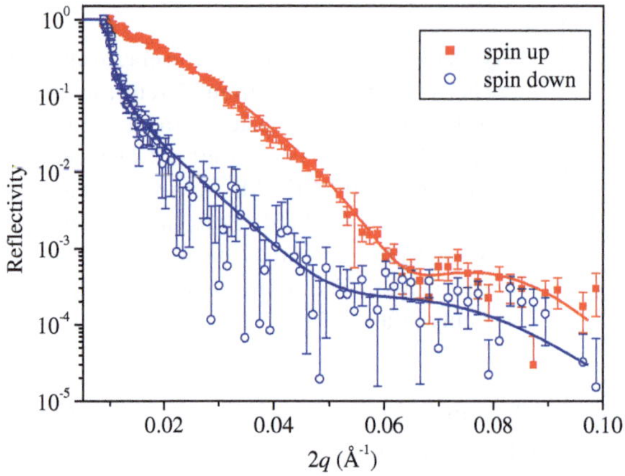

Fig. 7.16. Polarised neutron reflection of a 100 Å thick epitaxial bcc Fe(100) film deposited on InAs(001) and capped with a 30 Å Au film. Continuous lines are fit to the data and give a magnetic moment close to the bulk

determine the layer thickness of the films, and compared the x-ray results with the PNR results, with an overall good agreement between the two techniques.

Pasyuk et al. [7.129] have studied the magnetic moment of a 6 Å epitaxial bcc Fe(110) film buried between W(110) layers using PNR at room temperature and measured the temperature dependence of the magnetisation using SQUID magnetometry to determine the absolute ground state magnetic moment of the Fe film. The value obtained, 2.1 μ_B, slightly reduced compared to the bulk value (2.2 μ_B), agrees with theoretical calculations predicting slightly reduced moments for a monolayer of Fe(110) on W(110) due to hybridisation effects [7.130] but not with previous experimental results obtained by conversion-electron Mössbauer spectroscopy on a Fe/W(110) monolayer capped with Ag, where an enhancement of the magnetic moment was observed [7.131] (and where a slight reduction was also expected from the same theoretical study [7.130]).

Other substrates have been used for the epitaxial growth of thin Fe films. In particular, MgO(100) (where a bulk magnetic moment was found, irrespective of film thickness [7.132]), V(110) (where a reduction of the magnetic moment is observed, with the Fe film being uncapped and the measurements performed using an *in situ* PNR set-up [7.133, 134]), and InAs(001) (giving bulk like values for thick Fe films, see Fig. 7.16). Due to the importance that epitaxial growth of magnetic films on semiconductors may have in future magnetoelectronic devices, the magnetic characterisation of such films specially at the interface level is specially important and PNR is expected to contribute to the magnetic and structural characterisation of these structures. In all cases, PNR is showed to be an ideal technique when the measurement of the absolute value of magnetic moment of ultrathin magnetic films is required.

7.3.1.2 Magnetic Moment of Thin Co Films

The magnetic moment of thin Co films has also been extensively studied using PNR over the last decade. Co also exhibits several metastable crystalline phases, which can be experimentally stabilised by epitaxy. One system of great interest is the fcc phase of Co, which is obtained when Co is grown on fcc substrates with lattice constant close to the fcc equilibrium lattice value (as obtained from extrapolation from the high temperature fcc phase of Co down to low temperatures [7.136]). This stems from the fact that the moment of fcc Co (1.739 μ_B) is expected to be different from the equilibrium hcp value (1.708 μ_B) [7.137] and also ab initio calculations predict enhanced moments for some interfacial systems, such as Cu/Co [7.138–140] and Pd/Co [7.119, 141]. PNR results for the Cu/Co/Cu system have shown that the magnetic moment of Co is essentially bulk like [7.111, 113, 142–147], although a slight enhancement of the moment is not incompatible with the PNR data (enhanced Co/Cu(001) moments have been obtained from other experiments [7.148–150]). *In situ* SQUID magnetometry has been successfully used to measure the absolute moment of ultrathin fcc Co/Cu(001) films by Ney et al. [7.151–153] where they found an enhancement in the Co magnetic moment of 1.87(3) μ_B/atom

Table 7.4. Values for the magnetic moment of Co thin films for different interface materials (thickness in ML unless specified otherwise)

Sample	phase	μ^{Co} (μ_B)	T (K)	Ref.
42 Å Cu/10 Co/Cu(001)	fcc	1.8 ± 0.5	4	[7.111, 142]
40 Å Cu/18 Å Co/Cu(001)	fcc	1.8 ± 0.3	4	[7.113]
40 Å Cu/2 Co/Cu(001)	fcc	2.1 ± 0.3	4	"
80 Å Au/100 Å Co/GaAs(001)	bcc	1.40 ± 0.05 [a]	300	[7.167]
30 Å CoO/150 Å Co/GaAs(001) [b]	bcc [c]	1.6 [d]	300	[7.168, 169]
222 Å Ag/2 Co/222 Å Ag/GaAs(001)	fcc	2.10 ± 0.15	5	[7.114, 165, 166]
222 Å Ag/1 Co/222 Å Ag/GaAs(001)	fcc	2.15 ± 0.2	5	[7.165, 166]
125 Å Ag/2 Co/Ag(001)	fcc	2.05 ± 0.15	5	"
230 Å Ag/3 Co/230 Å Ag/GaAs(001)	fcc	1.65 ± 0.15	5	"
160 Å Ag/5 Co/Ag(001)	fcc	0.45 ± 0.1	5	"
70 Å Pd/21 Å Co/345 Å Pd/ 215 Å Au/Si(001)-H [e]	?	1.84		[7.51, 170]
1450 Å $Co_{81}Cr_{19}$/quartz	hcp	0.6 [f]		[7.173]
45 Å Cu/9 Å Co /50 Å Ni/Cu/Si(001)	fcc	1.70 ± 0.20	RT	[7.144, 145]
40 Å Cu/23 Å Co/10 Å Cu/ 53 Å Ni/Cu/Si(001)	fcc	1.57 ± 0.08	RT	[7.146]
300 Å $Ni_{80}Fe_{20}$/80 Å Co/ 180 Å Mo/Al_2O_3(1100)	hcp	1.5 ± 0.2		[7.174]
93 Å Cu/9 Å Co/830Å Cu/Si(001)	fcc	1.78 ± 0.14	RT	[7.147]
93 Å Cu/9 Å Co/62 Å Ni/830 Å Cu/Si(001)	fcc	1.71 ± 0.23	RT	[7.147]

[a] $\mu^{Co} = 1.7\ \mu_B$ (centre), $\mu^{Co} = 1.0\ \mu_B$ (interface).
[b] $p_{growth} \sim 10^{-8}$ Torr.
[c] Early growth stages.
[d] $\mu^{Co} = 1.7\ \mu_B$ (centre); $\mu^{Co} = 0.8\ \mu_B$ (interface).
[e] Also with Si(110)-H and SiO_2 substrates.
[f] $\mu_{bulk} = 0.5\ \mu_B$.

from the bulk value of 1.69(1) μ_B/atom for a 2 ML Co/Cu(001) film [7.151]. The effect of adding a Cu overlayer is to reduce the total magnetic moment [7.152].

Another system that has been studied extensively is the Co/Ni system, for which a strong perpendicular magnetic interface anisotropy has been predicted [7.154] for the (111) orientation, while for the (100) orientation this is overwhelmed by a strain and magnetostatic contribution from the Co layer [7.155–158]. In particular, the determination of the magnetic anisotropy constants requires the knowledge of the value of the saturation magnetisation. Lauhoff et al. [7.144] have performed PNR measurements on a Cu/9 Å Co/50 Å Ni/Cu/Si(001) structure to find that the magnetic moment of the Co layer, $\mu^{Co} = 1.70 \pm 0.20\ \mu_B$, was essentially bulk like. The PNR spectra for this structure are shown in Fig. 7.17. These data are representative of the data quality that can be achieved nowadays routinely at the CRISP reflectometer (U.K), where the measurements were performed. Typically such spectra require 8–10 h counting time and measurements are performed at 2–3 different angles of incidence (most of the counting time being spent at the larger q-values, for which the reflected intensities fall to very small values). For this particular system, the magnetic moment is predicted not to change significantly across the Co/Ni interface [7.93, 159], as the PNR results seem to confirm [7.143–147].

Body centred tetragonal (bct) Co has also been grown by epitaxy on Ag(001) [7.114, 160], GaAs(110) [7.161, 162] and GaAs(001) [7.163, 164] substrates. The magnetic moment of bct Co/Ag(001) using neutron reflection showed that for a 1 and

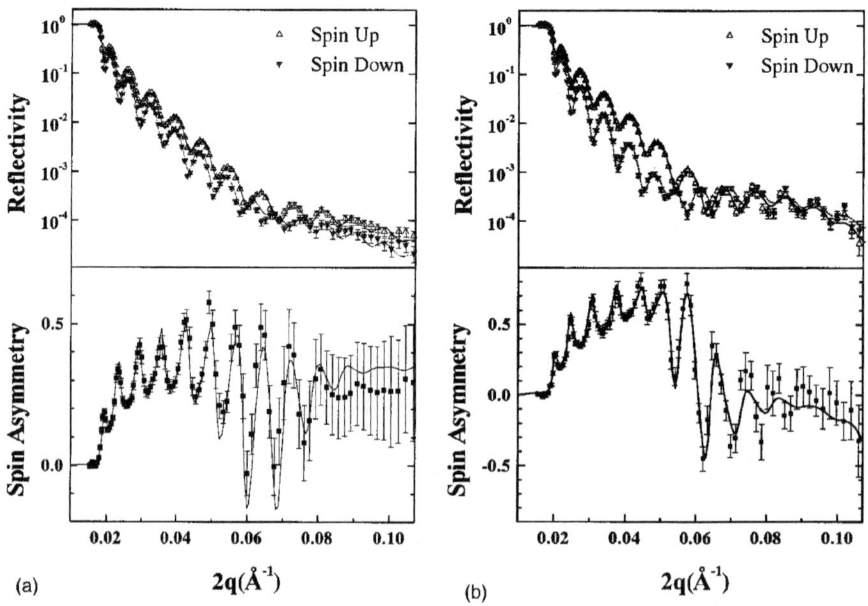

Fig. 7.17. Polarised neutron reflectivity and spin asymmetry data (symbols) and their best fits to data (continuous lines) for **(a)** a 45 Å Cu/50 Å Ni/686 Å Cu/Si(001) and **(b)** a 45 Å Cu/9Å Co/ 50 Å Ni/686 Å Cu/Si(001) structure (after [7.144])

2 ML Co film the magnetic moment is enhanced relative to the bulk value, but that the moment decreases rapidly as the thickness increases, an effect that was attributed to strain-induced disorder in the Co films [7.165, 166]. For the Co films grown on GaAs, strong interdiffusion at the interface decreases the average magnetic moment, which is found to be bulk like in the centre of the film [7.35, 167–169].

Studies of the magnetic moment of Co films sandwiched between Pd layers, using PNR magnetometry, have been reported by Pasyuk et al. [7.51, 170], where an enhanced moment of 1.84 μ_B is obtained, compared with the bulk value of 1.73 μ_B. Unfortunately, no information about the crystalline structure of the Co film is given nor is the temperature of measurement mentioned. The results seems to agree with numerical calculations that predict an enhanced moment for this system [7.141, 171, 172].

7.3.1.3 Magnetic Moment of Thin Ni Films

The study of the magnetic moment of thin Ni films, particularly for the Cu/Ni interface, has been studied also in some detail. The interest here resides in the unusual magnetic properties of the Ni/Cu(001) system, exhibiting dominant perpendicular magnetic anisotropy over a thickness range from 15 to 120 Å, which manifests itself in a state of perpendicular magnetisation at remanence. Reduced magnetic moments at the Cu/Ni interface have been predicted by numerical calculations (due to hybridisation effects) [7.140, 175–178], but experimentally a larger reduction in the moment of Ni has been observed for Cu/Si(001) substrates [7.149, 179–182], larger than that predicted by interface hybridisation effects and also not explainable by interdiffusion at the Ni/Cu interface. In particular, it is observed that the magnetic moment of Ni increases with thickness but in a thickness range where the interface

Table 7.5. Values for the magnetic moment of Ni thin films for different interface materials (thickness in ML unless specified otherwise)

Sample	phase	μ^{Ni} (μ_B)	T (K)	Ref.
Cu/Ni/Cu/Si(001)-H	fcc	[c]	RT	[7.179]
20 Å Cu/51 Å Ni/500 Å Cu/Si(001)	fcc	0.50 ± 0.02 [a]	RT	[7.188]
20 Å Cu/51 Å Ni/770 Å Cu/Si(001)	fcc	0.53 ± 0.02 [b]	RT	[7.188]
50 Å Cu/40 Å Ni/Cu/Si(001)	fcc	0.54 ± 0.03	RT	[7.183]
180 Å Cu/40 Å Ni/Cu/Si(001)	fcc	0.45 ± 0.03	RT	"
45 Å Cu/50 Å Ni/Cu/Si(001)	fcc	0.53 ± 0.03	RT	[7.144–146]
45 Å Cu/9 Å Co/50 Å Ni/Cu/Si(001)	fcc	0.57 ± 0.03	RT	"
45 Å Cu/23 Å Co/10 Å Cu/50 Å Ni/Cu/Si(001)	fcc	0.50 ± 0.04	RT	"
93 Å Cu/62 Å Ni/Cu/Si(001)	fcc	0.54 ± 0.03	RT	[7.147]
93 Å Cu/9 Å Co/62 Å Ni/Cu/Si(001)	fcc	0.50 ± 0.05	RT	"

[a] 0.55 ± 0.03 (XMCD).
[b] 0.58 ± 0.03 (XMCD).
[c] A variation in the Ni moment per atom with thickness is reported.

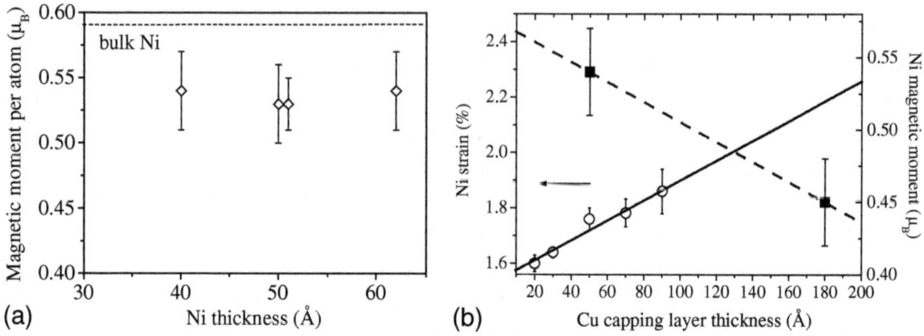

Fig. 7.18. (**a**) Variation of the Ni magnetic moment per atom as a function of the film thickness in Cu/Ni/Cu/Si(001) samples. (**b**) Variation of the Ni magnetic moment (from PNR) and of the strain in the Ni layer (from grazing incidence x-ray diffraction) for different values of the Cu overlayer thickness (measurements at room temperature). (After [7.183])

effects were expected not to contribute much to a moment variation [7.179–181]. This effect has been attributed to strain in the Ni layer, which has the effect of decreasing the magnetic moment per atom of Ni. This is also suggested by experiments where the Cu overlayer in a Cu/Ni/Cu(001) structure is varied, where the sample with the thicker Cu overlayer exhibits a smaller magnetic moment; this is concomitant with an increase of the Ni strain (observed in identical structures), see Fig. 7.18 [7.183].

A variation in the magnetic moment of Ni with strain has been predicted theoretically but is too small to account for the experimental results [7.175, 184]. Although Wu et al. [7.185] present a variation in the magnetic moment per Ni atom that is

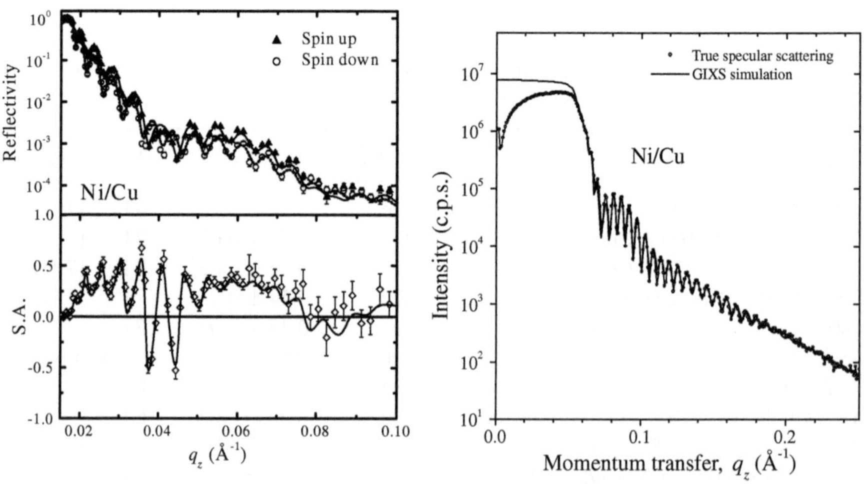

Fig. 7.19. Polarised neutron reflectivity and spin asymmetry data (*symbols*) and their best fits to data (*continuous lines*) for a 45 Å Cu/9Å Co/ 50 Å Ni/686 Å Cu/Si(001) structure. The value of the magnetic moment per Ni atom obtained from the fit to the data is $0.54 \pm 0.03\ \mu_B$

consistent with the variation reported in [7.183], the calculations refer to uniform expansions of the fct primitive cell along the *c*-axis, while in practice it is the in-plane lattice constant of Ni that is changed while the out of plane lattice parameter is free to adjust to the value that minimises the free energy of the system (usually, tending to keep constant the volume of the primitive cell). In Fig. 7.19a we show a recent PNR measurement on a Cu/60 Å Ni/Cu/Si(001) structure, where the structural parameters were compared to those obtained from x-ray reflectometry (Fig. 7.19b). An identical reduction in the magnetic moment of Ni films grown on Cu/Si(001) substrates has been recently reported independently by Gubbiotti et al. [7.186]. However, magnetometry studies on Ni/Cu(001) thin films using an *in situ* SQUID technique and XMCD has not corroborated this large decrease in the magnetic moment of Ni [7.187] with thickness, except for the decrease from the moment reduction at the Ni/Cu interface. In this case however, strain is reduced as a single crystal of Cu is used [7.187].

7.3.1.4 PNR on Other Magnetic Films

The previous three systems mentioned above account for most of the neutron reflectivity studies as far as the accurate determination of the magnetic moment of thin films is concerned, but other systems have also been studied with this technique, often with a view to determining the magnetisation profile across a relatively thick magnetic layer (e.g., Fe_3O_4 and γ-Fe_2O_3 thick films [7.189], strained thick Ni films [7.190], and NiFe films [7.191]). In these studies, the reflectivity data of the magnetic layer is fitted with a depth-dependent magnetic nuclear potential, which may be non-uniform due to interdiffusion between the substrate and the magnetic film, oxidation of the top surface or variations in stoichiometry due to the fabrication process [7.192, 193]. Hoffmann et al. [7.194] have investigated magnetic proximity effects in NiO/Pd with PNR with negative results. In this study, the nuclear scattering length density of the NiO layer was matched to that of Pd by a suitable mixture of Ni isotopes in order to enhance the PNR magnetic contrast.

Another example concerns the study of the magnetic moment of Cr in a Ag/Cr/Ag(001) sandwich in the submonolayer coverage regime, by Johnson et al. [7.195]. In this study, PNR measurements on a 0.33 ML Cr film yielded a magnetic moment per atom that was enhanced relative to the bulk Cr magnetic moment per atom (0.59 μ_B), while a thicker 3.3 ML film yielded no net magnetic moment (consistent with the antiferromagnetic phase of Cr).

Other systems have been successfully studied using PNR (or other neutron polarised techniques) namely systems with artificially depth dependent helicoidal magnetisation profiles, such as those obtained in exchange-spring magnets [7.189, 196] or growth induced initial magnetic configurations [7.192, 193].

A system which could prove to be of particular interest is the Fe/Co interface system. Fe and Co alloy to give an increased average magnetic moment up to 2.30 μ_B at 30% Co concentration, an effect caused by the progressive filling of the majority $3d$ spin band of the Fe [7.198]. Recent ab initio calculations have predicted for bcc

Fe/Co interface an enhancement of the magnetic moment of Fe from 2.23 μ_B (bulk) to 2.60 μ_B [7.93], an increase of ~18%.

7.3.2 Spin-valve Systems

The relevance of spin-valve systems hardly needs to be emphasised, so conspicuous have been their use in technological applications. There is therefore a strong incentive for understanding the physical mechanisms that govern the performance of a given spin-valve system. Due to the large number of parameters that determine the structure of such systems, a large amount of work remains to be done in terms of performance optimisation, involving systematic studies of the many material and physical parameters that define these structures. This is discussed in detail by Gurney et al. [7.197] and we consider here only the contribution PNR has made in the understanding of these systems. In fact, although much work has been devoted to the study of spin-valve systems (namely by magnetoresistance and conventional magnetometry), few techniques are able to determine the depth profile magnetisation of the magnetic layer stack, or whether they are in a state of uniform magnetisation. It is in addressing these questions that PNR has been employed.

For convenience, we discuss separately two classes of systems, multilayers, corresponding to the simplest spin-valve structures (2 to 5 magnetic layers) and superlattices, which consist of a periodic repetition of a basic magnetic/spacer bilayer. The former, due to their simpler structure, allow a study of the magnetic behaviour of each individual magnetic layer (as far as PNR is concerned). The latter are more difficult to study and uniform magnetic and structural properties of the different layers are often assumed, a simplification necessary for the interpretation of the data, but not always the case in reality. Another reason why this division is convenient is because these structures are studied in different scattering regimes, the low wavevector range close to the critical edge for the multilayer systems, and the high wavevector range (the diffraction limit) for the superlattice systems (this also has repercussions in the kind of information that can be obtained in these different systems/scattering regimes: structural/magnetic information for each individual layer in the first case, and structural/magnetic information for the superlattice as a whole in the second case).

The simplest spin-valve systems consist of two magnetic layers separated by a non-magnetic (metallic) spacer. Spin-valve systems derive their designation from their electric properties, which exhibit a high and low resistance states according to the relative orientation of the adjacent magnetic layers, thus acting as a spin-operated valve for the electric current (giant magnetoresistance effect, GMR [7.199, 200]). The magnetic layers may be of different materials and different thicknesses; it may be such that the coupling between the magnetic layers is antiferromagnetic at remanence (the coupling between two magnetic layer oscillates with the thickness of the non-magnetic spacer [7.201], an effect that has been described in terms of the RKKY model [7.202] or more generally, in terms of quantum interference due to confinement in ultrathin layers [7.203, 204]). The GMR effect varies roughly with the cosine of the relative angle of the magnetisation between adjacent magnetic layers for

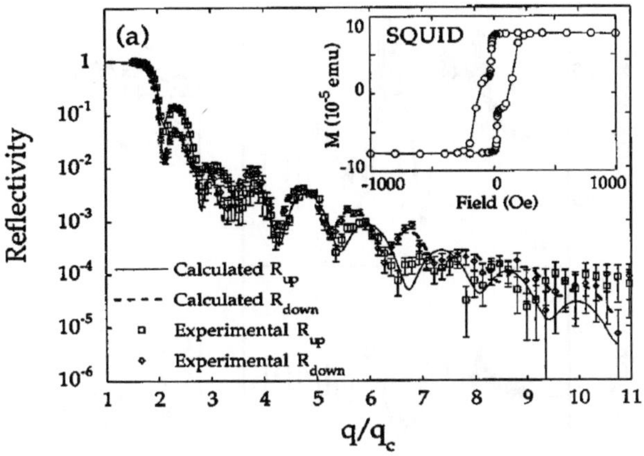

Fig. 7.20. PNR spectra for the Si(001)/Cu/FeNi/Cu/Co/Cu epitaxial trilayer structure held at 100 K with the layer magnetisations aligned parallel by an applied field of 3 kG. The lines correspond to fits to the data. The inset shows the magnetic hysteresis loop for the spin-valve structure obtained by SQUID magnetometry. (After [7.216])

the current in-plane geometry (CIP) [7.205–208] (but deviates from this behaviour for the current perpendicular to plane (CPP) geometry [7.209, 210] and in general a more complex behaviour is expected from theory [7.211–215]). The minimum in resistance is attained for the parallel configuration, and the maximum for the antiparallel configuration. We see therefore that in order to maximise the GMR effect, it is important to know what is the relative orientation of the magnetisation (which is determined by a competition between the intrinsic magnetocrystalline anisotropy axes, the exchange coupling between the adjacent layers, domain splitting at remanence and also pinning defects that can trap magnetic domains in local energy minima).

Polarised neutron reflection studies have been performed on some spin-valve systems in order to study some of these mechanisms. Bland et al. [7.216–218] have studied a single FeNi/Cu/Co trilayer spin-valve using PNR (in combination with other techniques) in order to obtain the magnetic moment profile across the interfaces of the trilayer, and also the orientation of the magnetisation for various values of the applied magnetic field. The spin-valve experiment is important since in general domain formation can occur, complicating the data analysis [7.12, 219]. In contrast to the case of multilayers, where it is assumed that equivalent layers have identical spin configurations, the magnetic moment of each individual layer can be controlled accurately in the single domain limit. The reflectivity data for the saturated state is shown in Fig. 7.20 while Fig. 7.21 shows the spin asymmetry spectra as a function of the relative orientation of each layer magnetisation. The soft NiFe layer acts as the free layer, and can be rotated towards the direction of the applied magnetic field for fields not strong enough to change the direction of the magnetisation of the Co layer from its easy axis. These data clearly illustrates the sensitivity of PNR to the magnetisation orientation of the magnetisation, even when polarisation analysis is not

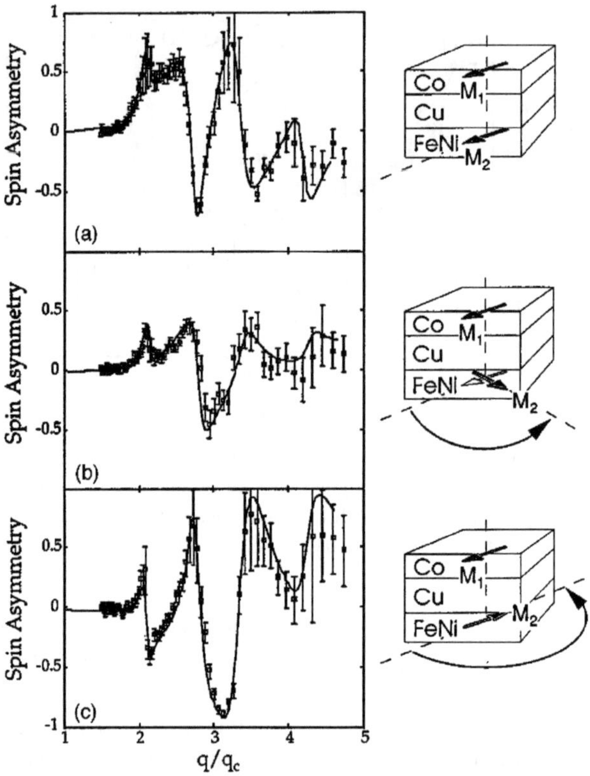

Fig. 7.21. The spin asymmetry spectra as a function of the relative layer magnetization orientation at room temperature for **(a)** parallel alignment and after the sample is rotated with respect to the applied field by approximately **(b)** 90° and **(c)** 180° causing the FeNi layer to rotate as shown in the schematic (right insets). The lines correspond to fits to the data. (After [7.216])

employed (the cost is that of knowing the angle of the magnetisation only within 180° relative to the direction of the quantisation axis but this is compensated by improved data statistics).

A number of systematic studies of the vector configuration in two double spin-valve systems has been reported by Samad et al. [7.220, 221] for a NiFe/Cu/Co/Cu/NiFe structure and by Choi et al. [7.222] for a Co/Cu/NiFe/Cu/Co structure. Both systems were grown epitaxially on Cu(001)/Si(001) buffer layers, and the orientation of the magnetisation in each layer as a function of the magnetic field in a *M-H* loop cycle was studied using PNR. Figure 7.22 (left) shows the data for the Co/Cu/NiFe/Cu/Co spin-valve in the saturated state, exemplifying the data quality that is possible to achieve in these kind of systems. The magnetic moments thus deduced for the Co and NiFe layers were similar to the bulk values, within experimental error. In Fig. 7.22 (right) the PNR spectra for a non-collinear configuration of the layer magnetisations is shown, for the same structure. In this case, after saturation along one field di-

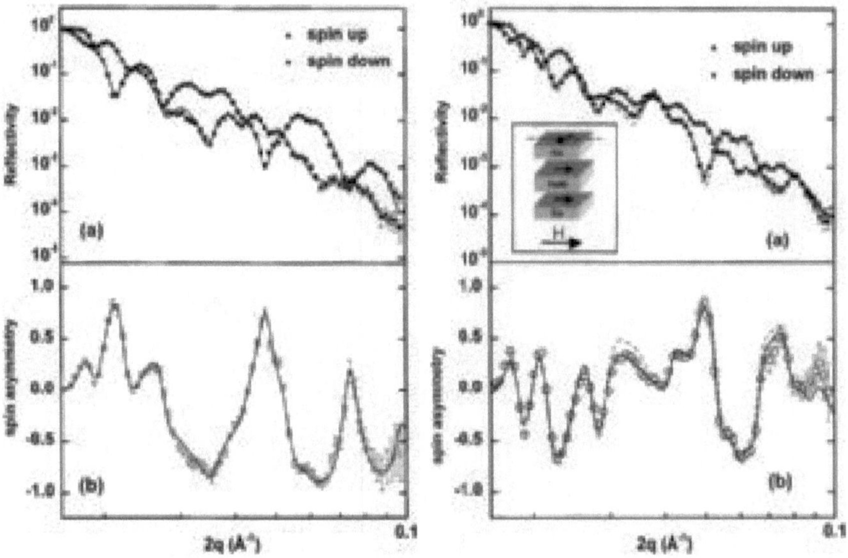

Fig. 7.22. PNR spectra for the Co/Cu/NiFe/Cu/Co double spin-valve for two magnetic states. *Left:* **(a)** Spin dependent reflectivity (R) and **(b)** spin asymmetry as a function of perpendicular momentum transfer for the saturated state (magnetisation aligned along the fcc Co [110] axis). *Right:* The field strength is set to +100 Oe after saturation in the negative direction, so that only the bottom Co layer flips toward the applied field. The magnetisation vectors are shown schematically in the inset. In all plots, the solid lines correspond to fits to the experimental data; the broken lines in the right panel are the simulated PNR spectrum corresponding to a multidomain configuration: magnetisation aligned along the negative field direction but with a reduced effective Co moment (After [7.222])

rection, the magnetic field is reversed until the magnetisation of the NiFe and that of the bottom Co layer (which exhibits a smaller switching field) flip towards the field direction. The magnetisation of the two bottom magnetic layers is aligned with the magnetic field, while the top Co layer makes an angle of 36° with respect to the negative direction of the applied field, as schematised in the figure inset. It is to be noticed that without spin polarisation analysis the mirror image configuration (i.e., that forming an angle of −36°) cannot be distinguished, but that a domain-split structure giving the same value of the magnetic moment $M_s \times \cos(36°)$ would give a different PNR spectrum, as shown by the simulated spectra in the figure (dashed line).

PNR data for the NiFe/Cu/Co/Cu/NiFe spin-valve is shown in Fig. 7.23 for different magnetic configurations, namely for the saturated ferromagnetic state and after successive rotation of 90° at an applied field of 50 Oe (sufficient to move the soft NiFe layers only). The PNR data shows that collinear configurations are not obtained in these cases, a result that is in agreement with the angular dependent CIP-GMR measurements and that was attributed to domain formation [7.220, 221].

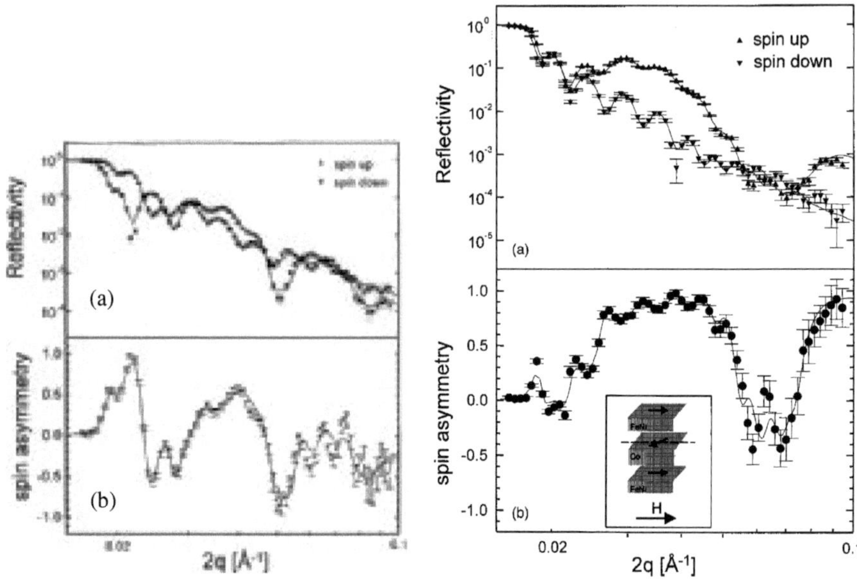

Fig. 7.23. PNR spectra for the NiFe/Cu/Co/Cu/NiFe double spin-valve for two magnetic states. *Left:* **(a)** Spin dependent reflectivity (R) and **(b)** spin asymmetry as a function of perpendicular momentum transfer for the saturated state (magnetisation aligned along the fcc Co [110] axis). *Right:* After saturation along the $[\bar{1}\bar{1}0]$ direction, a reversed applied field of 50 Oe is applied along the [110] direction. The magnetisation vectors are shown schematically in the *i*nset. In all plots, the solid lines correspond to fits to the experimental data (after [7.220])

The study of exchange bias (the unidirectional anisotropy induced in a ferromagnet in contact with an antiferromagnet) has also been studied with PNR. Parkin et al. [7.191] studied the magnetisation profile of an exchange biased NiFe/FeMn film using PNR to show that there is no evidence of a deviation from a uniform distribution of the magnetisation across the NiFe film, which has a sharp interface with the antiferromagnetic layer, therefore suggesting that non-uniform magnetisation states (such as planar magnetic domains and domain walls in the ferromagnet) and interface defects (such as intrusions of the FM into the AF or vice versa), are not responsible for the observed low exchange field values (compared with the values expected from a purely exchange coupling mechanism for the exchange anisotropy). The absence of magnetic domains in NiFe in a similar structure was also reported by Ball et al. [7.223] using PNR. However, for an exchange biased Fe_3O_4/NiO multilayer, PNR measurements for the two oppositely saturated magnetic states reveal different magnetic states which could be ascribed to the presence of a interfacial domain wall in the ferromagnetic layer. Felcher et al. [7.224] have considered the problem of determining the magnetic state of a NiFe exchange biased interface layer in contact with a NiCoO AF layer, in order to determine whether the coupling mechanism is through exchange coupling. For this purpose PNR spectra were taken at two points of the hysteresis curve corresponding to the two oppositely saturated states (at the same applied field value). The sum of the spin asymmetry for these two magnetic states should be non-zero in the case where an interface NiFe layer remains exchange coupled to the AF. The experimental results seem to suggest that this is the case, but the errors associated with the data do not allow a definitive conclusion. In fact, the magnetic configuration at the AF-FM interface is thought to be more complex, with local variations of the AF interface crystal plane (due to a distribution of grain orientations and grain boundaries across the film and due to film roughness) and possibly also the presence of domain walls in the AF film. As a consequence, only a relatively small number of uncompensated AF spins contributes to the exchange coupling with the FM spins, but other energy contributions may introduce other types of interactions between the AF and the FM layer, namely the "spin-flop" coupling [7.225, 226]. Velthuis et al. [7.227] have studied the effect of training (cycling of the *M-H* loop) in Co/CoO exchange biased structures using PNR, to show that after the first training cycle a spin-dependent diffuse scattering of the Yoneda type sets in, which appears due to the presence of magnetic domains oriented differently from the main magnetisation. In the same study the rotation of the magnetisation with an external magnetic field applied perpendicular to the unidirectional anisotropy axis was studied, to confirm that the magnetisation rotates coherently and uniformly throughout the Co layer thickness for small values of the magnetic field (it has been shown that an alternative method of measuring the exchange anisotropy is through small deviations of the magnetisation from the unidirectional anisotropy axis, without breaking the FM/AF exchange bias; this method yields values for the exchange anisotropy that is larger by a factor of 2 in Co/CoO bilayers than the conventional method of measuring the exchange bias field from the shift in the hysteresis loop [7.228, 229]).

Fitzsimmons et al. have used PNR in combination with SQUID and MR (magnetoresistance) to study the switching mechanisms in exchange biased Fe/MnF_2 and Fe/FeF_2 systems [7.230–234]. They found that Fe films grown on epitaxial Fe/FeF_2 and Fe/MnF_2(110) antiferromagnets (AF) with in-plane twin domains exhibit different switching behaviour for the two sides of the *M-H* characteristic (by coherent rotation in one case and domain wall motion in the other) when the samples are cooled with the applied field along the direction that bisects the anisotropy axis of the AF twin domains, and in this case the exchange bias fields are greatly enhanced [7.230, 231, 234]. This behaviour is not observed when the AF is polycrystalline or single crystal (all with the same out of plane (110) texture) [7.234]. This effect was attributed to an additional anisotropy axis in the ferromagnetic film introduced by the F-AF interface.

Graaf et al. [7.235] have studied the orientation of the layer magnetic moment in a NiFe/Cu/NiFe/FeMn exchange biased spin-valve (consisting of several repeats) as a function of the applied magnetic field to compare with a theoretical model developed by Rijks et al. [7.236] that estimates the magnetisation angle of the free NiFe layer as a function of several parameters, such as interlayer exchange coupling, exchange bias and Cu interlayer thickness. The experimental results (consisting of data for two values of the Cu spacer layer, and therefore for the interlayer coupling interaction) seem to agree with that model.

Velthuis et al. [7.237–239] have approached the exchange bias problem by fabricating an exchange biased system consisting of an exchange coupled Fe/Cr double superlattice, one with a AF interlayer coupling and the other with a FM interlayer coupling (this is achieved by suitably choosing the Cr spacing thickness). This system closely resembles an idealised exchange biased system with collinear spins and where the AF/FM interface problems are avoided (such as roughness and the effect of the AF material). The collinear configuration of the magnetic layers is confirmed by PNR, and the magnetic properties of this system compare well with the classical Meiklejohn-Bean model [7.240].

7.3.3 Experimental Results on Superlattice Systems

Superlattice systems have been extensively studied with PNR, mainly in the diffraction limit. For wavelengths comparable to those of the superlattice repetition period, a diffraction scattering process occurs, which leads to an enhanced peak in the scattered intensity. Since the magnetic potential is comparable to the nuclear potential, the magnetic superstructure also gives rise to peaks in the scattered intensity, namely for the antiferromagnetic configuration, where a half-order peak is observed (double the period of the superlattice, and therefore, half the wavevector value, compared to the position of the first order peak).

The Fe/Cr multilayer system has received much attention, due to the important role it played in the discovery of the GMR effect [7.199, 200] and of the antiferromagnetic exchange coupling between transition metal layers [7.241] (it is worth noting that this effect was first observed in Gd/Y multilayers using polarised neutron diffraction [7.242]). The complexities associated with this system (e.g., the influence of the structural parameters, such as interface roughness [7.243, 244]) also have the consequence that this system has been more investigated than others. The role played by the Cr layer, an antiferromagnet below $T_N = 311$ K (bulk Néel temperature) has also been investigated using polarised neutron scattering [7.245].

PNR offers a very direct method of confirming the parallel or antiparallel alignment of the magnetisation in a multilayer system, and the order of alignment of the magnetic layers can be easily monitored as a function of the magnetic field. Barthélémy et al. [7.247] have confirmed the existence of antiferromagnetic coupling in a Fe/Cr superlattice for Cr thicknesses thinner than 30 Å, by measuring the spin-flip intensity variation with the applied magnetic field at the antiferromagnetic scattering peak. While these structures were single-crystal grown by MBE, Parkin et al. [7.248] have studied polycrystalline sputtered superlattices of similar nominal structures, to show that they exhibit identical properties as the MBE samples, using PNR to identify the antiferromagnetic coupling and saturation fields. Bland et al. [7.21, 219] studied a Fe/Cr/Fe sandwich with antiferromagnetic coupling by PNR with polarisation analysis and determined the degree to which the magnetic alignment is fully antiparallel, to conclude that a multidomain structure may be present in the structure (see also [7.249]). The effect of annealing on the spin configuration of Fe/Cr superlattices has been studied by Hahn et al. [7.53, 250] using PNR, and

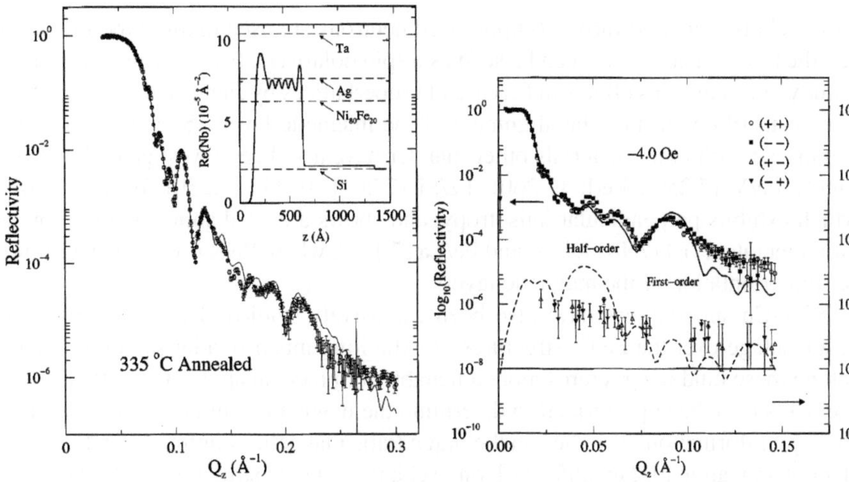

Fig. 7.24. *Left:* x-ray reflectivity of a [20 Å NiFe/40 Å Ag]$_4$ multilayer annealed at 335 °C. Circles correspond to the data and the solid line correspond to the fit. *Right:* corresponding polarised neutron reflectivity spectra, prepared in a field of −4 kOe after saturation in a 150 Oe field, in order to produce the demagnetising field. The circles and squares correspond to the non-spin-flip cross sections, and the triangles correspond to the spin-flip cross sections (shifted by a factor of 100 for clarity). The solid line corresponds to the fit to the data, and the dashed line represents the scattering expected for antiferromagnetically aligned NiFe layers with large in-plane domains assuming that the moments align perpendicular to the applied field (after [7.246])

related this effect with the magnetoresistance data to show that the GMR is proportional to the degree of antiferromagnetism, while Pechan et al. [7.251] have studied the magnetic profile as a function of structural disorder (by deposition at increasing Ar pressure) to find that the magnetic superlattice peak does not degrade appreciably (and therefore to conclude that the magnetic coherence length is significantly greater than that of the chemical superlattice structure). Schreyer et al. [7.252, 253] have reported an extensive study on the magnetic and structural properties of MBE grown Fe/Cr superlattices grown at different growth temperatures. They found that for samples grown at 250 °C a coupling angle of 50° is obtained from the PNR data (using polarisation analysis). This non-collinear coupling has been reported by later studies [7.254, 255] and while Adenwalla et al. [7.254] attribute it to the biquadratic interlayer coupling term, Schreyer et al. [7.245, 252, 253, 255] have explained this behaviour in terms of the proximity magnetism model [7.256] based on the antiferromagnetism of the Cr layer induced by the adjacent ferromagnetic layers. Temst et al. [7.257] report a low-field (∼230 Oe) spin-flop transition in Fe/Cr superlattices where, from an antiferromagnetic configuration, the moments antiparallel to the field align, at this critical field, to a direction perpendicular to the field direction, and are then progressively oriented towards the field direction as this is increased. This is seen as a sudden change in the spin-flip intensity at low fields and a gradual decrease as the field is increased further (unpolarised neutrons are used in this experiment; in fact, the first magnetic layer can be seen as a spin polariser). PNR has also been used to study exchange bias effects in Fe/Cr(211) superlattices by Jiang et al. [7.239, 258], as a means of confirming the alignment of the magnetic layers. Superlattices of Fe interspaced with other materials other than Cr have also been investigated by PNR, namely Fe/V [7.259], Fe/Ir [7.260], Fe/Si [7.261], Fe/Gd [7.250], Fe/Tb [7.262] (which exhibits perpendicular anisotropy, and shows a canted spin configuration at low temperatures), Fe/Nb [7.263] and Fe/La [7.193], where PNR is used to determine the spin orientation of the magnetic layers.

NiFe/Cu superlattices have also been extensively studied. These systems have the advantage over the Fe/Cr structures of exhibiting much smaller saturation fields making these kind of structures more amenable for practical applications. PNR measurements have been performed to determine the magnetic configuration, as well as structural information in studies on the biquadratic magnetic coupling [7.264], on the effect of annealing [7.246, 265, 266], on exchange coupled structures [7.235] (where the magnetic alignment is studied as a function of the applied field). Similar PNR studies have been reported for Co/Cu [7.17, 267–273], Co/Re [7.274, 275], Co/Pd [7.276], Co/Ru [7.277] and NiCo/Cu [7.278] multilayers.

7.4 Conclusions

We have illustrated the capabilities of polarised neutron reflection magnetometry for the accurate determination of the magnetic moment of ultrathin magnetic films (in the dynamical regime) and in determining the relative orientation of magnetic moment of buried layers in multilayer systems (in the diffraction regime). In particular, quantitative studies of the magnetic moment in single transition metal layers using PNR are demonstrated, with emphasis on the effect of the interface material on the magnetic moment. Such studies are the result of improvements both in the quality of the samples and in improvements in the neutron reflectometers (increased neutron flux and improved instrumentation) which has enabled on the one hand avoiding problems related with diffuse scattering (e.g., by using optically flat substrates) and on the other, improved data statistics. For the case of spin-valve structures we have shown how PNR can make vectorial magnetometry possible, even when polarisation analysis is not employed. It is shown that these systems can exhibit non-collinear magnetic configurations at low applied fields. Interface moments are key to devices, e.g., magnetic tunnel junctions, and therefore PNR may prove an important technique in future on account of its sensitivity to both magnetic and structural interface details. We also show that most of the studies have been limited to a small number of magnetic systems, suggesting that PNR has been under-used. It is our hope that the present review will increase the interest in this technique specifically when accurate determination of the magnetic moment and structure characterisation are required simultaneously.

References

7.1. S. S. Dhesi, H. A. Dürr, G. van ver Laan, E. Dudzik, and N. B. Brookes. *Phys. Rev. B*, 60:12852, 1999.
7.2. J. Störr. *Journal of Electron Spectroscopy and Related Phenomena*, 75:253, 1995.
7.3. J. Störr. *J. Magn. Magn. Mat.*, 200:470, 1999.
7.4. M. Tolan and W. Press. *Z. Kristallogr.*., 213:319, 1998.
7.5. G. P. Felcher, R. Felici, R. T. Kampwirth, and K. E. Gray. *J. Appl. Phys.*, 57:3789, 1985.
7.6. G. P. Felcher. *Phys. Rev. B*, 24:1595, 1981.
7.7. U. Gradmann. Magnetism in ultrathin transition metal films. In K. H. J. Buschow, editor, *Handbook of Magnetic Materials*, volume 7, page 1. Elsevier Science Publishers B. V., 1993.
7.8. J. F. Ankner and G. P. Felcher. *J. Magn. Magn. Mater.*, 200:741, 1999.
7.9. G. P. Felcher. *J. Appl. Phys.*, 87:5431, 2000.
7.10. H. Zabel, R. Siebrecht, and A. Schreyer. *Physica B*, 276-278:17, 2000.
7.11. J. A. C. Bland, J. Lee, S. Hope, G. Lauhoff, J. Penfold, and D. Bucknall. *J. Magn. Magn. Mat.*, 165:46, 1997.
7.12. J. A. C. Bland. *J. Vac. Sci. Technol. A.*, 15:1759, 1998.
7.13. C. F. Majkrzak. *Physica B*, 221:342, 1996.
7.14. H. Zabel. *Appl. Phys. A*, 58:159, 1994.
7.15. H. Zabel, K. Theis–Bröhl. *J. Phys.: Condens. Matter*, 15:S505, 2003.
7.16. J. Penfold. *Physica B*, 173:1, 1991.
7.17. D. E. Joyce, C. A. Faunce, P. J. Grundy, B. D. Fulthorpe, T. P. A. Hase, I. Pape, and B. K. Tanner. *Phys. Rev. B*, 58:5594, 1998.
7.18. G. P. Felcher, R. O. Hilleke, R. K. Crawford, J. Haumann, R. Kleb, and G. Ostrowski. *Rev. Sci. Instrum.*, 58:609, 1987.
7.19. S. J. Blundell and J. A. C. Bland. *Phys. Rev. B*, 46:3391, 1992.
7.20. S. J. Blundell and J. A. C. Bland. *J. Mag. Mag. Mater.*, 121:185, 1993.
7.21. J. A. C. Bland, R. D. Bateson, A. D. Johnson, S. J. Blundell, V. S. Speriosu, S. Metin, B. A. Gurney, and J. Penfold. *J. Magn. Magn. Mat.*, 123:320, 1993.
7.22. L. Koester. Neutron scattering lengths and fundamental neutron interactions. In G. Höhler, editor, *Neutron Physics*, volume 80 of *Springer Tracts in Modern Physics*. Springer-Verlag, 1977.
7.23. E. Fermi. *Ric. Scientifica*, 7:13, 1936. See: E. Fermi, Collected Papers, Ed: E. Segré, (University of Chicago Press, 1962) p. 980.
7.24. H. A. Bethe. *Rev. Mod. Phys.*, 9:69, 1937.
7.25. R. Golub, D. Richardson, and S. K. Lamoreaux. *Ultra-Cold Neutrons*. Adam Hilger, 1991.
7.26. V. F. Sears. *Neutron Optics*. Oxford University Press, 1989.
7.27. H. Dachs. Principles of neutron diffraction. In H. Dachs, editor, *Neutron Diffraction*, page 1. Springer-Verlag, 1978.
7.28. S. W. Lovesey. *Theory of neutron scattering from condensed matter*, volume 1. Oxford Universi Press, Oxford, 1984.
7.29. L. Dobrzynski and K. Blinowski. *Neutrons and Solid State Physics*. Ellis Horwood, 1994.
7.30. X.-L. Zhou and S.-H. Shen. *Physics Reports*, 257:223, 1995.
7.31. J. M. F. Gunn. Introductory theory of neutron scattering. In R. J. Newport, B. D. Rainford, and R. Cywinski, editors, *Neutron Scattering at a Pulsed Source*, pages 1–107. Adam Hilger, Bristol, 1988.

7.32. O. Halpen and M. H. Johnson. *Phys. Rev.*, 55:898, 1939.
7.33. R. P. Hunt. *J. Appl. Phys.*, 38:1652, 1967.
7.34. S. D. Bader and J. L. Erskine. Magneto-optical effects in ultrathin magnetic structures. In B. Heinrich and J. A. C. Bland, editors, *Ultrathin magnetic structures*, volume II, page 297. Springer-Verlag, Berlin Heidelberg, 1994.
7.35. S. J. Blundell, M. Gester, J. A. C. Bland, H. J. Lauter, V. V. Pasyuk, and A. V. Petrenko. *Phys. Rev. B*, 751:9395, 1995.
7.36. V. F. Sears. Neutron scattering lengths and cross sections. In Sköld and Price [7.279], page 521.
7.37. A. G. Klein and S. A. Werner. *Rep. Prog. Phys.*, 46:259, 1983.
7.38. J. Penfold. Neutron optics. In R. J. Newport, B. D. Rainforf, and R. Cywinski, editors, *Neutron Scattering at a Pulsed Source*, page 287. Adam Hilger, 1988.
7.39. S. J. Blundel. *Spin-dependent transport in artificial structures*. PhD thesis, University of Cambridge, 1993.
7.40. J. A. C. Bland. Polarised neutron reflection. In J. A. C. Bland and B. Heinrich, editors, *Ultrathin magnetic structures*, volume I, page 305. Springer-Verlag, Berlin Heidelberg.
7.41. C. Fermon, F. Ott, and A. Menelle. Neutron reflectometry. In J. Daillant and A. Gibaud, editors, *X-Ray and Neutron Reflectivity: Principles and Applications*, page 163. Springer, Berlin, 1999.
7.42. A. Galindo and P. Pascual. *Quantum Mechanics I.* Springer-Verlag, 2nd edition, 1989.
7.43. A. Messiah. *Quantum mechanics I.* North-Holland, 1961.
7.44. E. Fermi and W. H. Zinn. *Phys. Rev.*, 70:103, 1946.
7.45. E. Fermi and L. Marshall. *Phys. Rev.*, 71:666, 1947.
7.46. J. B. Hayter, J. Penfold, and W. G. Williams. *Nature*, 262:569, 1976.
7.47. J. R. Lu and R. K. Thomas. *Nucl. Instr. Meth. Phys. Res. A*, 354:149, 1995.
7.48. P. S. Pershan. *Phys. Rev. E*, 50:2369, 1994.
7.49. J. Kasper, H. Leeb, and R. Lipperheide. *Phys. Rev. Lett.*, 80:2614, 1998.
7.50. P. Krams, F. Lauks, R. L. Stamps, B. Hillebrands, and G. Güntherodt. *Phys. Rev. B*, 52:10827, 1995.
7.51. V. Pasyuk, H. J. Lauter, M. T. Johnson, F. J. A. den Broader, E. Janssen, J. A. C. Bland, and A. V. Petrenko. *Appl. Surf. Sci.*, 65-66:118, 1993.
7.52. J. Penfold, J. Webster, and D. G. Bucknall. *The use of polarisation analysis in polarised neutron reflection studies*. Science and Engineering Research Council, RAL-93-018 edition, 1993.
7.53. W. Hahn, M. Loewenhaupt, G. P. Felcher, Y. Y. Huang, and S. S. P. Parkin. *J. Appl. Phys.*, 75:3564, 1993.
7.54. N. K. Pleshanov. *Z. Phys. B*, 94:233, 1994.
7.55. G. P. Felcher, S. Adenwalla, V. O. De Haan, and A. A. Van Well. *Nature*, 377:409, 1995.
7.56. D. A. Korneev, V. I. Bodnarchuk, and V. K. Ignatovich. *JETP Lett.*, 63:944, 1996.
7.57. H. Fredrikze, T. Rekveldt, A. van Well, Y. Nikitenko, and V. Syromyatnikov. *Physica B*, 248:157, 1998.
7.58. R. W. E. van de Kruijs, H. Fredrikze, M. Th. Rekveldt, A. A. van Well, Yu. V. Nikitenko, and V. G. Syromyatnikov. *Physica B*, 283:189, 2000.
7.59. C. Fermon. *Physica B*, 213-214:910, 1995.
7.60. A. Rühm, B. P. Toperverg, H. Dosch *Phys. Rev. B*, 60:16,073, 1999.
7.61. M. Wormington, I. Pape, T. P. A. Hase, B. K. Tanner, and D. K. Bowen. *Phil. Mag. Lett.*, 74:211, 1996.

7.62. L. Névot and P. Croce. *Revue Phys. Appl.*, 15:761, 1980.
7.63. R. A. Cowley and T. W. Ryan. *J. Phys. D*, 20:61, 1987.
7.64. S. K. Sinha, E. B. Sirota, S. Garoff, and H. B. Stanley. *Phys. Rev. B*, 38:22977, 1988.
7.65. R. Pynn. *Phys. Rev. B*, 45:602, 1992.
7.66. F. de Bergevin. Interaction of X-rays (and neutrons) with matter. In J. Daillant and A. Gibaud, editors, *X-Ray and Neutron Reflectivity: Principles and Applications*, page 3. Springer-Verlag, Berlin, 1999.
7.67. F. de Bergevin, J. Daillant, A. Gibaud, and A. Sentenac. The treatment of roughness in specular reflectivity. In J. Daillant and A. Gibaud, editors, *X-Ray and Neutron Reflectivity: Principles and Applications*, page 116. Springer-Verlag, Berlin, 1999.
7.68. A. Steyerl. *Z. Physik*, 254:169, 1972.
7.69. R. D. Bateson, G. W. Ford, J. A. C. Bland, H. J. Lauter, B. Heinrich, and Z. Celinski. *J. Magn. Magn. Mat.*, 121:189, 1993.
7.70. J. Penfold and R. K. Thomas. *J. Phys. Condens. Matter*, 2:1369, 1990.
7.71. G. P. Felcher. *Physica B*, 267-268:154, 1999.
7.72. R. Felici, J. Penfold, R. C. Ward, and W. G. Williams. *Nucl. Instr. and Meth. A*, 260:309, 1987.
7.73. A. Schreyer, R. Siebrecht, U. Englisch, U. Pietsch, and H. Zabel. *Physica B*, 248:349, 1998.
7.74. A. Schreyer, R. Siebrecht, U. Englisch, U. Pietsch, and H. Zabel. *Physica B*, 241-243:34169, 1998.
7.75. Institut Laue-Langevin, Scientific Coordination Office. *The ILL Yellow Book*, 2001.
7.76. B. Nickel, A. Rühm, W. Donner, J. Major, H. Dosh, A. Schreyer, H. Zabel, and H. Humblot. *Rev. Sci. Instrum.*, 72:163, 2001.
7.77. H. Doschand K. Al Usta, A. Lied, W. Drexel, and J. Peisl. *Rev. Sci. Instrum.*, 63:5533, 1992.
7.78. C. Fermon, F. Ott, H. Glätti, and G. Saux. *C. R. Acad. Sci. Paris IIB*, 322:479, 1996.
7.79. D. Clemens, P. Gross, P. Keller, N. Schlumpf, and M. Könnecke. *Physica B*, 276-278:140, 2000.
7.80. V.-O. de Haan, J. de Blois, P. van der Ende, H. Fredrikze, A. van der Graaf, M. N. Schipper, A. A. van Well, and J. van der Zanden. *Nucl. Instr. and Meth. A*, 362:434, 1995.
7.81. A. A. van Well, V.-O. de Haan, and H. Fredrikze. *Physica B*, 198:217, 1994.
7.82. F. Mezei, R. Golub, F. Klose, and H. Toews. *Physica B*, 213-214:898, 1995.
7.83. M. Stamm, S. Hüttenbach, and G. Reiter. *Physica B*, 173:11, 1991.
7.84. U. Rücker, B. Alefeld, W. Bergs, E. Kentzinger, and Th. Brückel. *Physica B*, 276-278:95, 2000.
7.85. Los Alamos Neutron Science Center. *Instrument highlight: ASTERIX.*
7.86. M. Vedpathak, S. Basu, and S. K. Kulkarni. *Appl. Surf. Sci.*, 115:317, 1997.
7.87. G. S. Lodha, S. Basu, A. Gupta, S. Pandita, and R. V. Nandedkar. *Phys. Stat. Sol. (a)*, 163:415, 1997.
7.88. Council for the Central Laboratory of the Research Councils. *ISIS 98, the ISIS facility Annual Report 1997-98*, 1998.
7.89. J. M. Carpenter and W. B. Yelon. Neutron sources. In Sköld and Price [7.279], page 99.
7.90. V. Nunez, A. T. Boothroyd, J. Penfold, S. Langridge, D. G. Bucknall, P. Boni, D. Clemens, and M. Senthil Kumar. *Physica B*, 241-243:148, 1998.
7.91. D. G. Bucknall and S. Langridge. *CRISP Instrument Manual.* Council for the Central Laboratory of the Research Council, Technical Report RAL-TR-97-022 edition, 1997.

7.92. S. Blügel and P. H. Dederichs. *Europhys. Lett.*, 9:597, 1989.
7.93. A. M. N. Niklasson, B. Johansson, and H. L. Skriver. *Phys. Rev. B*, 59:6373, 1999.
7.94. D. Bagayoko and J. Callaway. *Phys. Rev. B*, 28:5419, 1983.
7.95. A. J. Freeman, C. L. Fu, T. Oguchi, and M. Weinert. Electronic and magnetic structure of solid surfaces. In R. Feder, editor, *Polarized electrons in surface physics*, page 3. World Scientific, Singapore, 1985.
7.96. V. L. Moruzzi, P. M. Marcus, K. Schwarz, and P. Mohn. *Phys. Rev. B*, 34:1784, 1986.
7.97. P. M. Marcus and V. L. Moruzzi. *J. Appl. Phys.*, 63:4045, 1988.
7.98. Z. Celinski, B. Heinrich, and J. F. Cochran. *J. Appl. Phys.*, 73:5966, 1993.
7.99. J. A. C. Bland, C. Daboo, B. Heinrich, Z. Celinski, and R. D. Bateson. *Phys. Rev. B*, 51:258, 1995.
7.100. D. A. Steigerwald, I. Jacob, and Jr. W. F. Egelhoff. *Surface Science*, 202:472, 1988.
7.101. D. P. Pappas, C. R. Brundle, and H. Hopster. *Phys. Rev. B*, 45:8169, 1992.
7.102. P. Xhonneux and E. Courtens. *Phys. Rev. B*, 46:556, 1992.
7.103. M. T. Kief and Jr. W. F. Egelhoff. *Phys. Rev. B*, 47:10785, 1993.
7.104. F. J. Himpsel. *Phys. Rev. Lett.*, 67:2363, 1991.
7.105. M. Wuttig and J. Thomassen. *Surf. Sci.*, 282:237, 1993.
7.106. J. Thomassen, B. Feldmann, and M. Wuttig. *Surface Science*, 264:406, 1992.
7.107. M. Wuttig, B. Feldmann, J. Thomassen, F. May, H. Zillgen, A. Brodde, H. Hannemann, and H. Neddermeyer. *Surface Science*, 291:14, 1993.
7.108. D. Li, M. Freitag, J. Pearson, Z. Q. Qiu, and S. D. Bader. *Phys. Rev. Lett.*, 72:3112, 1994.
7.109. J. Shen, Z. Gai, J. Kirschner. *Surface Science Reports*, 52:163, 2004.
7.110. B. Heinrich, Z. Celinski, J. F. Cochran, A. S. Arrot, and K. Myrtle. *J. Appl. Phys.*, 70:5769, 1991.
7.111. J. A. C. Bland, D. Pescia, and R. F. Willis. *Phys. Rev. Lett.*, 358:1244, 1987.
7.112. J. A. C. Bland, D. Pescia, R. F. Willis, and O. Schaërpf. *Physica Scripta*, 35:528, 1987.
7.113. R. F. Willis, J. A. C. Bland, and W. Schwarzacher. *J. Appl. Phys.*, 63:4051, 1988.
7.114. J. A. C. Bland, A. D. Johnson, C. Norris, and H. J. Lauter. *J. Appl. Phys.*, 67:5397, 1990.
7.115. J. A. C. Bland, A. D. Johnson, H. J. Lauter, R. D. Bateson, S. J. Blundell, C. Shackleton, and J. Penfold. *J. Magn. Magn. Mat.*, 93:513, 1991.
7.116. J. A. C. Bland, R. D. Bateson, B. Heinrich, Z. Celinski, and H. J. Lauter. *J. Magn. Magn. Mat.*, 104-107:1909, 1992.
7.117. J. A. C. Bland, R. D. Bateson, A. D. Johnson, B. Heinrich, Z. Celinski, and H. J. Lauter. *J. Magn. Magn. Mat.*, 93:331, 1991.
7.118. J. A. C. Bland, C. Daboo, B. Heinrich, Z. Celinski, E. E. Fullerton, K. Ounadjela, and D. Stoeffler. *J. Magn. Magn. Mat.*, 148:85, 1995.
7.119. E. E. Fullerton, Stoeffler, K. Ounadjela, B. Heinrich, Z. Celinski, and J. A. C. Bland. *Phys. Rev. B*, 51:6364, 1995.
7.120. S. Ohnishi, M. Weinert, and A. J. Freeman. *Phys. Rev. B*, 30:36, 1984.
7.121. F. J. A. den Broeder, H. C. Donkersloot, H. J. G. Draaisma, and W. J. M. de Jonge. *J. App. Phys.*, 61:4317, 1987.
7.122. Z. Celinski, B. Heinrich, J. F. Cochran, W. B. Muir, A. S. Arrott, and J. Kirschner. *Phys. Rev. Lett.*, 65:1156, 1990.
7.123. J. R. Childress, R. Kergoat, O. Durand, J.-M. George, P. Galtier, J. Miltat, and A. Schuhl. *J. Magn. Magn. Mat.*, 130:13, 1994.
7.124. J. Vogel, A. Fontaine, V. Cross, F. Petroff, J.-P. Kappler, and G. Krill. *Phys. Rev. B*, 55:3663, 1997.

7.125. C. L. Fu and A. J. Freeman. *Phys. Rev. B*, 35:925, 1987.

7.126. S. Müller, P. Bayer, C. Reischl, K. Heinz, B. Feldmann, H. Zilgen, and M. Wuttig. *Phys. Rev. Lett.*, 74:765, 1995.

7.127. Z. Q. Qiu and S. D. Bader. *J. Magn. Magn. Mater.*, 200:664, 1999.

7.128. Y. Li, C. Polaczyk anf F. Klose, J. Kapoor, H. Maletta, F. Mezei, and D. Riegel. *Phys. Rev. B*, 53:5541, 1996.

7.129. V. Pasyuk, O.F.K. Mc Grath, H. J. Lauter, A. Petrenko, A. Liénard, and D. Givord. *J. Magn. Magn. Mat.*, 148:38, 1995.

7.130. S. C. Hong, A. J. Freeman, and C. L. Fu. *Phys. Rev. B*, 38:12156, 1988.

7.131. H. J. Elmers, G. Liu, and U. Gradmann. *Phys. Rev. Lett.*, 63:566, 1989.

7.132. Y. Y. Huang, C. Liu, and G. P. Felcher. *Phys. Rev. B*, 47:183, 1993.

7.133. T. Nawrath, H. Fritzsche, and H. Maletta. *J. Magn. Magn. Mat.*, 212:337, 2000.

7.134. T. Nawrath, H. Fritzsche, F. Klose, J. Nowikow, and H. Maletta. *Phys. Rev. B*, 60:9525, 1999.

7.135. H. Fritzsche, T. Nawrath, H. Maletta, H. Lauter. *Physica*, 241–243:707, 1998.

7.136. W. B. Pearson. *Lattice spacings and structures of metals and alloys*, volume 2. Pergamon Press, Oxford, 1967.

7.137. S. Chikazumi. *Physics of Ferromagnetism*. Clarendon Press, Oxford, 2nd edition, 1997.

7.138. C. Li, A. J. Freeman, and C. L. Fu. *J. Magn. Magn. Mater.*, 83:51, 1990.

7.139. M. Alden, S. Mirbt, H.L. Skriver, N.M. Rosengaard, and B. Johansson. *Phys. Rev. B*, 46:6303, 1992.

7.140. O. Hjortstam, J. Trygg, J.M. Wills, B. Johnansson, and O. Eriksson. *Phys. Rev. B*, 53:9204, 1996.

7.141. S. Blügel, B. Drittler, R. Zeller, and P. H. Dederichs. *Appl. Phys. A*, 49:547, 1989.

7.142. D. Pescia, R. F. Willis, and J. A. C. Bland. *Surface Science*, 189-190:724, 1987.

7.143. G. Lauhoff, J. Lee, J. A. C. Bland, J. Ph. Schillé, and G. van der Laan. *J. Magn. Magn. Mater.*, 177-181:1253, 1998.

7.144. G. Lauhoff, J. A. C. Bland, J. Lee, S. Langridge, and J. Penfold. *Phys. Rev. B*, 60:4087, 1999.

7.145. G. Lauhoff, J. Lee, J. A. C. Bland, S. Langridge, and J. Penfold. *J. Magn. Magn. Mater.*, 198-199:331, 1999.

7.146. G. Lauhoff, A. Hirohata, J. Lee, J. A. C. Bland, S. Langridge, and J. Penfold. *J. Phys. Condens. Matter*, page 6707, 1999.

7.147. C. A. F. Vaz, G. Lauhoff, J. A. C. Bland, S. Langridge, D. Bucknall, J. Penfold, J. Clarke, S. K. Halder, and B. K. Tanner. Interface dependent magnetic moments in Cu/Co,Ni/Cu/Si(001) epitaxial structures. *Unpublished*.

7.148. M. Tischer, O. Hjortstam, D. Arvanitis, J.H. Dunn, F. May, K. Baberschke, J. Trygg, J.M. Willis, B. Johansson, and O. Eriksson. *Phys. Rev. Lett.*, 75:1602, 1995.

7.149. P. Srivastava, F. Wilhelm, A. Ney, M. Farle, H. Wende, N. Haack, G. Ceballos, and K. Baberschke. *Phys. Rev. B*, 58:5701, 1998.

7.150. D. Schmitz, C. Charton, A. Scroll, C. Carbone, and W. Eberhardt. *Phys. Rev. B*, 59:4327, 1999.

7.151. A. Ney, P. Poulopoulos, M. Farle, and K. Baberschke. *Phys. Rev. B*, 62:11336, 2000.

7.152. A. Ney, P. Poulopoulos, and K. Baberschke. *Europhys. Lett.*, 54:820, 2001.

7.153. A. Ney, P. Poulopoulos, F. Wilhelm, A. Scherz, M. Farle, and K. Baberschke. *J. Magn. Magn. Mat.*, 226-230:1570, 2001.

7.154. G. H. O. Daalderop, P. J. Kelly, and F. J. A. den Broeder. *Phys. Rev. Lett.*, 68:682, 1992.

7.155. M. T. Johnson, J. J. de Vries, N. W. E. McGee, J. aan de Stegge, and F. J. den Broeder. *Phys. Rev. Lett.*, 69:3575, 1992.
7.156. M. T. Johnson, F. J. A. der Broeder, J. J. de Vries, N. W. E. McGee, R. Jungblut, and J. aan de Stegge. *J. Magn. Magn. Mater.*, 121:494, 1993.
7.157. J. Lee, G. Lauhoff, and J. A. C. Bland. *Phys. Rev. B*, 56:R5728, 1997.
7.158. C. A. F. Vaz and J. A. C. Bland. *Phys. Rev. B*, 61:3098, 2000.
7.159. Ž. V. Šljivančanin and F. R. Vukajlović. *J. Phys. Condens. Matter*, 10:8679, 1998.
7.160. H. Li and B. P. Tonner. *Phys. Rev. B*, 40:10241, 1989.
7.161. G. A. Prinz. *Phys. Rev. Lett.*, 54:1051, 1985.
7.162. Y. U. Idzerda, W. T. Elam, B. T. Jonker, and G. A. Prinz. *Phys. Rev. Lett.*, 62:2480, 1989.
7.163. F. Xu, J. J. Joyce, M. W. Ruckman, H.-W. Chen, F. Boscherini, D. M. Hill, S. A. Chambers, and J. H. Weaver. *Phys. Rev. B*, 35:2375, 1987.
7.164. Y. B. Xu, E. T. M. Kernohan, D. J. Freeland, A. Ercole, M. Tselepi, and J. A. C. Bland. *Phys. Rev. B*, 58:890, 1998.
7.165. J. A. C. Bland, A. D. Johnson, R. D. Bateson, and H. T. Lauter. *J. Magn. Magn. Mat.*, 104-107:1798, 1992.
7.166. J. A. C. Bland, C. Daboo, G. A. Gehring, B. Kaplan, A. J. R. Ives, R. J. Hicken, and A. D. Johnson. *J. Phys. Condens. Matter*, 7:6467, 1995.
7.167. J. A. C. Bland, R. D. Bateson, P. C. Riedi, R. G. Graham, H. J. Lauter, J. Penfold, and C. Shackleton. *J. Appl. Phys.*, 69:4989, 1991.
7.168. J. A. C. Bland, S. J. Blundell, M. Gester, R. D. Bateson, J. Singleton, U. J. Cox, C. A. Lucas, W. C. K. Poon, , and J. Penfold. *J. Magn. Magn. Mat.*, 115:359, 1992.
7.169. S. J. Blundell, M. Gester, J. A. C. Bland, C. Daboo, E. Gu, M. J. Baird, and A. J. R. Ives. *J. Appl. Phys.*, 73:5948, 1993.
7.170. V. Pasyuk, H. J. Lauter, M. T. Johnson, F. J. A. den Broader, E. Janssen, J. A. C. Bland, A. V. Petrenko, and J. M. Gay. *J. Magn. Magn. Mat.*, 121:180, 1993.
7.171. S. Blügel, M. Weinert, and P. H. Dederichs. *Phys. Rev. Lett.*, 60:1077, 1988.
7.172. R. Wu, C. Li, and A. J. Freeman. *J. Magn. Magn. Mat.*, 99:71, 1991.
7.173. H. Fredrikze, P. de Haan, J. C. Lodder, and M. Th. Rekveldt. *J. Magn. Magn. Mat.*, 120:369, 1993.
7.174. C.-H. Lee, K.-L. Yu, M.-H. Lee, J. A. C. Huang, and G. P. Felcher. *J. Magn. Magn. Mat.*, 209:110, 2000.
7.175. G. Y. Guo. *J. Magn. Magn. Mater.*, 176:97, 1997.
7.176. V. dos Santos and C. A. Kuhnen. *Thin Solid Films*, 350:258, 1999.
7.177. Z. Yang, V. I. Gavrilenko, and R. Wu. *Surf. Sci.*, 447:212, 2000.
7.178. Z. Yang and R. Wu. *Phys. Rev. B*, 63:064413, 2001.
7.179. S. Hope, J. Lee, P. Rosenbusch, G. Lauhoff, J. A. C. Bland, A. Ercole, D. Bucknall, J. Penfold, H. J. Lauter, V. Lauter, and R. Cubitt. *Phys. Rev. B*, 55:11422, 1997.
7.180. J. Lee, G. Lauhoff, M. Tselepi, S. Hope, P. Rosenbusch, J.A.C. Bland, H. A. Durr, G. van der Laan, J. Ph. Schillé, and J.A. D. Matthew. *Phys. Rev. B*, 55:15103, 1997.
7.181. J. Lee, G. Lauhoff, S. Hope, C. Daboo, J. A. C. Bland, J. Ph. Schillé, G. van der Laan, and J. Penfold. *J. App. Phys.*, 81:3893, 1997.
7.182. M. Farle. *Rep. Prog. Phys.*, 61:755, 1998.
7.183. C. A. F. Vaz, G. Lauhoff, J. A. C. Bland, B. D. Fulthorpe T. P. A. Hase, B. K. Tanner, S. Langridge, and J. Penfold. *J. Magn. Magn. Mater.*, 226-230:1618, 2001.
7.184. H. Jansen. Unpublished.
7.185. R. Wu, L. Chen, and A. J. Freeman. *J. Appl. Phys.*, 81:4417, 1997.
7.186. G. Gubbiotti, G. Carlotti, M. Ciria, and R. C. O'Handley. *IEEE Trans. Magn.*, 38:2649, 2002

7.187. A. Ney, A. Scherz, P. Poulopoulos, K. Lenz, H. Wende, K. Baberschke, F. Wilhelm, and N. B. Brookes. *Phys. Rev. B*, 65:024411, 2002.
7.188. J. Lee, G. Lauhoff, C. Fermon, S. Hope, J. A. C. Bland, J. Ph. Schillé, G. van der Laan, C. Chappert, and P. Beauvillain. *J. Phys. Condens. Matter*, 9:L137, 1997.
7.189. S. S. P. Parkin, R. Sigsbee, R. Felici, and G. P. Felcher. *Appl. Phys. Lett.*, 48:604, 1986.
7.190. F. Ott and C. Fermon. *J. Magn. Magn. Mat.*, 165:475, 1997.
7.191. S. S. Parkin, V. R. Deline, R. O. Hilleke, and G. P. Felcher. *Phys. Rev. B*, 42:10583, 1990.
7.192. S. Manguin, C. Bellouard, and H. Fritzsche. *Physica B*, 276-278:558, 2000.
7.193. W. Lohstroh, M. Münzenberg, W. Felsch, H. Fritzsche, H. Maletta, R. Goyette, and G. P. Felcher. *J. Appl. Phys.*, 85:5873, 1999.
7.194. A. Hoffmann, M. R. Fitzsimmons, J. A. Dura, and C. F. Majkrzak. *Phys. Rev. B*, 65:024428, 2001.
7.195. A. D. Johnson, J. A. C. Bland, C. Norris, and H. Lauter. *J. Phys. C: Solid State Phys.*, 21:L899, 1988.
7.196. K. V. O'Donovan, J. A. Borchers, C. F. Majkrzak, O. Hellwig, and E. E. Fullerton. *Phys. Rev. Lett.*, 88:067201, 2002.
7.197. B. Gurney, M. Carey, C. Tsang, M. Williams, S. Parkin, R. Fontana, Jr., E. Grochowski, M. Pinarbasi, T. Lin, D. Mauri. Spin valve giant magnetoresistance sensor materials for hard disk drives. In J. A. C. Bland, B. Heinrich, editors, *Ultrathin Magnetic Structures IV.*, Springer-Verlag, forthcoming.
7.198. P. James, O. Eriksson, B. Johansson, and I. A. Abrikosov. *Phys. Rev. B*, 59:419, 1999.
7.199. M. N. Baibich, J. M. Broto, A. Fert, F. Nguyen Van Dau, F. Petroff, P. Etienne, G. Creuzet, A. Friederich, and J. Chazelas. *Phys. Rev. Lett.*, 61:2472, 1988.
7.200. G. Binasch, P. Grünberg, F. Saurenbach, and W. Zinn. *Phys. Rev. B*, 39:4828, 1989.
7.201. M. D. Stiles. *J. Magn. Magn. Mat.*, 200:322, 1999.
7.202. P. Bruno and C. Chappert. *Phys. Rev. B*, 46:261, 1992.
7.203. P. Bruno. *Phys. Rev. B*, 52:411, 1995.
7.204. P. Bruno. *J. Phys. Condens. Matter*, 11:9403, 1999.
7.205. A. Chaiken, G. A. Prinz, and J. J. Krebs. *J. Appl. Phys.*, 67:4892, 1990.
7.206. B. Dieny, V. S. Speriosu, S. S. P. Parkin, B. A. Gurney, D. R. Wilhoit, and D. Mauri. *Phys. Rev. B*, 43:1297, 1991.
7.207. L. B. Steren, A. Barthélémy, J. L. Duvail, A. Fert, R. Morel, F. Petroff, P. Holody, and P. A. Schroeder. *Phys. Rev. B*, 51:292, 1995.
7.208. S. Mao, M. Plumer, A. Mack, Z. Zhijun, and E. Murdock. *J. Appl. Phys.*, 85:5033, 1999.
7.209. B. Dieny, C. Cowache, A. Nossov, P. Dauguet, J. Chaussy, and P. Gandit. *J. Appl. Phys.*, 79:6370, 1996.
7.210. P. Dauguet, P. Gandit, and J. Chaussy. *J. Appl. Phys.*, 79:5823, 1996.
7.211. A. Vedyayev, B. Dieny, N. Ryzhanova, J. B. Genin, and C. Cowache. *Europhys. Lett.*, 25:465, 1994.
7.212. K. Wang, S. Zhang, and P. M. Levy. *Phys. Rev. B*, 54:11965, 1996.
7.213. J. Barnás, O. Baksalary, and A. Fert. *Phys. Rev. B*, 56:6079, 1997.
7.214. R. H. Brown, D. M. C. Nicholson, W. H. Butler, X.-G. Zhang, W. A. Shelton, T. C. Schulthess, and J. M. MacLaren. *Phys. Rev. B*, 58:11146, 1998.
7.215. D. Huertas-Hernando, G. E. W. Bauer, and Yu. V. Nazarov. *J. Magn. Magn. Mat.*, 240:174, 2002.
7.216. J. A. C. Bland, C. Daboo, M. Patel, T. Fujimoto, and J. Penfold. *Phys. Rev. B*, 57:10272, 1998.

7.217. M. Patel, T. Fujimoto, A. Ercole, C. Daboo, and J. A. C. Bland. *J. Magn. Magn. Mater.*, 156:53, 1996.
7.218. T. Fujimoto, M. Patel, C. Dadoo, E. Gu, and J. A. C. Bland. *J. Magn. Magn. Mater.*, 156:365, 1996.
7.219. J. A. C. Bland, H. T. Leung, S. J. Blundell, V. S. Speriosu, S. Metin, B. A. Gurney, and J. Penfold. *J. Appl. Phys.*, 79:6295, 1996.
7.220. A. Samad, B. C. Choi, S. Langridge, J. Penfold, and J. A. C. Bland. *Phys. Rev. B*, 60:7304, 1999.
7.221. B. C. Choi, A. Samad, W. Y. Lee, S. Langridge, J. Penfold, and J. A. C. Bland. *IEEE Trans. Magn.*, 35:3847, 1999.
7.222. B. C. Choi, A. Samad, C. A. F. Vaz, J. A. C. Bland, S. Langridge, and J. Penfold. *Appl. Phys. Lett.*, 77:892, 2000.
7.223. A. R. Ball, H. Fredrikze, P. J. van der Zaag, R. Jungblut, A. Reinders, A. van der Graaf, and M. Th. Rekveldt. *J. Magn. Magn. Mat.*, 148:46, 1995.
7.224. G. P. Felcher, Y. Y. Huang, M. Carey, and A. Berkowitz. *J. Magn. Magn. Mat.*, 121:105, 1993.
7.225. N. C. Koon. *Phys. Rev. Lett.*, 78:4865, 1997.
7.226. M. D. Stiles and R. D. McMichael. *Phys. Rev. B*, 59:3722, 1999.
7.227. S. G. E. te Velthuis, A. Berger, G. P. Felcher, B. K. Hill, and E. Dan Dalberg. *J. Appl. Phys.*, 87:5046, 2000.
7.228. B. H. Miller and E. Dan Dahlberg. *Appl. Phys. Lett.*, 69:3932, 1996.
7.229. V. Ström, B. J. Jönsson, K. V. Rao, and D. Dahlberg. *J. Appl. Phys.*, 81:5003, 1997.
7.230. M. R. Fitzsimmons, P. Yashar, C. Leighton, I. K. Schuller, J. Nogués, C. F. Majkrzak, and J. A. Dura. *Phys. Rev. Lett.*, 84:3986, 2000.
7.231. C. Leighton, M. R. Fitzsimmons, P. Yashar, A. Hoffmann, J. Nogués, J. A. Dura, C. F. Majkrzak, and I. K. Schuller. *Phys. Rev. Lett.*, 86:4394, 2001.
7.232. M. R. Fitzsimmons, C. Leighton, A. Hoffmann, P. Yashar, J. Nogués, K. Liu, C. F. Majkrzak, J. A. Dura, H. Fritzsche, and I. K. Schuller. *Phys. Rev. B*, 64:104415, 2001.
7.233. C. Leighton, M. R. Fitzsimmons, A. Hoffmann, J. A. Dura, C. F. Majkrzak, M. S. Lund, and I. K. Schuller. *Phys. Rev. B*, 65:064403, 2002.
7.234. M. R. Fitzsimmons, C. Leighton, J. Nogués, A. Hoffmann, K. Liu, C. F. Majkrzak, J. A. Dura, J. R. Groves, R. W. Springer, P. N. Arendt, V. Leiner, H. Lauter, and I. K. Schuller. *Phys. Rev. B*, 65:134436, 2002.
7.235. A. van der Graaf, A. R. Ball, and J. C. S. Kools. *J. Magn. Magn. Mater.*, 165:479, 1997.
7.236. Th. G. S. M. Rijks, R. Coehoorn, J. T. F. Daemen, and W. J. M. de Jonge. *J. Appl. Phys.*, 76:1092, 1994.
7.237. S. G. E. te Velthuis, G. P. Felcher, J. S. Jiang, A. Inomata, C. S. Nelson, A. Berger, and S. D. Bader. *Appl. Phys. Lett.*, 75:4174, 1999.
7.238. S. G. E. te Velthuis, J. S. Jiang, and G. P. Felcher. *Appl. Phys. Lett.*, 77:2222, 2000.
7.239. J. S. Jiang, G. P. Felcher, A. Inomata, R. Goyette, C. Nelson, and S. D. Bader. *Phys. Rev. B*, 61:9653, 2000.
7.240. W. H. Meiklejohn. *J. Appl. Phys.*, 33:1328, 1962.
7.241. P. Grünberg, R. Schreiber, Y. Pang, M. B. Brodsky, and H. Sowers. *Phys. Rev. Lett.*, 57:2442, 1986.
7.242. C. F. Majkrzak, J. W. Cable, J. Kwo, M. Hong, D. B. McWhan, Y. Yafet, J. V. Waszczak, and C. Vettier. *Phys. Rev. Lett.*, 56:2700, 1986.
7.243. M. Takeda, Y. Endoh, H. Yasuda, K. Yamada, A. Kamijo, J. Muzuki. *J. Phys. Soc. Japan*, 62:3015, 1993.

7.244. H. Lauter, V. Lauter-Pasyuk, B. Toperverg, L. Romashev, M. Milyaev, T. Krinitsina, E. Kravtov, V. Ustinov, A. Petrenko, V. Aksenov. *J. Magn. Magn. Mat.*, 258–259:338, 2003.
7.245. A. Schreyer, T. Schmitte, R. Siebrecht, P. Bödeker, H. Zabel, S. H. Lee, R. W. Erwin, C. F. Majkrzak, J. Kwo, and M. Hong. *J. Appl. Phys.*, 87:5443, 2000.
7.246. J. A. Borchers, P. M. Gehring, R. W. Erwin, J. F. Ankner, C. F. Majkrzak, T. L. Hylton, K. R. Koffey, M. A. Parker, and J. K. Howard. *Phys. Rev. B*, 54:9870, 1996.
7.247. A. Barthélémy, A. Fert, M. N. Baibich, S. Hadjoudj, F. Petroff, P. Etienne, R. Cabanel, S. Lequien, F. Nguyen Van Dau, and G. Creuzet. *J. Appl. Phys.*, 67:6295908, 1990.
7.248. S. S. P. Parkin, A. Mansour, and G. P. Felcher. *Appl. Phys. Lett.*, 58:6291473, 1991.
7.249. N. Hosoito, K. Mibu, S. Araki, T. Shinjo, S. Itoh, Y. Endoh. *J. Phys. Soc. Japan*, 61:300, 1992
7.250. M. Loewenhaupt, W. Hahn, Y. Y. Huang, G. P. Felcher, and S. S. P. Parkin. *J. Magn. Magn. Mat.*, 121:173, 1993.
7.251. M. J. Pechan, J. F. Ankner, C. F. Majkrzak, D. M. Kelly, and I. K. Schuller. *J. Appl. Phys.*, 75:6178, 1994.
7.252. A. Schreyer, J. F. Ankner, Th. Zeidler, H. Zabel, M. Schäfer, J. A. Wolf, P. Grünberg, and C. F. Majkrzak. *Phys. Rev. B*, 52:16066, 1995.
7.253. A. Schreyer, J. F. Ankner, Th. Zeidler, H. Zabel, C. F. Majkrzak, M. Schäfer, and P. Grünberg. *Europhys. Lett.*, 32:595, 1995.
7.254. S. Adenwalla, G. P. Felcher andE. E. Fullerton, and S. D. Bader. *Phys. Rev. B*, 53:2474, 1996.
7.255. J. F. Ankner, H. Kaiser, A. Schreyer, Th. Zeidler, H. Zabel, M. Schäfer, and P. Grünberg. *J. Appl. Phys.*, 581:3765, 1997.
7.256. J. C. Slonczewski. *J. Magn. Magn. Mat.*, 150:13, 1995.
7.257. K. Temst, E. Kunnen, V. V. Moshchalkov, H. Maletta, H. Fritzsche, and Y. Bruynseraede. *Physica B*, 276-278:684, 2000.
7.258. J. S. Jiang, G. P. Felcher, A. Inomata, R. Goyette, C. Nelson, and S. D. Bader. *J. Vac. Sci. Technol. A*, 18:1264, 2000.
7.259. V. L. Aksenov, Yu. V. Nikitenko, V. V. Proglyado, M. A. Andreeva, B. Kalska, L. Häggström, R Wäppling. *J. Magn. Magn. Mat.*, 258–259:332, 2003
7.260. S. Andrieu, M. Piecuch, H. Fischer, J. F. Bobo, F. Bertram, Ph. Bauer, and M. Hennion. *J. Magn. Magn. Mat.*, 121:30, 1993.
7.261. A. van der Graaf, M. Valkier, J. Khohepp, and F. J. A. den Broeder. *J. Magn. Magn. Mat.*, 165:157, 1997.
7.262. J. Tappert, F. Klose, Ch. Rehm, W. S. Kim, R. A. Brand, H. Maletta, and W. Keune. *J. Magn. Magn. Mater.*, 156:58, 1996.
7.263. J. E. Mattson, E. E. Fullerton, C. H. Sowers, Y. Y. Huang, G. P. Felcher, and S. D. Bader. *J. Appl. Phys.*, 73:77, 1993.
7.264. B. Rodmacq, K. Dumesnil, P. Manguin, and M. Hennion. *Phys. Rev. B*, 48:3556, 1993.
7.265. J. A. Borchers, P. M. Gehring, R. W. Erwin, J. F. Ankner, C. F. Majkrzak, T. L. Hylton, K. R. Koffey, M. A. Parker, and J. K. Howard. *J. Appl. Phys.*, 79:4762, 1996.
7.266. J. A. Borchers, P. M. Gehring, C. F. Majkrzak, A. M. Zeltser, N. Smith, and J. F. Ankner. *J. Appl. Phys.*, 81:3771, 1997.
7.267. A. Schreyer, K. Bröhl, J. F. Ankner, C. F. Majkrzak, Th. Zeidler, P. Bödeker, N. Metoki, and H. Zabel. *Phys. Rev. B*, 47:15334, 1993.
7.268. W. Schwarzecher, W. Allison, J. Penfold, C. Shackleton, C. D. England, W. R. Bennett, J. R. Dutcher, and C. M. Falco. *J. Appl. Phys.*, 69:4040, 1991.

7.269. D. E. Joyce, S. I. Campbell, P. R. T. Pugh, and P. J. Grundy. *Physica B*, 248:152, 1998.
7.270. N. D. Telling, S. Langridge, and C. C. Tang. *J. Magn. Magn. Mat.*, 198-199:692, 1999.
7.271. J. A. Borchers, J. A. Dura, J. Unguris, D. Tulchinsky ans M. H. Kelly, C. F. Majkrzak, S. Y. Hsu, R. Loloee, W. P. Pratt, Jr., and J. Bass. *Phys. Rev. Lett.*, 82:2796, 1999.
7.272. J. A. Borchers, J. A. Dura, C. F. Majkrzak, S. Y. Hsu, R. Lolee, W. P. Pratt, and J. Bass. *Physica B*, 283:162, 2000.
7.273. J. Unguris, D. Tulchinsky ans M. H. Kelly, J. A. Borchers, J. A. Dura, C. F. Majkrzak, S. Y. Hsu, R. Loloee, W. P. Pratt, Jr., and J. Bass. *J. Appl. Phys.*, 87:6639, 2000.
7.274. Z. Tun, W. J. L. Buyers, I. P. Swainson, M. Sutton, and R. W. Cochrane. *J. Appl. Phys.*, 76:7075, 1994.
7.275. T. Charlton, D. Lederman, S. M. Yusuf, and G. P. Felcher. *J. Appl. Phys.*, 85:4436, 1999.
7.276. J. A. Borchers, J. F. Ankner, C. F. Majkrzak, B. N. Engel, M. H. Wiedmann, R. A. Van Leeuwen, and C. M. Falco. *J. Appl. Phys.*, 75:6498, 1994.
7.277. Y. Y. Huang, G. P. Felcher, and S. S. P. Parkin. *J. Magn. Magn. Mat.*, 899:L31, 1991.
7.278. M. Mao, S. H. Nguyen, B. D. Gaulin, Z. Tun, X. Bian, Z. Altounian, and J. O. Ström-Olsen. *J. Appl. Phys.*, 79:4769, 1996.
7.279. K. Sköld and D. L. Price, editors. volume 23-A of *Methods of Experimental Physics*. Academic Press, Inc., 1986.

8

X-ray Scattering Studies of Ultrathin Metallic Structures

E.E. Fullerton and S.K. Sinha

8.1 Introduction

The study of magnetic thin films and multilayers has been driven by the variety of novel physics and phenomena that can be studied in such structures as outlined in these volumes. In general, much of the new physics comes from combining materials (e.g. ferromagnetic, antiferromagnetic, paramagnetic, nonmagnetic) on nanometer length scales. With reduced layer thicknesses, the interfaces increasingly dominate the magnetic response of the system. The interfaces can perturb the system in a variety of important ways including interfacial strains that alter the atomic structure within the layer, reduced coordination of the atoms at the interface producing unique properties of the interfacial atoms compared to the bulk and coupling of the layers either directly across the interface or indirectly via an interlayer. In general, these properties coexist and depend sensitively on the structural integrity of the thin films or multilayers. Since these structures are typically made by vapor deposition and are not in thermodynamic equilibrium, structural characterization of the atomic and interfacial structure is a prerequisite for a complete understanding of the magnetic properties.

The use of neutron and x-ray scattering as a tool to study the structure and magnetization of the films and multilayers is by now well established. Such scattering techniques are non-destructive and provide quantitative statistically-averaged structural information ranging in scale from the atomic spacing up to the coherence length of the x-ray source ($\sim$ microns). This allows x-ray scattering to address several issues, such as the crystalline structure within the films and the morphology of the interfaces. One of the most important aspects of the latter is the issue of roughness, both structural and magnetic. Ordinary charge scattering (observable even in in-house x-ray facilities such as x-ray tube or rotating anode sources) can be used to study crystalline structure and structural roughness at interfaces, using techniques such as x-ray specular reflectivity and wide-angle x-ray diffraction [8.1–4]. Some applications of such techniques to magnetic multilayer films have been reviewed in an earlier volume of this series [8.5].

Over the last decade or so it has been realized that x-rays can also be used to study magnetism in solids by utilizing the magnetic cross section for x-rays, which is nevertheless small compared to the ordinary charge or Thompson scattering cross section. The discovery of a large resonant enhancement of the magnetic cross section for x-rays scattering at the L-edges of transition metal and rare earth atoms and at the M-edges of actinide atoms has made x-ray studies of magnetism in thin films and multilayers readily feasible at synchrotron sources, where intense x-ray beams which can be tuned to the resonant energies are available [8.6, 7].

Much of the initial magnetic x-ray studies have been carried out on bulk antiferromagnetic crystals, where peaks distinct from the much larger Bragg peaks due to ordinary charge scattering may be studied relatively easily [8.8, 9]. In cases where the two types of scattering are superimposed, distinguishing magnetic scattering from charge scattering requires polarization techniques, such as polarization analysis of the scattered beam (to pick out scattering in the $\sigma \rightarrow \sigma$ vs. $\sigma \rightarrow \pi$ polarization channels). Alternatively, one may make use of circularly polarized x-ray beams where the charge scattering may, in principle, be subtracted out by changing the sense of circular polarization of the incident beam (or equivalently, the magnetization direction in the sample). The use of these techniques to study magnetism in films and surface magnetism is a growing but still relatively unexplored field.

In this chapter we will discuss the basic concepts of x-ray scattering from thin films and multilayers with a focus on determining both the average structure as well as the deviations from the average (i.e. interfacial roughness and disorder). This is generally achieved by measuring both the specular and diffuse scattering and fitting the scattered intensities to models that have been generalized to include disorder. We will show how this approach can be extended to study magnetic order and roughness. The scattering from thin films and multilayers is generally separated into reflectivity and wide-angle measurements. Reflectivity scans refer to scattering regimes where the magnitude of the scattering wavevector transfer $q \ll 2\pi a$ where a is the lattice constant of the constituent material. In this regime one can neglect short length scale fluctuations due to the discreteness of the atoms or atomic scale-disorder and consider both the total electron density and the magnetization as uniform inside the medium contained within the "structural" and "magnetic" surfaces/interfaces respectively. Wide-angle diffraction occurs for $q \simeq 2\pi a$ where the scattering becomes sensitive to the atomic order (both structural and magnetic) which needs to be included in any modeling. Since these two techniques probe different length scales they often provide complementary information about the structure. We will first describe reflectivity measurements focusing on determining both the structural and magnetic roughness. We will then discuss methods for determining the atomic structure and disorder from the wide-angle diffraction measurements.

8.2 Reflectivity Measurements of Interfacial Structure

8.2.1 Interfacial Roughness

The effect of roughness at interfaces has been recognized as playing an important role in affecting important magnetic and magneto-transport properties of thin films, including magnetic anisotropy, magnetic proximity effects, giant magnetoresistance (GMR), exchange bias, dipolar interactions in magnetic devices, etc. but the relationship between these properties and the interface structure is still poorly understood [8.7]. Such interfaces are never chemically abrupt or completely smooth or flat as shown schematically in Fig. 8.1. Both the magnitude and the spatial frequency of the interface roughness play an extremely important role in affecting the way in which magnetic anisotropy and exchange can determine the ordering of magnetic moment near an interface. Since x-ray scattering provides a statistically averaged measure of the surface roughness, it proves useful to have general statistical descriptions of the surfaces. The morphology of a surface (or interface) may be characterized by the statistical height-height correlation function $C(\boldsymbol{r})$

$$C(\boldsymbol{r}) = \langle \delta z(0) \delta z(\boldsymbol{r}) \rangle \tag{8.1}$$

where $\boldsymbol{r}$ represents an in-plane separation vector and δz(0) and $\delta z(\boldsymbol{r})$ represent height fluctuation from the average interface at the origin and $\boldsymbol{r}$, respectively (Fig. 8.1). This correlation function is the Fourier transform of the two-dimensional power spectral density (PSD) of the surface given by

$$\mathrm{PSD}(\boldsymbol{f}) = \frac{1}{A} \left| \int_A z(\boldsymbol{r}) \exp\left(\mathrm{i}2\pi \boldsymbol{f} \cdot \boldsymbol{r}\right) \mathrm{d}\boldsymbol{r} \right|^2 \quad [A = \text{surface area}] \tag{8.2}$$

where $\boldsymbol{f}$ is a vector of the spatial frequency of the roughness. Thus most surfaces can be characterized by the corresponding $C(\boldsymbol{r})$ in real space (or PSD($\boldsymbol{f}$) in frequency space) [10, 11]. A rather general and convenient way of representing such a correlation function is in the modified self-affine form:

$$C(r) = \sigma^2 \exp\left[(-r/\xi)^{2h}\right] \tag{8.3}$$

where $r = |\boldsymbol{r}|$, σ^2 represents a mean-square roughness for the interface, ξ is the correlation length for the roughness and h $(0 < h < 1)$ is the roughness exponent [8.10, 11]. Although not appropriate to describe surfaces with lateral periodicity such a form is able to describe a wide variety of interfacial structures. Low values of h give a jagged morphology while large h results in a smoothly varying morphology. For small values of r, the roughness can be mapped onto a fractal surface with the fractal dimension $D = 3 - h$.

If one extends such a description to a multilayer structure where there are multiple interfaces (Fig. 8.1), then a description of this structure needs a description that include how the roughness of ith interface is correlated with roughness in jth interface. This can also be characterized by a correlation function

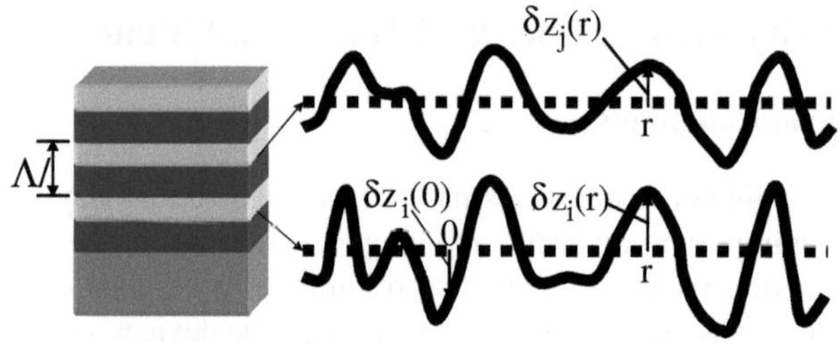

Fig. 8.1. Schematic representation of a multilayer structure on a substrate with a periodicity of Λ. An expanded view of the ith and jth interfaces where the average interface position (dashed line) is given by $\bar{z}_i$ and $\bar{z}_j$ and the local deviation from the average position at a point $\boldsymbol{r}'$ given by $\delta z_i(\boldsymbol{r}')$ and $\delta z_j(\boldsymbol{r}')$, respectively

$$C_{ij}(\boldsymbol{r}) = \langle \delta z_i(0) \delta z_j(\boldsymbol{r}) \rangle \tag{8.4}$$

where $\delta z_i(0)$ corresponds to a height fluctuation at the origin in the ith layer and $\delta z_j(\boldsymbol{r})$ is a height fluctuation in the jth layer at position $\boldsymbol{r}$. If fluctuations in the ith or jth interface are independent then $C_{ij}(\boldsymbol{r}) = 0$. However, if the fluctuations in the ith and jth layers are conformal, that is perfectly replicated in adjacent layers during growth then $C_{ij}(\boldsymbol{r}) = C(r)$, the correlation function of the individual layers as given in (8.3). In most multilayer thin film structures there is partial correlation of the roughness in adjacent interfaces (as shown schematically in Fig. 8.1) with the correlation being stronger for longer length scale fluctuations (for examples see [8.12–17]). It is often observed that the correlation function between interfaces decays as a function of the separation of these interfaces [8.13–16] and can be described with a perpendicular correlation length $\xi_\perp$, having a similar in-plane form as (8.3). In addition to being correlated the roughness can also be cumulative. That is the roughness of each interface adds to the roughness of subsequent layers and the value of σ increases from interface to interface as the multilayer grows.

Up to this point we have only considered the nature of the chemical (or structural) interface. One needs to similarly describe the magnetic interface. An interface between a ferromagnetic and non-magnetic layer can sometime produce a magnetic "dead layer" at the interface, as in the case of surface magnetism where the order parameter often decreases faster with increasing temperature than the bulk order parameter. This also can result from hybridization of the electron states or interdiffusion of the interfacial atoms that suppresses the moments. A conceptually simple way to model such effects is to visualize a "magnetic interface" between the two layers that may be distinct from the actual chemical interface, but may also be highly correlated with the latter (Fig. 8.2). The relevant lateral length scale being considered here may be from 5-nm to microns, which is the range amenable to study with reflectivity. (Magnetic moments, which are disordered on shorter length scales, will simply appear as a loss of magnetization). In addition there may be domain formation within the ferromagnetic layers, which can also be studied as described below. The concept of the magnetic interface leads naturally to the concept of "magnetic roughness", which is distinct from the "chemical roughness" [8.18–22] and may be specified

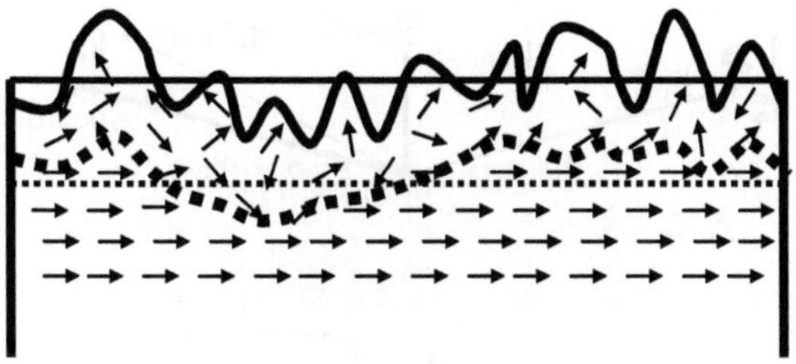

Fig. 8.2. Schematic representation of a possible magnetic interface or surface with a 'dead' layer. The structural interface is shown by the solid line while the magnetic interface (the boundary between the magnetic and nonmagnetic region) is shown by the dashed lines. The magnetic interface has both a different average position as well as a different lateral height-height correlation function

in terms of a similar type of magnetic correlation function $C_{mm}(r)$, which can be characterized by a root-mean-square value for the height fluctuations about the average magnetization σ_m, a magnetic roughness correlation length ξ_m and a roughness exponent h_m. Work on thin films of cobalt indicates that it is the magnetic roughness parameters (in particular the correlation length ξ_m) rather than the chemical roughness that determine the coercive field in these films [8.19]. The behavior of electron transport across magnetic layers and in particular, spin-dependent transport is clearly sensitive to both the chemical and magnetic roughness at the interfaces. Attempts to correlate the GMR effect in Fe/Cr multilayers with the chemical roughness often show contradictory results [8.23–26] which may resolved from an understanding of the correlation between the magnetic and chemical roughness.

8.2.2 Reflectivity Measurements

As is now well established both the specular reflectivity and the off-specular or diffuse scattering from the interface are excellent non-destructive ways of probing these roughness parameters. The general scattering geometry is shown in Fig. 8.3 where the incoming x-ray beam is at grazing angle of incidence θ_i. For a specular reflectivity experiment the scattered beam will reflect off the interfaces in the structure at an angle $\theta_f = \theta_i$. For such measurements, the scattering vector $\boldsymbol{q} = \vec{k}_f - \vec{k}_i$ is normal to the surface with magnitude $q = 4\pi \sin\theta/\lambda$ where λ is the x-ray wavelength. The scattered intensity is monitored as a function of q by scanning θ and keeping the total angle of scattering $= 2\theta$, so that θ_f always equals θ_i. Such a scan is often referred to as a "$\theta - 2\theta$" scan. Since $q_x = q_y = 0$ the scattering intensity is only sensitive to variations of the laterally averaged structure normal to the surface and provides information on the layer thicknesses and the root-mean-square interfacial width σ. For a multilayer with period Λ (Fig. 8.1 and 8.3a) the scattering from the different interfaces will interfere, resulting in Bragg peaks expected at $q_z = 2\pi n/\Lambda$ where n is an integer as shown schematically in Fig. 8.3b. The measured positions of the Bragg peaks are shifted slightly from the expected position when the refractive index of the material is taken into account [8.27].

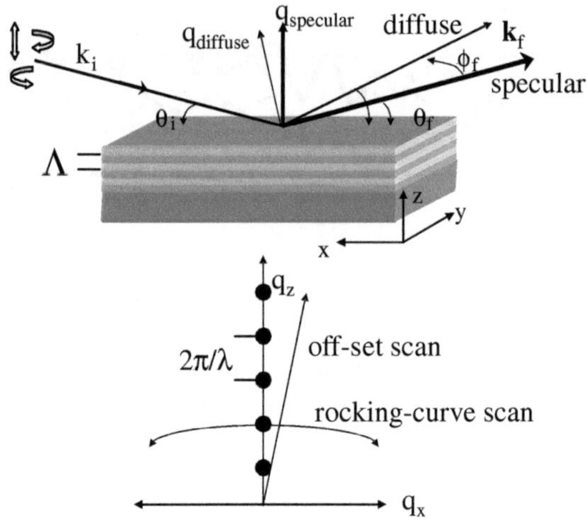

Fig. 8.3. Scattering geometry for a reflectivity or diffraction measurement from a multilayer structure (*top*) with the corresponding scattering geometry in reciprocal space (*bottom*). The incoming x-ray have either linear or circular polarization with an incoming angle θ_i and wavevector $\vec{k}_i$. The specularly scattered x-rays scatter at $\theta_f = \theta_i$ such that the scattering wavevector q is normal to surface. The multilayer periodicity results in Bragg peaks separated by $2\pi/\Lambda$ along q_z as shown by circles. For rough interfaces the diffuse scattering occurs at $\theta_f \neq \theta_i$.and $\phi_f \neq 0$ such that the scattered wavevector has a finite in-plane component. Offset and rocking curve scans (described in text) that probe the diffuse scattering are shown

For a rough interface there is diffuse intensity that is scattered into a cone about the specular beam such that $\theta_f \neq \theta_i$. and $\phi_f \neq 0$. However, some diffuse scattering occurs along the specular scattering direction and needs to be subtracted from the measured intensity to obtain the true specular scattering. If q_x or q_y is non-zero, the scattered intensity becomes sensitive to the lateral morphology of the interface where full details of the interface roughness are manifest in the diffuse intensity. The most convenient approach to obtain the diffuse scattering is by using an area detector that collects both the specular and diffuse scattering. In absence of an area detector, typical diffuse intensity scans are 'rocking curve' scans where the incoming beam and detector are fixed and the sample is rocked. For such a scan the magnitude of the wave vector is fixed while the in-plane component q_x ($q_y = 0$) of the wavevector increases as the sample is rocked out of the specular condition. Often, the experimental set-up utilizes detector slits which are quite wide open in the direction normal to the plane of scattering and thus what is measured is the integral of the diffuse scattering over q_y.

A related scan known as a grazing incidence scattering maintains θ_i and θ_f fixed and scans the detector along ϕ_f (In general θ_i and θ_f are both kept small) [8.28]. For such a scan the q_z component of the wavevector remain fixed and the in-plane component increases with increasing ϕ and the diffuse scattering at a given q_z is mapped out. Although a more difficult scan than the rocking-curve scan this approach allows a larger in-plane wavevector to be obtained. Another common scan is the 'offset scan' which is equivalent to a specular reflectivity scan but with the normal of the

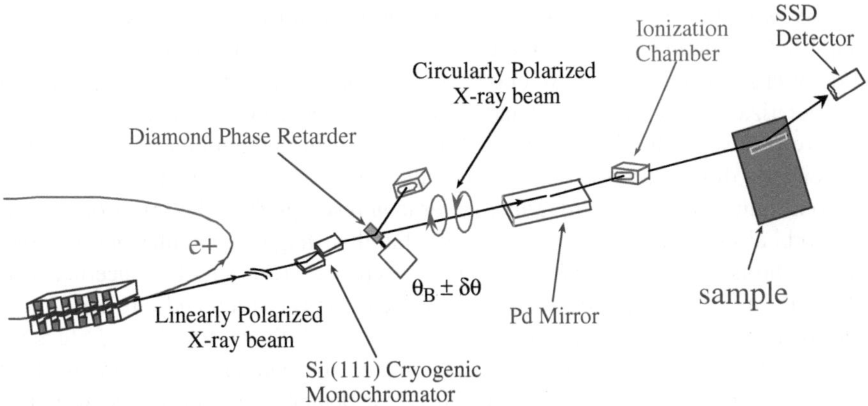

Fig. 8.4. Schematic representation of a synchrotron beamline for structural and magnetic scattering measurements as described in [8.20]. The components included a monochromator to select the energy and a phase retarder to select the polarization

film surface offset by a small angle δ so that the intensity is probed in reciprocal space along a line making a small angle to the q_z axis (Fig. 8.3). This scan probes the q_z dependence of the diffuse scattering and is often used to estimate the diffuse scattering under the specular scattering since the diffuse scattering usually has a much weaker dependence on q_x near the specular condition. This intensity can then be subtracted from the specular reflectivity scans to obtain the true specular scattering. From such scans a rather general description of the specular and diffuse scattering is obtained.

To probe magnetic scattering one also needs to control the polarization of the incoming x-ray beam and the magnetic state of the sample. The polarization of the x-ray beam from a synchrotron is in general linear and horizontal. It may be converted to circular polarized radiation of either sense by suitable use of an x-ray quarter wave plate. Alternatively, on can study scattering from the linearly polarized incident beam by looking for a rotated plane of polarization of the scattered beam using a polarization analyzer crystal or optic (e.g. to study $\sigma \rightarrow \pi$ scattering processes). For soft x-rays this has required the development of specialized optical elements [8.29]. A schematic diagram of a typical synchrotron set up for magnetic scattering is shown in Fig. 8.4.

8.2.3 Scattering Formalism

8.2.3.1 Atomic Scattering Amplitudes

The x-ray scattering amplitude from a magnetic atom in the dipole approximation has been derived by Hannon et al. [8.6] and is given by:

$$f = f_0\,(\boldsymbol{e}_\mathrm{f}\cdot\boldsymbol{e}_\mathrm{i}) + \frac{3\lambda}{8\pi}\,[F_{11}+F_{1-1}]\left(\boldsymbol{e}_\mathrm{f}^*\cdot\boldsymbol{e}_\mathrm{i}\right) - \boldsymbol{i}\,[F_{11}-F_{1-1}]\left(\boldsymbol{e}_\mathrm{f}^*\times\boldsymbol{e}_\mathrm{i}\right)\cdot\hat{\boldsymbol{m}} + [2F_{10}-F_{11}-F_{1-1}]\left(\boldsymbol{e}_\mathrm{f}^*\cdot\hat{\boldsymbol{m}}\right)\left(\boldsymbol{e}_\mathrm{i}\cdot\hat{\boldsymbol{m}}\right) \tag{8.5}$$

where f_0 is the usual non-magnetic charge form factor or the Thomson scattering amplitude (including anomalous terms from x-ray edges other than the one under consideration), λ is the x-ray wavelength, $\boldsymbol{e}_\mathrm{i}$, $\boldsymbol{e}_\mathrm{f}$ and $\hat{\boldsymbol{m}}$ are the unit vectors representing the polarizations of the incident and scattered photons and the magnetization direction respectively, and the F_{LM} are the resonant scattering amplitudes (with $L = 1$ for electric dipolar transitions) which contain resonant denominators and thus can become quite large. The third term is linear in $\hat{\boldsymbol{m}}$ resulting from the anisotropy in spin and orbital moments and gives rise to the familiar magnetic circular birefringence and dichroism of ferromagnetic materials. Here we are primarily concerned with the real (or scattering) part. Because the d states on transition and rare earth metal atoms show significant polarization when magnetized, dipolar transitions, which occur mainly from core p-states, corresponding to the photons with energies at the L edges will be sensitive to magnetism [for a detailed measure of the scattering cross sections of Fe see [8.30]]. Working at core levels has the additional advantage that the magnetic scattering becomes element specific since the core levels for different elements are well separated in energy [8.7]. The final term is second order in $\hat{\boldsymbol{m}}$ resulting from an anisotropy in charge related to the magnetization and gives a linear birefringence and linear dichroism. Since this final term probes the *axis* of magnetization, it provides sensitivity to antiferromagnetic as well as ferromagnetic order [8.31].

The expression for the differential cross-section in the Born Approximation may be obtained by summing (8.5) over all atoms in the system with the appropriate phase factors and taking the square of the modulus. In this expression cross terms involving products of the charge-scattering-like terms (first two terms in (8.5)) with the third term will occur, provided both are finite at that particular point in reciprocal space (the last term is usually small and may often be neglected). We can define a scattering amplitude density for charge (involving the first two terms in (8.5) times the number density of the corresponding atoms) and similarly the magnetic scattering amplitude density involving the last two terms in (8.5). It may be noted that the symmetry of the third term is such that it predominantly scatters in the $\sigma \rightarrow \pi$ or $\pi \rightarrow \sigma$ channels, causing a rotation of the plane of polarization of the scattering photons. If we now consider circular polarization incident x-rays, where $\boldsymbol{e}_\mathrm{i} = \sqrt{1/2}\left(\boldsymbol{e}_x \pm \boldsymbol{e}_y\right)$, $\boldsymbol{e}_x$ and $\boldsymbol{e}_y$ being two mutually orthogonal unit vectors transverse to the incident direction of propagation, and denote by I^+ and I^- the scattered intensities obtained with each sense of circular polarization (corresponding to the $+/-$ signs in the expression for e_i above), it can be easily shown that neglecting the last term in (8.2), $(I^+ - I^-)$ depends only on the cross-terms involving charge and magnetic scattering, i.e. on the interference between them. Thus the much larger purely charge scattering (and also the purely magnetic scattering) disappears from the expression for $I^+ - I^-$ [8.32]. Inspection of (8.5) shows the same result occurs by reversing the direction of magnetization of the sample (again neglecting the final term quadratic in m), a method which is sometimes more convenient to use than reversing the circular polarization of the beam. Conversely, scattering with linear polarization which is a linear superposition of right and left circular polarization $I^+ + I^-$ is sensitive to the sum of the charge and magnetic scattering where the magnetic-charge cross term is suppressed [8.33]. The charge and magnetic terms can further be separated by polarization analysis as discussed above.

8.2.3.2 Interfacial scattering

The cross section for scattering can, in general, be written as:

$$\frac{\mathrm{d}\sigma}{\mathrm{d}\Omega} = \frac{1}{16\pi^2} \left|\langle k_{\mathrm{i}\mu}\right| T \left|k_{\mathrm{f}\nu}\rangle\right|^2 \tag{8.6}$$

where we assume that the scattering is elastic, from photon state $\boldsymbol{k}_{\mathrm{i}\mu}$ (with incident photon wavevector $\boldsymbol{k}_{\mathrm{i}}$ and polarization state μ) to photon state $\boldsymbol{k}_{\mathrm{f}\nu}$ (similarly defined) and the T-matrix element can be written down in various approximations. The simplest is the so-called kinematic or Born approximation where $\langle k_{\mathrm{i}\mu}\right| T \left|k_{\mathrm{f}\nu}\rangle$ is given by the matrix element of the scattering amplitude (f in (8.5)) times the density) evaluated between plane wave photon states. If we are interested in the scattering at small $\boldsymbol{q}$ we can assume a uniform scattering amplitude density (which includes both the charge and magnetic scattering terms given in (8.5)) except at an interface between materials of different scattering amplitude densities. Therefore, scattering only occurs at the surface or interface between layers. Then it may be shown that the above matrix element may be written as a sum over all interfaces in system in the form

$$\begin{aligned} \langle k_{\mathrm{i}\mu} | T | k_{\mathrm{f}\nu}\rangle =& i \sum_j \frac{(\Delta f)^j_{\mu\nu}}{q_z} \exp\left[-iq_z\bar{z}_j\right] \\ &\times \iint \mathrm{d}x\,\mathrm{d}y \exp\left[-iq_z\delta_{z_j}(x,y)\right] \exp\left[-i(q_x x + q_y y\right] \end{aligned} \tag{8.7}$$

where the z-axis is normal to the average interfaces which are all assumed to be parallel as in a thin film or multilayer (Fig. 8.3), j denotes the interface number, the average position of the jth interface is denoted as $\bar{z}_j$, the fluctuation $\delta z_j(x, y)$ is the height fluctuation of the jth interface from its average value at a lateral position (x, y), $(\Delta f)^j_{\mu\nu}$ is the change in scattering amplitude density (evaluated between polarization states $\mu\nu$) across the jth interface.

In (8.7) the sum over j includes all interfaces (structural as well as magnetic, although their average heights often coincide except when magnetic dead layers are present). It is often convenient to write the cross section in terms of the scattering function $S_{\mu\nu}(\boldsymbol{q})$ from polarization state μ to ν at wavevector $\boldsymbol{q}$. Inserting (8.7) into (8.6) and taking the appropriate statistical averages, it may be shown that

$$\begin{aligned} S_{\mu\nu}(\boldsymbol{q}) =& \frac{A}{q_z^2} \sum_{i,j} \left[(\Delta f)^{*i}_{\mu\nu} (\Delta f)^j_{\mu\nu}\right] \exp\left[-\frac{1}{2} q_z^2 \left(\sigma_i^2 + \sigma_j^2\right)\right] \exp\left[-\mathrm{i}q_z\left(\bar{z}_i - \bar{z}_j\right)\right] \\ &\times \iint \mathrm{d}x\,\mathrm{d}y \exp\left[q_z^2 C_{ij}(x,y)\right] \exp\left[-\mathrm{i}\left(q_x x + q_y y\right)\right] \end{aligned} \tag{8.8}$$

where σ_i^2 is the appropriate mean square roughness (structure or magnetic) for interface denoted by i, and C_{ij} (x, y) denotes the statistical height-height correlation function from the appropriate interfaces depending on the subscript i, j. In making this derivation, one generally makes the assumption of a Gaussian statistical distribution of height fluctuations. However other profiles can be calculated [8.34, 35]. In general, $C_{ij}(\boldsymbol{r})$ (where $\boldsymbol{r} \equiv (x, y)$, the lateral separation) will tend to zero as $\boldsymbol{r} \to \infty$, and hence the exponential in the integral in (8.8) will tend to unity. We can explicitly separate out this contribution, as it will yield a delta function in (q_x, q_y), which corresponds to the specular reflection. We thus get for the specular contribution:

$$S_{\mu\nu}^{\text{specular}}(\boldsymbol{q}) = \frac{16\pi^2 A}{q_z^2} \sum_{i,j} [(\Delta f)_{\mu\nu}^{*i} (\Delta f)_{\mu\nu}^{j}] \times \exp\left[-iq_z\left(\bar{z}_i - \bar{z}_j\right)\right] \exp\left[-\frac{1}{2} q_z^2 \left(\sigma_i^2 + \sigma_j^2\right)\right] \delta(q_x)\delta(q_y) \quad (8.9)$$

and for the off-specular or diffuse contribution

$$S_{\mu\nu}^{\text{diffuse}}(q) = \frac{A}{q_z^2} \sum_{i,j} [(\Delta f)_{\mu\nu}^{*i} (\Delta f)_{\mu\nu}^{j}] \exp\left[-iq_z\left(\bar{z}_i - \bar{z}_j\right)\right] \exp\left[-\frac{1}{2} q_z^2 \left(\sigma_i^2 + \sigma_j^2\right)\right] \times \iint \mathrm{d}\boldsymbol{r} \exp\left[-\mathrm{i}\boldsymbol{q}_{//} \cdot \boldsymbol{r}\right] \left(\exp\left|q_z^2 C_{ij}(\boldsymbol{r})\right| - 1\right) \quad (8.10)$$

where $\boldsymbol{q}_{//}$ is the in-plane component of $\boldsymbol{q}$. From (8.9) and the delta-function form of $S_{\mu\nu}^{\text{specular}}(\boldsymbol{q})$, we may, using standard methods, convert it to an expression for the specular reflectivity R:

$$R_{\mu\nu}(\boldsymbol{q}) = \frac{16\pi^2}{q_z^4} \sum_{i,j} [(\Delta f)_{\mu\nu}^{*i} (\Delta f)_{\mu\nu}^{j}] \exp\left[-\frac{1}{2} q_z^2 \left(\sigma_i^2 + \sigma_j^2\right)\right] \times \exp\left[-iq_z\left(\bar{z}_i - \bar{z}_j\right)\right]. \quad (8.11)$$

Let us now consider some special cases of (8.10) and (8.11). In the absence of the magnetic terms in (8.5), the second term, which has the same form as the Thomson scattering, can be combined with the first to give the total effective scattering amplitude $\tilde{f}_0$. For structural interfaces only therefore we obtain:

$$R_{\mu\nu}(\boldsymbol{q}) = \frac{16\pi^2}{q_z^4} \sum_{i,j} \left|\boldsymbol{e}_\mu^* \cdot \boldsymbol{e}_\nu\right|^2 \left(\Delta \tilde{f}_0\right)_i^* \left(\Delta \tilde{f}_0\right)_j \exp\left[-\frac{1}{2} q_z^2 \left(\sigma_{S,i}^2 + \sigma_{S,j}^2\right)\right] \times \exp\left[-iq_z\left(\bar{z}_i - \bar{z}_j\right)\right] \quad (8.12)$$

$$S_{\mu\nu}^{\text{diffuse}} = \frac{A}{q_z^2} \sum_{i,j} \left|\boldsymbol{e}_\mu^* \cdot \boldsymbol{e}_\nu\right|^2 \left(\Delta \tilde{f}_0\right)_i^* \left(\Delta \tilde{f}_0\right)_j \exp\left[-\frac{1}{2} q_z^2 \left(\sigma_{S,i}^2 + \sigma_{S,j}^2\right)\right] \times \exp\left[-iq_z\left(\bar{z}_i - \bar{z}_j\right)\right] \iint \mathrm{d}\boldsymbol{r} \exp\left[-\mathrm{i}\boldsymbol{q}_{//} \cdot \boldsymbol{r}\right] \left(\exp\left|q_z^2 C_{ij}^S(\boldsymbol{r})\right| - 1\right) \quad (8.13)$$

where $(\Delta \tilde{f}_0)_j$ is the discontinuity of the scattering amplitude density across the ith interface, $\sigma^2_{S,i}$ is the corresponding mean square structural roughness and C^S_{ij} is the structural roughness height-height correlation function between interfaces i and j, often expressible in the form given in (8.3) [8.35]. In the case of charge scattering, equations of the form (8.12) and (8.13) have been used successfully to describe the scattering from rough surfaces, thin films and multilayers, allowing modeling of the structural roughness of each interface in terms of the parameters σ, h and in-plane and out-of-plane correlation lengths ξ and $\xi_\perp$, respectively. Equations (8.12) and (8.13) can be readily extended to have more specific function for C^S_{ij} [8.36] such as obtain for self-assembled structures in multilayers [8.37].

From the above equations it may be noted that (a) charge scattering cannot yield intensity in channels where the photon polarization is flipped by 90°, e.g. $\sigma \to \pi$, or $\pi \to \sigma$; (b) the roughness of each interface enters into the specular reflectivity only through σ the root-mean-square roughness whereas the expression for diffuse scattering involves all the roughness parameters including the correlation lengths and (c) peaks in the specular reflectivity appear due to the periodicity of bilayers in the multilayer arising from the sum over the phase factor $\exp\lfloor -iq_z\left(\bar{z}_i - \bar{z}_j\right)\rfloor$, which are known as the multilayer Bragg peaks. These multilayer Bragg peaks are mirrored in peaks (or rather ridges) in the diffuse scattering at the same q_z values, provided the correlation function $C_{ij}(\boldsymbol{r})$ is non-zero for significant separations of the interfaces i and j along the z-axis, i.e. provided that the out-of-plane roughness correlation length is not too small. Because of this, there are examples of multilayers with large conformal roughness where 'Bragg' peaks are observed in the diffuse scattering but not readily observed in the specular scan [8.17].

The above models, applied to x-ray scattering from surfaces [8.38], thin films [8.39] or multilayers where the magnetic scattering can be ignored, have been widely used to extract details of the roughness at the interfaces [8.13–16, 35, 40–48]. As may be seen from (8.12) and (8.13), the specular reflectivity may be used to yield the values of the mean-square structural roughness of the ith interface $\sigma^2_{s,i}$, while the off-specular or diffuse scattering may be used to obtain the correlation function $C_{ij}(\boldsymbol{r})$ which depend on the roughness exponent, the in-plane roughness correlation length ξ, and the correlation *between* interfaces which depends on the out-of-plane correlation length $\xi_\perp$. With these parameters determined, then one can then make a quantitative comparison to various growth models. The evolution and correlations of the interfacial roughness with film thickness can be calculated and compared with the measured values and provide insight into the underlying growth mechanisms [8.13, 41–47].

Figure 8.5 shows an example of the specular reflectivity and diffuse (rocking curve) scans for a GaAs/AlAs multilayer that have been analyzed in terms of this kind of model [8.48]. The specular scan shows the expected multilayer Bragg peaks that are separated by $\Delta q = 2\pi/\Lambda$. For the transverse or diffuse scan, the specular scattering is seen as the sharp peak at $q_x = 0$ with the diffuse scattering seen as broad scattering at finite q_x (the peaks in the diffuse scattering arise from dynamical effects as discussed below). For single surfaces, in cases where detailed profiles has been also studied by scanning tunneling microscopy (STM) or atomic force microscopy (AFM) and the height-height correlation function reconstructed, the values for the

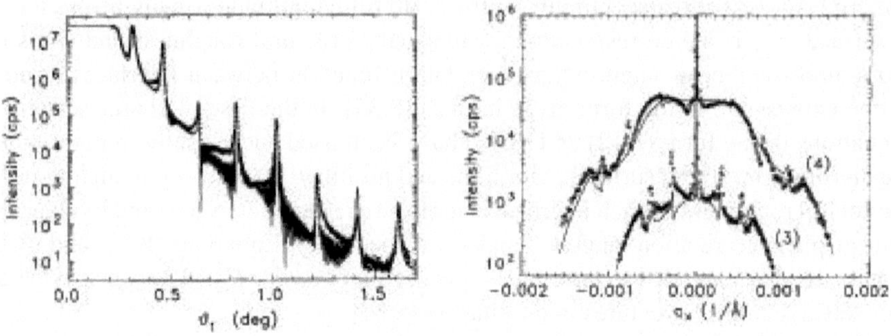

Fig. 8.5. Specular (*left*) and transverse (*right*) scans for a GaAl/AlAs superlattice from [8.48]. The peaks in the specular scan arise from the multilayer periodicity. The transverse scans are scans of q_x where the q_z values correspond to the 3rd and 4th Bragg peak of the specular scan

structural roughness parameters σ, ξ, and h are generally in good agreement with values obtained from scattering studies [8.37, 38, 42, 43, 46]. It is to be noted that the scanning microscopy techniques cannot be used to study multiple or buried interfaces which however pose no problem for scattering methods. Limited comparisons with transmission electron microscopy (TEM) have which can access buried interfaces has also shown quantitative agreement with x-ray scattering [8.16, 26, 49].

As mentioned above, for many experiments with slits wide in the direction normal to the plane of scattering, one measures $S_{\mu\nu}^{\text{diffuse}}(\boldsymbol{q})$ integrated over the in-plane component of $\boldsymbol{q}$, i.e. q_y. In this case it is easy to see from (8.13), that instead of the 2-D Fourier transform of $\left(\exp\lfloor q_z^2 C_{ij}(\boldsymbol{r})\rfloor - 1\right)$, one measures only its 1-D Fourier transform (with respect to x). An interesting variation to the above discussion arises when there is appreciable interdiffusion across an interface yielding a 'graded' rather than 'abrupt' interface. In such a case, $\sigma_{s,i}^2$ measured from specular reflectivity is really the sum:

$$\sigma_{s,i}^2 = \sigma_i^2(\text{ID}) + \sigma_i^2(\text{R}) \tag{8.14}$$

where $\sigma_i^2(\text{ID})$ represents the smearing of the interface due to interdiffusion and $\sigma_i^2(\text{R})$ represents that due to roughness. However, the diffuse scattering still represents scattering from roughness only, except that (8.13) has to be multiplied by "form factors" for each interface representing the Fourier-transform of the graded density across the interface in the z direction. Thus a combined measurement of the integrated diffuse scattering and specular reflectivity can be used to separate the two components [8.14].

At this point we should stress that the kinematic approximation is only valid when the scattering is weak and breaks down in the vicinity of total reflection or strong Bragg reflections. In such a case, for specular reflectivity one can develop a full dynamical theory by using well-known iterative or matrix methods as are used for discussing the optical properties of compound films. Such methods are usually formulated without including roughness at the interfaces, which can then be included as an extra Debye-Waller-like factor multiplying the reflectivity of each interface or by fine slicing of the graded (x, y)-averaged scattering amplitude density representing the rough interface . To extend this approach to achieve a more accurate calculation of the diffuse scattering, one may use the so-called Distorted Wave Born Approximation (DWBA) [8.11].

In this approximation, the matrix element $\langle k_{i\mu} | T | k_{f\nu} \rangle$ is the scattering matrix element between the appropriate states for a smooth surface plus the matrix element of the difference between the scattering amplitude density of the rough surface and that of the smooth surface evaluated between the actual eigenstates of the smooth surface problem (i.e. incident plus reflected and transmitted beams). Details of this method for a single interface are given in [8.11] and the multiple interfaces in [8.35, 48, 50]. The main results of the DWBA for the diffuse scattering from a single interface with no magnetic component is given by:

$$S_{\mu\nu}^{\text{diffuse}}(\boldsymbol{q}) = \frac{A}{16\pi^2} \left| \boldsymbol{e}_\mu^* \cdot \boldsymbol{e}_\nu \right|^2 \left| k_0^2 (1 - n^2) \right|^2 \left| T(\overrightarrow{\boldsymbol{k}_\mathrm{i}}) \right|^2 \left| T(\overrightarrow{\boldsymbol{k}_\mathrm{f}}) \right|^2 S(\boldsymbol{q}_\mathrm{t}) \tag{8.15}$$

where k_0 is the magnitude of the incident wave vector, n is the refractive index of the medium, A is the illuminated surface area, $T(\overrightarrow{\boldsymbol{k}_\mathrm{i}})$ is the Fresnel transmission coefficient for a smooth interface for incident wavevector $\overrightarrow{\boldsymbol{k}_\mathrm{i}}$ and similarly for $T(\overrightarrow{\boldsymbol{k}_\mathrm{f}})$, $\boldsymbol{q}_\mathrm{t}$ is the wavevector transfer in the medium as opposed to free space, i.e. allowing for for refractive effects. Note that $\boldsymbol{q}_\mathrm{t}$ may be complex or even purely imaginary (as is the case when the wave in the medium is purely evanescent). $S(\boldsymbol{q}_\mathrm{t})$ is given by:

$$S(\boldsymbol{q}_\mathrm{t}) = \frac{\exp\left[-\frac{1}{2}\sigma_S^2 (q_{z,t}^2 + q_{z,t}^{*2})\right]}{\left| q_{z,t} \right|^2} \iint \mathrm{d}^2\boldsymbol{r} \exp\left[-\mathrm{i} q_{//} \cdot \boldsymbol{r}\right] \\ \times \left(\exp[|q_{z,t}|^2 C_j^S(\boldsymbol{r})] - 1 \right) . \tag{8.16}$$

For multiple interfaces we need to know the transmission and reflection coefficients for specular reflection at each interface (obtained from the corresponding specular calculations) and one obtains:

$$S_{\mu\nu}^{\text{diffuse}}(\boldsymbol{q}) = \frac{A}{16\pi^2} \left| \boldsymbol{e}_\mu^* \cdot \boldsymbol{e}_\nu \right|^2 \sum_{i,j=1}^{N} \left[k_0^2 (n_i^2 - n_{i+1}^2) \right] \left[k_0^2 (n_j^2 - n_{j+1}^2) \right] \\ \times \sum_{m,n=0}^{3} \frac{1}{(q_{z,m}^{i+1})(q_{z,m}^{j+1})^*} \exp\left[-\frac{1}{2} \sigma_j^2 (q_{z,m}^{i+1})^2 \right] \exp\left[-\frac{1}{2} \sigma_k^2 (q_{z,n}^{j+1})^{*2} \right] \\ \times \iint \mathrm{d}^2\boldsymbol{r} \exp[-\mathrm{i} q_{//} \cdot \boldsymbol{r}] \left(\exp[(q_{z,m}^{i+1})(q_{z,n}^{j+1})^* C_{ij}^S(\boldsymbol{r})] - 1 \right) \tag{8.17}$$

where i, j denote the layers in the multilayer running from 1 to N (N defining the substrate), n_i is the refractive index for the ith layer and $q_{z,m}^i$ ($m = 0$–3) are the z components in layer i defined as follows

$$\overrightarrow{q}_0^i = \overrightarrow{k}_2^i - \overrightarrow{k}_1^i ; \quad \overrightarrow{q}_1^i = \overrightarrow{k}_2^i \overrightarrow{k}_1^i ; \quad \overrightarrow{q}_2^i = \overrightarrow{k}_2^i \overrightarrow{k}_1^i ; \quad \overrightarrow{q}_3^i = \overrightarrow{k}_2^i \overrightarrow{k}_1^i ; \tag{8.18}$$

In (8.13), $\vec{k}_1^{\,i}$, $\vec{k}_2^{\,i}$ are the wavevectors of the transmitted and reflected waves in layer i, while $\vec{k}_1'^{\,i}$, $\vec{k}_2'^{\,i}$ represent the wavevectors of the corresponding time-inverted states.

In general, the DWBA provides a good description of the diffuse scattering and explains additional structure in the diffuse scattering not accounted for in the Born Approximation. This additional scattering appears as peaks or dips in transverse scans i.e. rocking curves or scans through the specular peaks with only the in-plane component of $\boldsymbol{q}$ being varied (see Fig. 8.5 for example). These features occur when either of the angles which the incident or outgoing beams makes with the surface (θ_i or θ_f respectively in Fig. 8.3) is at the critical angle for the total reflection (in which case the peaks are know as "Yoneda wings" [8.51]) or is at an angle for multilayer Bragg reflection [8.52]. The physical explanation of the Yoneda wing peaks is that the incident and reflected beams at the surface are in phase at the critical angle for total reflection, resulting in a 2-fold increase in the magnitude of the electric field at the surface and thus a fourfold increase in the diffusely scattered intensity. At angles where θ_i or θ_f are set for multilayer Bragg reflections, the X-ray wavefield in the multilayer has the form of standing waves with the nodes or antinodes moving rapidly through the interfaces as a function of θ_i or θ_f, thus giving rise to sharp structure in the diffuse intensity (see Fig. 8.5).

8.2.3.3 Magnetic Scattering

Let us go back to the Born approximation expressions and now include the resonant magnetic scattering. This means that the sum over i, j in (8.10) and (8.11) sum over both the chemical *and* magnetic scattering. The polarization states of the photon are taken into account in the factors $(\Delta f)^i_{\mu\nu}$. In evaluating the latter, it is often convenient to neglect the term in (8.5) which is much smaller than the others and rewrite (8.5) in the form

$$f_{\mu\nu} = A[\vec{e}^{\,*}(\nu)\cdot\vec{e}(\mu)] - \mathrm{i}B[\vec{e}^{\,*}(\nu)\times\vec{e}(\mu)]\cdot\hat{\boldsymbol{m}} \tag{8.19}$$

where

$$A = f_0 + \frac{3\lambda}{8\pi}[F_{11}+F_{1-1}] \quad \text{and} \quad B = \frac{3\lambda}{8\pi}[F_{11}-F_{1-1}]. \tag{8.20}$$

We can then write the explicit expressions for the scattering for important special cases. For $\sigma \to \pi$ scattering process (where the incident photon polarization $\vec{e}(\mu)$ is parallel to the surface and the scattering photon polarization is rotated by 90°), there will obviously be no contribution from the first term (i.e. no charge scattering) and we obtain:

$$S^{\text{diffuse}}_{\mu\nu}(\boldsymbol{q}) = \frac{A}{q_z^2}\sum_{i,j=m}\left[(\Delta g_1)^{*i}(\Delta g_1)^j\right]\exp\left[-\mathrm{i}q_z(\bar{z}_i-\bar{z}_j)\right]\exp\left[-\frac{1}{2}q_z^2(\sigma^2_{m,i}+\sigma^2_{m,j})\right]$$
$$\times\iint \mathrm{d}^2\boldsymbol{r}\exp\left[-\mathrm{i}q_{//}\cdot\boldsymbol{r}\right]\left(\exp\left[q_z^2-C^M_{ij}(\boldsymbol{r})\right]-1\right) \tag{8.21}$$

and

$$R_{\sigma\pi}(\boldsymbol{q}) = \frac{16\pi^2}{q_z^4} \sum_{i,j} \left[(\Delta g_1)^{*i} (\Delta g_1)^j \right] \exp\left[-\mathrm{i} q_z (\bar{z}_i - \bar{z}_j) \right] \times \exp\left[-\frac{1}{2} q_z^2 (\sigma_{m,i}^2 + \sigma_{m,j}^2) \right] \tag{8.22}$$

where $i, j = m$ in the sum defines a sum over magnetic interfaces only. $(\Delta g_1)^i$ is the *discontinuity* in the quantity $[n_\mathrm{m} B(\hat{\boldsymbol{k}}_\mathrm{f} \cdot \hat{\boldsymbol{m}})]$ going across the ith interface in the positive z direction, where n_m is the number density of resonant magnetic atoms, B is defined in (8.20), $\hat{k}_\mathrm{f}$ and $\hat{m}$ are unit vectors in the direction of $\overrightarrow{k_\mathrm{f}}$ and the average magnetization in the layer, respectively. $\sigma_{m,i}^2$ is the mean square *magnetic* roughness of the ith interface and $C_{ij}(\boldsymbol{r})$ is the correlation function defined in (8.4) for deviations from the *magnetic* interface i, j. From the definition of $(\Delta g_1)^i$, it can be seen that components of the magnetization perpendicular to the plane of scattering will not contribute to (8.21) and (8.22) and that for grazing angles of incidence and scattering, the biggest contribution comes from the components of the magnetization in the plane of the surface and also in the plane of the scattering, i.e. along the x-axis (Fig. 8.3). Similar expressions can be written down for $S_{\pi\sigma}^{\mathrm{diffuse}}(\boldsymbol{q})$ and $R_{\pi\sigma}(\boldsymbol{q})$ except the $(\Delta g_1)^i$ now stands for the discontinuity in the quantity $[n_\mathrm{m} B(\hat{\boldsymbol{k}}_\mathrm{f} \cdot \hat{\boldsymbol{m}})]$ across the ith interface.

For circularly polarized incident photons without using any polarization analysis of the scattered radiation, we obtain for the difference in the cross-sections or reflectivities for (+) and (−) circular polarization, the expression

$$\Delta S^{\mathrm{diffuse}}(\boldsymbol{q}) = \frac{A}{q_z^2} \sum_{i=s,j=m} \left[(\Delta g_2)^{*i} (\Delta g_3)^j \right] \exp\left[-\mathrm{i} q_z (\bar{z}_i - \bar{z}_j) \right] \times \exp\left[-\frac{1}{2} q_z^2 (\sigma_{s,i}^2 + \sigma_{m,j}^2) \right] \iint \mathrm{d}^2\boldsymbol{r} \exp\left[-\mathrm{i}\boldsymbol{q}_{//} \cdot \boldsymbol{r} \right] \times \left(\exp\left[q_z^2 - C_{ij}^{S,M}(\boldsymbol{r}) \right] - 1 \right) \quad + \text{complex conjugate} \tag{8.23}$$

and

$$\Delta R(\boldsymbol{q}) = \frac{16\pi^2}{q_z^4} \sum_{i=s,j=m} \left[(\Delta g_2)^{*i} (\Delta g_3)^j \right] \exp\left[-\mathrm{i} q_z (\bar{z}_i - \bar{z}_j) \right] \times \exp\left[-\frac{1}{2} q_z^2 (\sigma_{s,i}^2 + \sigma_{m,j}^2) \right] \quad + \text{complex conjugate} \tag{8.24}$$

where the sum over $i = s$, $j = m$ denotes i is to be summed over only structural interfaces and j is to be summed over only magnetic interfaces. $(\Delta g_2)^i$ is the discontinuity in the quantity $[N_\mathrm{nr} + n_\mathrm{m} A]$ across the ith interface going in the positive z direction where N_nr is the number density of non-resonant atoms times their scattering factors and A in defined in (8.20). $(\Delta g_3)^i$ is the discontinuity in the quantity $[n_\mathrm{m} B\{(\hat{\boldsymbol{k}}_\mathrm{f} \cdot \hat{\boldsymbol{m}}) + \cos(\theta_\mathrm{i} + \theta_\mathrm{f})(\hat{\boldsymbol{k}}_\mathrm{i} \cdot \hat{\boldsymbol{m}})\}]$ across the jth interface. From the above expressions, we see that pure charge scattering is again absent from $\Delta S^{\mathrm{diffuse}}(\boldsymbol{q})$ and

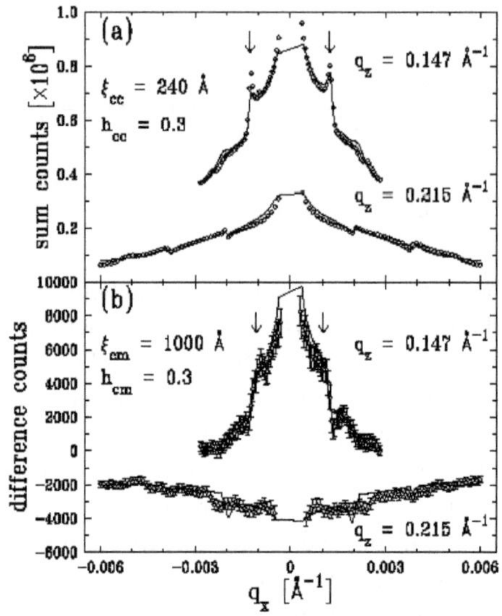

Fig. 8.6. Resonant diffuse scattering from a [Gd(53.2 Å)/Fe(36.4 Å)]$_{15}$ multilayer with the photon energy to the Gd L_3 edge at 7245 eV [8.54]. Measured sum [$(I_+ + I_-)$, (**a**)] and difference [$(I_+ - I_-)$, (**b**)] of opposite photon helicity rocking curve data (circles) at the second ($q_z = 0.147\,\text{Å}^{-1}$) and the third ($q_z = 0.215\,\text{Å}^{-1}$) multilayer Bragg peaks. The lines represent the dynamical calculations in the DWBA and explain well the anomalous scattering features indicated by the arrows. The fit allows one to extract the interfacial roughness and the correlation between the charge and magnetic roughness

$\Delta R(\boldsymbol{q})$, and that $\Delta S^{\text{diffuse}}(\boldsymbol{q})$ depends only on the cross-correlation function $C_{ij}^{S,M}(\boldsymbol{r})$ between the *structure* and *magnetic* roughness fluctuations, i.e. if there was no correlation between the magnetic and structural interfaces this term would vanish [8.32]. The approach can be extended to include magnetic scattering with the DWBA (the magnetic analog to (8.13)) as described in [8.53] and [8.54] and is applied to quantify the diffuse magnetic scattering from a Gd/Fe multilayer in Fig. 8.6.

From the above discussion it is clear that magnetic x-ray scattering provides considerable opportunity to directly characterize the magnetism in thin films and at surfaces and interfaces [8.20, 21, 30–33]. In addition to characterizing the magnetic roughness, it can also characterize domain structures that can be viewed as a type of magnetic roughness and incorporated into the definition of the magnetic correlation function [8.31–33, 55]. However, it also points out the need to control the polarization of both the incoming and scattered radiation. Techniques for doing this are increasingly common at synchrotron facilities and open the opportunities for applying these techniques to a variety of thin film and nanostructured systems. The above discussion also points to the need to understand the source of the scattering (magnetic, charge, or interference of the two). This can, in part be achieved by analysis of the polarization as well as the energy dependence of the scattering. The energy dependence of the charge and magnetic scattering factors in (8.2) typically has different spectral features [8.30, 33]. Understanding the energy dependence of the scattering is particularly important since the measurements are performed at resonant edges to enhance the magnetic scattering where there are other magneto-optical and absorption effects that are also quite strong and can also affect the measured scattering curves [8.30].

8.3 Wide-angle Diffraction Measurements of Layered Structures

8.3.1 Introduction

Wide-angle diffraction measurements are performed in similar scattering geometries but at higher q values ($q \simeq 2\pi/a$) than for reflectivity measurements and the scattering is sensitive to the atomic order (both structural and magnetic). This sensitivity providing complementary information on the structure often at very different length scales when compared to reflectivity measurements which results in important differences in interpreting wide-angle diffraction scans. At higher q values the scattering arises from all the atoms in the structure and not just the interfaces as encountered in reflectivity measurements. The atomic structure within the layers and how the atomic structure propagates across the interface are both important to the scattered intensity (in contrast to the reflectivity which is insensitive to the structure within the layer). A trivial example that highlights this difference is comparing a multilayers made up of amorphous layers compared to one consisting of crystalline layers. Assuming similar interface morphologies, the two multilayers would have indistinguishable reflectivity scans whereas the amorphous multilayer would not contribute any wide-angle diffraction since it only has long-range chemical order from the layering but no long-range crystalline order. A similar effect occurs in multilayers where one of the layers is amorphous [8.12] or an amorphous interfacial compound forms at the interface [8.56]. Again, the long-range crystalline order is disrupted and the wide-angle diffraction is only sensitive to order within the layers but not to the multilayer structure.

For a multilayer consisting of crystalline layers as shown schematically in Fig. 8.7 the scattering will become sensitive to the multilayer structure [8.57] where the scattering averages coherently only over regions set by the crystalline coherence length of the sample. This length is often less than the coherence of the x-ray beam where the coherence is limited by grain boundaries for a polycrystalline films, defects within the layers, disorder introduced at the interfaces or fluctuation at the interfaces. In general this coherence length is significantly less than that that probed in reflectivity measurements or imaging techniques such as atomic force or transmission electron microscopy [8.58, 59]. For a polycrystalline multilayer with lateral grain sizes of $\sim$ 10–20 nm, this sets the lateral coherence lengths whereas reflectivity measurements are not sensitive to the granular structures and can measure grain-to-grain variations in the interface position. This latter measurement of the interfacial roughness may differ from that determined from wide-angle diffraction.

Another difference in comparing wide-angle and reflectivity measurements is that most thin film and multilayer samples are characterized by a distribution in the angle

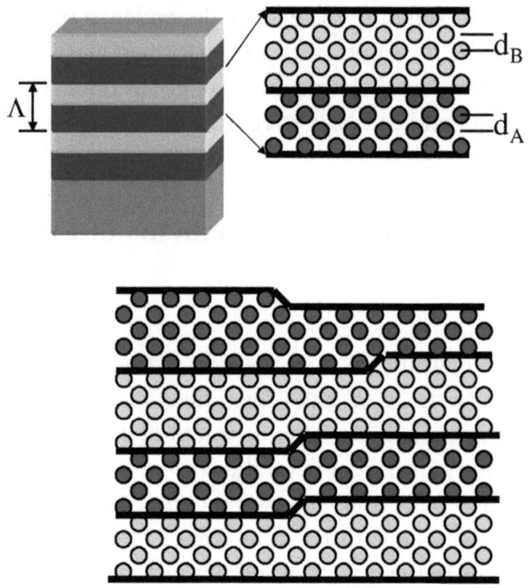

Fig. 8.7. Schematic representation of a multilayer structure on a substrate with a periodicity of Λ. An expanded view of a single bilyer (one period) assuming crystalline layers with no disorder. (*below*) Schematic representation of a two multilayer periods with random thickness fluctuations

the atomic planes make with the surface and described as a mosaic crystal [8.60]. This coupled with the limited lateral coherence of the atomic structure results, in general, that the scattering cannot be separated into a specular and diffuse components and the scattering is treated as the integrated intensity (specular + diffuse). As such, the in-plane length scale of the interfacial roughness is not accessible through $q_{//}$ resolved diffuse scans as done in reflectivity measurement. Therefore most wide-angle diffraction models are one-dimensional and ensemble average lateral variation in the interfaces and layer structures. Thus, in most cases an equivalent measure to the height-height correlation function determined in diffuse reflectivity scans is not obtainable from a wide-angle diffraction measurement [see [8.61] and [8.62] for examples where this is not the case].

8.3.2 Wide Angle Diffraction Measurements

The scattering geometries used for wide-angle diffraction are the same as those described in Fig. 8.3 but the scattering is not nicely separated into specular and diffuse components. The scattering geometries are chosen such the scattering wavevector $\boldsymbol{q}$ is commensurate with a crystallographic direction of the lattice. The most common scan has the $\boldsymbol{q}$ normal to surface and probes the crystalline and multilayer order normal to the layers (again resulting in multilayer Bragg peaks). A rocking curve scan about this direction gives the mosaic spread of the crystallites that make up the sample. The other common scan is the grazing incident x-ray scan (GIXS) where θ_i

and θ_f are $\sim$ few degrees and are just above the angle for total external reflection. ϕ_f and the sample are rotated (with θ_i and θ_f fixed) such that scattering wavevector is nearly in plane and thus probes different in-plane directions of the samples. In this geometry the scattering measures the in-plane lattice spacing of the two layers independently avoiding any interference between the layers. From the positions of the diffraction peaks, the in-plane strain of the layers and the epitaxial relations of the sample can be directly determined.

8.3.3 Scattering Formalism

We will focus on the scattering normal to surface of a multilayer although the results are easily extended to thin films. In addition, we will describe the scattering in the Born Approximations since most magnetic thin films and multilayers do not have sufficient crystalline perfection that multiple scattering needs be considered. This assumption holds as long as one avoids regions of large reflectivity such as critical angles or substrate reflections. We will first discuss the expected scattering intensity without disorder and then extend this formalism to discuss this inclusion of disorder.

8.3.3.1 Without Disorder

As was discussed earlier the differential cross-section in the Born Approximation is obtained by summing (8.1) over all atoms in the system with the appropriate phase factors and taking the square of the modulus. For scattering normal to surface ($q_x = q_y = 0$) this sum reduces to a sum over the atomic planes normal to the surface [8.57]:

$$S(q_z) = \left| \sum_j f_j \exp\left(\mathrm{i} q_z z_j\right) \right|^2 \tag{8.25}$$

where z_j is now the position of the jth atomic layer (compared to the jth interface in the reflectivity modeling) and f_j is the atomic scattering factors (8.5) averaged laterally over the layers [8.57]. If one includes the model for a multilayer assuming that each bilayer (layer A and B of the multilayer) repeats with the periodicity of Λ (shown schematically in the top of Fig. 8.7) then (8.25) reduces to

$$S(q_z) = \left| \sum_n^M F_n \exp\left(\mathrm{i} q_z n \Lambda\right) \right|^2 \tag{8.26}$$

where M is the number of bilayer repeats and F_n is the scattering amplitude of a single bilayer given by

$$F_n = \sum_i f_i \exp\left(\mathrm{i} q_z z_i\right) \tag{8.27}$$

where the summation is over a single bilayer period. If one assumes that each bilayer repeats exactly then (8.26) reduces to the simple equation:

$$S(q_z) = |F|^2 \frac{1 - \exp(\mathrm{i}q_z M\Lambda)}{1 - \exp(\mathrm{i}q_z \Lambda)} \tag{8.28}$$

where $|F|^2$ is the square of the scattering amplitude of a single bilayer and the second term is the interference term that for large M results in Bragg peaks at $q_z = 2\pi n/\Lambda$ just as observed in reflectivity measurements. If we define the average lattice spacing $\bar{d} = \Lambda/N$ where N is the average number of atomic planes in a unit cell (single multilayer bilayer) then the Bragg peaks can be indexed to [8.63]

$$q_z = \frac{2\pi}{\bar{d}} \pm \frac{2\pi n}{\Lambda}. \tag{8.29}$$

An example is shown in Fig. 8.8 for a textured Mo/Ni multilayer that displays a series of equally-spaced diffraction peaks whose separation is determined by $\Delta q = 2\pi/\Lambda$. Thus for a multilayer the peak positions are determined solely by $\bar{d}$ and the period Λ (the peak position do not depend on the lattice constant of either material but on the average lattice constant). The atomic structure within the bilayer determines the peak intensity through $|F|^2$. The out-of-plane coherence length can be estimated from the line width. By fitting the measured peaks intensities to model bilayer structures one can extract the average bilayer structure. This type of structural characterization is commonly used for determining the structure of bulk crystals using the Rietveld refinement procedure [8.64]. The structure of a single unit cell is modeled and the relative intensities of the diffraction peaks are determined from the structure factor of the unit cell and the line shapes are fit to a model-independent line shape.

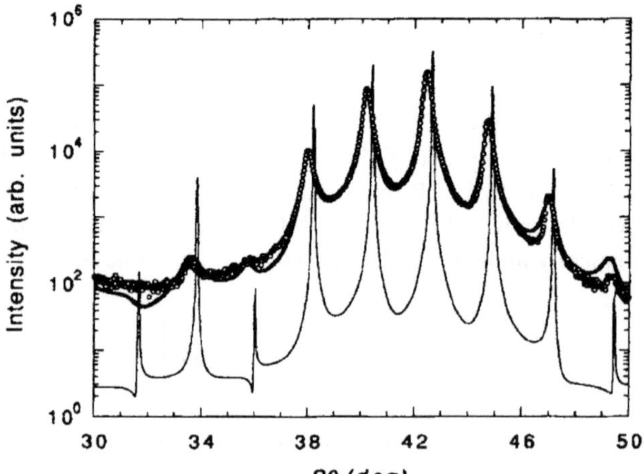

Fig. 8.8. Wide-angle diffraction scan from a textured Mo/Ni multilayer (circles) [8.63]. The thin solid line is a calculation assuming no disorder and the bulk structure, the thick solid line shows the refined structure

This approach can also be used to estimate the average structural properties of the multilayer where the peak positions are determined from (8.29) and peak intensity by (8.28). If one assumes that the bilayer consists of layers A and B which are composed of N_A and N_B atomic layers with lattice spacing d_A and d_B and scattering amplitudes f_A and f_B, respectively then F is given by:

$$F = \sum_{n=0}^{N_A-1} f_A \exp\left(iq_z n d_A\right) + \exp\left(iq_z t_A\right) \sum_{m=0}^{N_B-1} f_B \exp\left(iq_z \left(m d_B\right)\right) \tag{8.30}$$

where t_A is the thickness of layer A. One can easily include the effects of strain by varying the lattice spacing within the unit cell (assumed constant in (8.30)) or include interdiffusion by taking a weighted average of f_A and f_B as well as d_A and d_B for the atomic planes near the interface to mimic the compositional gradient across the interface [8.65–67]. Where this approach is lacking is that it does not include the effects of disorder which requires not only determining the average structure but the statistical deviation about the average structure.

8.3.3.2 With Disorder

Including the role of disorder in modeling the wide-angle scattering has a long history where a variety of approaches have been employed [8.63, 68–75]. The difficulty is that there are many types of disorder that are cumulative that lead to a loss of long-range order such as deviations from layer-by-layer growth, misfit dislocations, or interface fluctuations from lattice mismatches. To incorporate these effects most modeling has been based on the approach of Hendricks and Teller [8.76] that treats disorder as a random sequencing of layers where either the scattering power of the layers or the relative phase between layers varies randomly as one moves vertically through the structure. For instance if there are variations of the layer thickness (either from variations in the growth rates or roughness of the layers), then the number of atomic planes that characterize a layer will vary (at least locally) from one layer to the next. This is shown schematically in Fig. 8.7 where there are random steps in the layers that result in local variations in the layer thicknesses and the local periodicity Λ. To include these effects the integrated scattering intensity $S^{\mathrm{int}}(q)$ (specular plus diffuse) needs to calculated as pointed out in [8.68] and is given by ensemble averaging the intensity of all the possible sequences of different layer thicknesses:

$$S^{\mathrm{int}}(q_z) = \left\langle \left| \sum_{n}^{M} F_n \exp\left(iq_z \sum^{n-1} \Lambda_n \right) \right|^2 \right\rangle \tag{8.31}$$

where F_n and Λ_n are the structure factor and bilayer period of the nth bilayer, respectively. Equation (8.31) calculates the scattering intensity from each possible sequence and then averages these intensities. The specular intensity alone would be given by averaging the scattering amplitudes of each sequence and then squaring the average amplitude to calculate the scattered intensity:

$$S^{\text{specular}}(q_z) = \left| \left\langle \sum_{n}^{M} F_n \exp\left(\mathrm{i} q_z \sum^{n-1} \Lambda_n \right) \right\rangle \right|^2 . \tag{8.32}$$

In (8.31) and (8.32) both the scattering amplitude F_n and period Λ_n is allowed to vary from layer to layer and the brackets $\langle\ \rangle$ represent the ensemble average over all possible sequences. The diffuse scattering alone is given by $S^{\text{int}} - S^{\text{specular}}$. In general, the average can be done numerically where various structures are generated randomly and the ensemble averaged intensity calculated. However, for most data analysis an analytical expression proves invaluable. This can be achieved under the assumption that fluctuations from the average layer structure are cumulative (i.e. a phase error in one layer perturbs all subsequent layers) but are statistically independent for each layer. That is, the probability of a given F_n or Λ_n does not depend on F_{n-1} or Λ_{n-1}. Although this may seem rather constraining it has proven applicable to a wide variety of multilayers systems (there are examples where these assumptions clearly don't hold [8.77]). Under these assumptions it can be shown that (8.31) and (8.32) reduces to [8.63, 73, 78]:

$$S^{\text{int}}(q_z) = M\left\langle F^* F \right\rangle + 2\mathrm{Re}\left[\langle F\rangle \Phi\Psi/T\right] \tag{8.33}$$

$$S^{\text{specular}}(q_z) = \left| \frac{1 - T^M}{1 - T} \langle F\rangle \right|^2 \tag{8.34}$$

where Re refers to the real component within the brackets and $\langle F\rangle$ and $\langle F^* F\rangle$ are the ensemble average over all possible bilayer scattering amplitudes and intensities, respectively. The remaining terms are given by:

$$\begin{aligned} T &= \langle \exp(\mathrm{i} q_z \Lambda) \rangle \\ \Phi &= \left\langle \exp(\mathrm{i} q_z \Lambda)\, F^* \right\rangle \\ \Psi &= \frac{M - (M+1)T + T^{M+1}}{(1-T)^2} - M \end{aligned} \tag{8.35}$$

where the brackets in T and Φ ensemble average over all possible bilayers. Thus, the ensemble averages of all possible multilayer sequences can be calculated from the ensemble averaged bilayer properties ($\langle F\rangle$, $\langle F^* F\rangle$, T, and Φ) and (8.33) and (8.34).

This approach describes averaging different regions of the sample that scatter incoherently to achieve the intensity. One can extend this approach and assume that some fluctuations occur within the lateral coherence of the system where the scattering amplitude is averaged. The final intensity average (8.31) averages the square of these locally averaged scattering amplitudes. Again assuming statistically-independent fluctuations it can be shown that the scattered intensity reduces to a weighted sum of (8.31) and (8.32) where the relative weight depends on the number of fluctuations within the coherent region and includes the roughness of the substrate [8.75, 79].

These expressions can be made explicit for a multilayer where the unit cell has layers A and B that can be described by a scattering factors F_A and F_B and thicknesses t_A and t_B, respectively. The ensemble averaged parameters become:

$$\langle F \rangle = \langle F_A \rangle + T_A \langle F_B \rangle$$
$$\langle F^* F \rangle = \langle F_A{}^* F_A \rangle + \langle F_B{}^* F_B \rangle + 2\mathrm{Re}\left[\Phi_A \langle F_B \rangle\right] \tag{8.36}$$
$$\Phi = T_B \Phi_A + \Phi_B$$
$$T = T_A\, T_B \tag{8.37}$$

where the fluctuation are averaged over the individual layers A and B where $\Phi_A = \langle \exp\left(iq_z t_A\right) F_A^* \rangle$ and $T_A = \langle \exp\left(iq_z t_A\right) \rangle$ with similar terms for layer B.

The type of roughness will depend on the multilayer structure but can be generally separated into continuous and discrete contributions. Discrete roughness is associated with variations in the layer thickness by an integer number of atomic layers resulting from steps or non-uniform growth modes (Fig. 8.7). For this case the layer thickness varies by discrete values set by the lattice spacing. To ensemble average these fluctuations require calculating the parameters F_A, $F_A{}^* F_A$, $\exp(iq_z t_A) F_A{}^*$, and $\exp(iq_z t_A)$ for each possible number of atomic planes N_{Aj} and add them weighted by there probability of occurrence (and similarly for layer B):

$$\langle F_A \rangle = \sum_j P(N_{Aj}) F_A(N_{Aj})$$
$$\langle F_A{}^* F_A \rangle = \sum_j P(N_{Aj}) F_A(N_{Aj}) F_A^*(N_{Aj}) \tag{8.38}$$
$$\Phi_A = \sum_j P(N_{Aj}) \exp(iqt_{Aj}) F_A^*(N_{Aj})$$
$$T_A = \sum_j P(N_{Aj}) \exp(iqt_{Aj}) \tag{8.39}$$

where $P(N_{Aj})$ is the normalized probability of a layer A with N_{Aj} number of atomic planes and t_{Aj} is the corresponding layer thickness, and the sum is over all possible values of N_{Aj}. $F_A(N_{Aj})$ is the scattering amplitude calculated for layer A for N_{Aj} number of atomic planes (first term in (8.30)) which can include the effects of strain (varying lattice spacing) or interdiffusion (varying scattering amplitude and lattice spacing). There are similar terms for layer B.

Continuous roughness describes atomic level roughness that can vary continuously and can arise from dislocation, and incommensurate lattice mismatch interface, and is also used to describe thickness variation of an amorphous layer where the thickness is not determined by the lattice spacing [8.68]. A simple way to include continuous roughness is to assume that the layer thickness of the jth layer t_{Aj} (or equivalently the interfacial lattice spacing between layers) fluctuates about the average thickness by δt_{Aj}. Assuming δt_{Aj} is characterized by a Gaussian distribution of width c and averaging the appropriate parameters by integrating over all possible δt_{Aj}

one arrives at an additional term $\exp(-q^2c^2/2)$ that is multiplied to T_A and Φ_A. This term limits the phase coherence in the multilayer structure and tends to broaden all the diffraction peaks. It can be easily shown that for c values which approach the lattice spacing of the material (since $q \sim 2\pi/a$), long-range coherence is lost. This has been used to explain why long-range coherence is not observed in crystalline/amorphous multilayers or multilayers that form amorphous interfacial layers [8.68]. In contrast the role of discrete roughness is depends on the properties of the constituent materials and can broaden some diffraction peak will other are relatively narrow as was first pointed out by Hendricks and Teller [8.76].

These different roughness signatures allow them to be independently determined from the diffraction scans. Fitting the scattering profile (both peak heights and line widths) to general multilayer models refines both the average structure (lattice spacings, layer thicknesses, interdiffusion) and well as the fluctuations about the average structure (i.e. roughness). Such an approach is described in [8.63] for multilayers and [8.74] for thin films. Examples are shown in Figs. 8.8 and 8.9 for textured Mo/Ni multilayers. The calculated intensities in Fig. 8.8 were calculated using (8.28) and (8.30) assuming the bulk lattice constant and fit to (8.33) allowing both discrete and continuous disorder and lattice strains. The refined structure is able to reproduce the both the measured intensities as well as the peak profiles. The results of the fitting determines the average lattice spacings and the fluctuations in the thicknesses (however, without any in-plane length scale of the fluctuations). Comparisons to independent determinations of the structure have shown that this approach provides a quantitative measure of the structure [8.63, 80–82]. This same formalism has been also been used for analysis of magnetic neutron scattering [8.83] using neutron cross-sections instead of the atomic x-ray cross sections given in (8.2).

Shown in Fig. 8.9 are wide angle x-ray diffraction data and fits of a multilayer similar to that shown in Fig. 8.8 (measured both about the Mo(110)/Ni(111) and Mo(220)/Ni(222) reflections). In the (a) panels the multilayer was grown in the conventional manner. In the (b) and (c) panels the thickness of the Ni and Mo layer, respectively, was allowed to fluctuate during the growth process. As described originally by Hendricks and Teller, this type of disorder broadens some peaks more than other [8.70]. In particular, the variation in the Ni layer thickness ((b) panels) cause increased broadening of the peak that are more weighted to the Mo lattice position while fluctuations in the Mo layer thicknesses ((c) panels) results in broadening of the peaks more associated with the Ni layer. This rather counter intuitive finding results from the fact that variations in the Ni layer thickness has the effect of altering the interference between adjacent Mo layers and tends to broaden these peaks. This example also points out the importance of including disorder when modeling diffraction data. The multilayer in panels (a)–(c) have the same average structure (lattice constants, average thicknesses etc) but have varying peak intensities resulting from the disorder. The solid lines are fitted curves that reproduce the amount of fluctuations in the layer thickness and confirm that they have the same average structure.

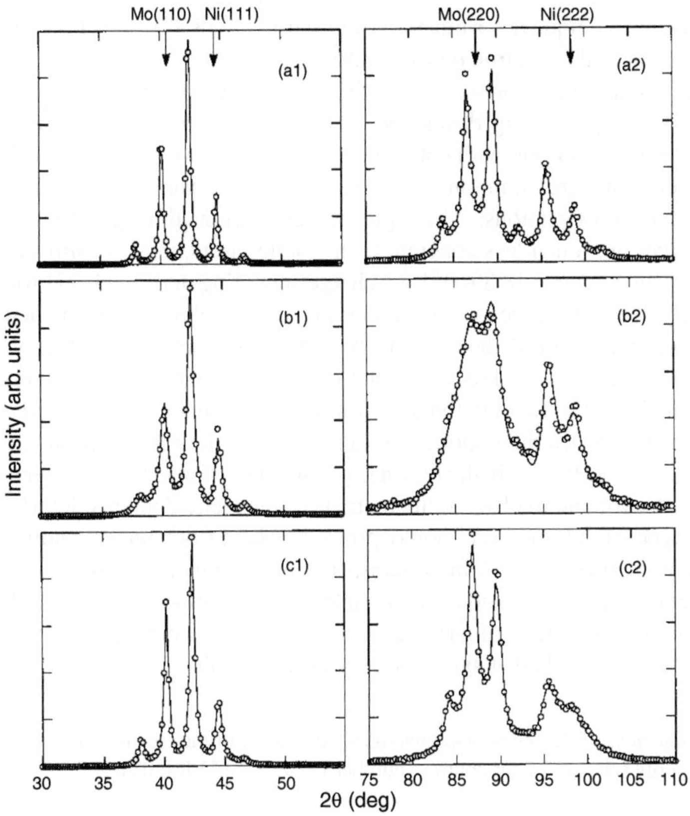

Fig. 8.9. Wide-angle diffraction scans from textured Mo/Ni multilayers (circles) and fits (line) [8.63]. The arrows indicate the expected position of the Mo and Ni peaks. The sample in (**a**) is the nominal structure. The sample in (**b**) and (**c**) were deposited with fluctuations in the Ni and Mo layer thicknesses, respectively

8.4 Outlook

Both reflectivity and off-specular diffuse scattering at small angles and diffraction at wide angles are becoming standard techniques for characterizing thin film and multilayer structures with there applications being helped by the availability of programs for fitting of the scattering profiles [8.63, 84, 85]. It appears that extending these approaches to magnetic scattering, coupled with other more direct imaging methods, such as x-ray absorption microscopes and photoemission electron microscope [8.7] should gives additional information on the magnetic order and the correlations of magnetic and structural interfacial roughness. Resonant scattering in addition to providing magnetic contrast also give additional structural information by changing (or enhancing) the contrast between layers. This ability to alter the contrast allows the study of films that typically have low contrast for x-rays [8.86], provide element specificity and a unique determination of the structure [8.87], and allow for a Fourier reconstruction of the electron density [8.88].

The previous discussion focused on the statistically averaged properties. However there are techniques being developed that provide direct structural and magnetic information via scattering with coherent x-rays also known as speckle [8.31, 89–97]. Such experiments are common in the visible because lasers provide a remarkable coherent source of photons. Although the best current third-generation undulator x-ray sources are much less coherent than visible lasers, they are still sufficient for many experiments [e.g. [8.89–92] for charge scattering and [8.93–97] for magnetic scattering]. The coherence is achieved by placing a pinhole in the incident beam to limit the spatial extent of the beam to be within the lateral coherence of the x-ray beam ($\sim$ tens of microns). For an undulator synchrotron source one obtains typically a coherent flux of $\sim 10^{12}$ photons/sec through the pinhole. The coherence of the x-ray source provides the opportunity to reconstruct the real space image from the speckle pattern [8.91], study the return point memory of magnetic domains with field history [8.93, 96] and to observe fluctuations of magnetic domains [8.94].

This type of information should prove crucial to our understanding of many magnetic properties of thin film systems, such as giant magnetoresistance, exchange bias, magnetic anisotropy and coercive fields. Such measurements are still relatively in their infancy at the time of writing, but we can expect a big increase in the number of magnetic films studied at synchrotron sources over the next decade.

Acknowledgement. S.K. Sinha acknowledges the support of the Department of Energy office of basic energy sciences under contract number DE-FG02-03ER46084.

References

8.1. A. Segmüller and M. Murakami, *Thin films from free atoms and particles,* edited by K.J. Klabunde, (Academic Press, London, 1985), p. 344.
8.2. H. Zabel, Appl. Phys. A **58**, 159 (1994).
8.3. P.F. Fewster, Rep. Prog. Phys. **59**, 1339 (1996).
8.4. E. Chason and T.M. Mayer, Critical Rev. in Solid State and Materials Sciences **22**, 1 (1997).
8.5. R. Clarke and F.J. Lamelas, *Ultrathin Magnetic Structures I* ed. by J.A.C. Bland and B. Heinrich (Springer-Verlag, Berlin, 1994) p. 264.
8.6. J.P. Hannon, G.T. Trammell, M. Blume, and D. Gibbs, Phys. Rev. Lett. **61**, 1245 (1988); Rev. Lett. **62,** 2644 (1989).
8.7. J.B. Kortright, D.D. Awschalom, J. Stöhr, S.D. Bader, Y.U. Idzerda, S.S.P. Parkin, I.K. Schuller, and H.-C. Siegmann, J. Magn. and Magn. Mater. **207**, 7, 1999.
8.8. D. Gibbs, D.R. Harshman, E.D. Isaacs, D.B. McWhan, D. Mills, and C. Vettier, Phys. Rev. Lett. **61** 1241 (1988).
8.9. D. Gibbs, J.P. Hill and C. Vettier, Phys.Stat.Sol. B **215**, 667 (1999).
8.10. B.B. Mandelbrodt, *The Fractal Geometry of Nature*, (Freeman, New York, 1982).
8.11. S.K. Sinha, E.B. Sirota, S. Garoff, and H.B. Stanley, Phys. Rev. B **38**, 2297 (1988).
8.12. E. Fullerton, J. Pearson, C.H. Sowers, S.D. Bader, X.Z. Wu, and S. K Sinha Phys. Rev. B **48**, 17 432 (1993).

8.13. R. Paniago, H. Homma, P.C. Chow, S.C. Moss, S.S.P. Parkin, and D. Cookson, Phys. Rev. B **52**, 17052 (1995).
8.14. D.E. Savage, J. Kleiner, N. Schimke, Y.-H. Phang, T. Jankowski, J. Jacobs, R. Kariotis, and M.G. Lagally, J. Appl. Phys. **69**, 1411 (1991).
8.15. S.A. Stepanov, E.A. Kondrashkina, M. Schmidbauer, R. Köhler, J.-U. Pfeiffer, T. Jach, A.Yu. Souvorov, Phys. Rev. B **54**, 8150 (1996).
8.16. M. Chládek, V. Valvoda, C. Dorner, V. Holý and J. Grim, Appl. Phys. Lett **69**, 1318 (1996).
8.17. D. Weller, L. Folks, M. Best, E.E. Fullerton, B.D. Terris, G.J. Kusinski, K.M. Krishnan, and G. Thomas, J. Appl. Phys. **89**, 7525 (2001).
8.18. M.J. Pechan, J.F. Ankner, C.F. Majkrazk, D.M. Kelly, nad I.K. Schuller, J. Appl. Phys. **75**, 6178 (1994).
8.19. J.W. Freeland, K. Bussmann and Y.U. Idzerda and C.-C. Kao, Phys. Rev. **B 60**, R9923 (1999).
8.20. C.S. Nelson , G. Srajer, J.C. Lang, C.T. Venkataraman, and S.K. Sinha , H. Hashizume, N. Ishimatsu, Midori, N. Hosoito, Phys. Rev. **B 60**, 12234 (1999).
8.21. J.F. MacKay, C. Teichert, D.E. Savage, and M.G. Lagally, Phys. Rev. Lett. **77**, 3925 (1996).
8.22. R.M. Osgood III, S.K. Sinha, J.W. Freeland, Y.U. Idzerda, and S.D. Bader, J. Magn. Magn. Mater. **198-199**, 698 (1999).
8.23. E.E. Fullerton, D.M. Kelly, J. Guimpel, I.K. Schuller, Y. Bruynseraede, Phys. Rev. Lett. **68**, 859 (1992).
8.24. N.M. Rensing, A.P. Payne, and B.M. Clemens, J. Magn. Magn. Mater. **121**, 436 (1993).
8.25. R. Schad, P. Belien, G. Verganck, K. Temst, V.V. Moshchalkov, Y. Bruynseraede, D. Bahr, J. Falta, J. Dekoster, and G. Langouche, Europhys. Lett. **44**, 379 (1998).
8.26. M.C. Cyrille, S. Kim, M.E. Gomez, J. Santamaria, C. Leighton, K.M. Krishnan, and I.K. Schuller, Phys. Rev. B **62**, 15079 (2000).
8.27. P.F. Miceli, D.A. Neumann and H. Zabel, Appl. Phys. Lett. **48**, 24 (1986).
8.28. T. Salditt, T.H. Metzger, J. Peisl and G. Goerigk, J. Phys. D: Appl. Phys. **28** A236 (1995).
8.29. J.B. Kortright, M. Rice, S.K. Kim, C.C. Walton and T. Warwick, J. Magn. Magn. Mater. **191**, 79 (1999).
8.30. J.B. Kortright and S.K. Kim, Phys. Rev. B **62**, 12216 (2000).
8.31. A. Rahmim, S. Tixier, T. Tiedje, S. Eisebitt, M. Lörgen, R. Scherer, W. Eberhardt, J. Lüning, A. Scholl, Phys. Rev. B **65**, 235421 (2002).
8.32. R.M. Osgood III, S.K. Sinha, J.W. Freeland, Y.U. Idzerda, and S.D. Bader, J. Appl. Phys. **85**, 4619 (1999).
8.33. J.B. Kortright, S.-K. Kim, G.P. Denbeaux, G. Zeltzer, K. Takano, and E.E. Fullerton, Phys. Rev. B **64**, 092401 (2001).
8.34. D.G. Stearns, J. Appl. Phys. **65**, 491 (1989).
8.35. S.K. Sinha, J. Phys. (France) III **4**, 1543 (1994).
8.36. G. Palasantzas and J. Krim, Phys. Rev. B **48**, 2873 (1993).
8.37. M. Meduòa, V. Holý, T. Roch, J. Stangl, G. Bauer, J. Zhu, K. Brunner, and G. Abstreiter, J. Appl. Phys. **89**, 4839 (2001).
8.38. W. Weber and B. Lengeler, Phys. Rev. B **46**, 7953 (1992).
8.39. H. You, R.P. Chiarello, H.K. Kim, and K.G. Vandervoort, Phys. Rev. Lett. **70**, 2900 (1993).
8.40. M.K. Sanyal, S.K. Ginha, A. Gibaud, S.K. Satija, C.F. Majkrzak, and H. Homma, *Surface X-ray and Neutron Scattering,* H. Zabel and I.K. Robinson Eds. (Springer-Verlag, Berlin, Heidelberg, 1992) 91.

8.41. T. Salditt, T.H. Metzger, and J. Peisl, Phys. Rev. Lett. **73**, 2228 (1994).
8.42. T. Salditt, T.H. Metzger, J. Peisl, B. Reinker, M. Moske, and K Samwer, Europhys. Lett. **32**, 331 (1995).
8.43. R. Paniago, P.C. Chow, R. Forrest, and S.C. Moss, Physica B **248**, 39 (1990).
8.44. T. Salditt, D. Lott, T.H. Metzger, J. Peisl, R. Fischer, J. Zweck, P. Høghøj, O. Schärpf, and G. Vignaud, Europhys. Lett. **36**, 565 (1996).
8.45. T. Salditt, D. Lott, T.H. Metzger, J. Peisl, G. Vignaud, P. Høghøj, O. Schärpf, P. Hinzc and R. Lauer, Phys. Rev. B **54**, 5860 (1996).
8.46. Z. Kovats, T. Salditt, T.H. Metzger, J. Peisl, T. Stimpel, H. Lorenz, J.O. Chu, K. Ismail, J. Phys. D: Applied Physics **32**, 359 (1999).
8.47. J.M. Freitag and B.M. Clemens, J. Appl. Phys. **89**, 1101 (2001).
8.48. V. Holý and T. Baumbach, Phys. Rev. B **49**, 10668 (1994).
8.49. M. Chládek, V. Valvoda, C. Dorner, W. Ernst, J. Magn. Magn. Mater. **172**, 209 (1997).
8.50. D.K.G. de Boer, Phys. Rev. B **53**, 6048 (1996).
8.51. Y. Yoneda, Phys. Rev. **131**, 2010 (1963).
8.52. J.B. Kortright, J. Appl. Phys. **70**, 3620 (1991).
8.53. D.R. Lee, S.K. Sinha, D. Haskel, Y. Choi, J.C. Lang, S.A. Stepanov and G. Srajer, Phys. Rev. B **68**, 224409 (2003).
8.54. D.R. Lee, S.K. Sinha, C.S. Nelson, J.C. Lang, C.T. Venkataraman, G. Srajer, and R.M. Osgood III, Phys. Rev. B **68**, 224410 (2003).
8.55. H.A. Dürr, E. Dudzik, S.S. Dhesi, J.B. Goedkoop, G. van der Laan, M. Belakhovsky, C. Mocuta, A. Marty, and Y. Samson, Science **284**, 2166 (1999).
8.56. B.M. Clemens and J.G. Gay, Phys. Rev. B **35**, 9337 (1987).
8.57. I.K. Schuller, Phys. Rev. Lett. **44**, 1597 (1980).
8.58. K. Temst, M.J. de Groot, N. Koeman, and R. Griessen, Appl. Phys. Lett. **67**, 3429 (1995).
8.59. I. Heyvaert, K. Temst, C. Van Haesendonch, and Y. Bruynseraede, J. Vac. Sci. Technol. B **14**, 1121 (1996).
8.60. C.G. Darwin, Philos. Mag. **43**, 800 (1922).
8.61. P.F. Miceli and C.J. Palmstrøm, Phys. Rev. B **51**, 5506 (1995).
8.62. D.F. McMorrow, P.P. Swaddling, R.A. Cowley, R.C.C. Ward and M.R. Wells, J. Phys.: Condens. Matter **8**, 6553 (1996).
8.63. E.E. Fullerton, Ivan K. Schuller, H. Vanderstraeten, Y. Bruynseraede, Phys. Rev. B **45**, 9292 (1992).
8.64. Advances in the Rietveld Methods, ed. R.A. Yound (Oxford University Press, Oxford, 1992).
8.65. V.S. Speriosu and J.T. Vreeland, J. Appl. Phys. **56**, 1591 (1984).
8.66. J. Kwo, E.M. Gyorgy, D.B. McWhan, F.J. DiSalvo, C. Vettier, and J.E. Bower, Phys. Rev. Lett. **55**, 1402 (1985).
8.67. M.B. Stearns, C.H. Lee, and T.L. Groy, Phys. Rev. B **40**, 8256 (1989).
8.68. W. Sevenhans, M. Gijs, Y. Bruynseraede, H. Homma and I.K. Schuller, Phys. Rev. B **34**, 5965 (1986).
8.69. D. Chrzan and P. Dutta, J. Appl. Phys. **59**, 1504 (1986).
8.70. J.-P. Locquet, D. Neerinck, L. Stockman, Y. Bruynseraede, and I.K. Schuller, Phys. Rev. **39**, 13338 (1989).
8.71. M.A. Hollanders and B.J. Thijsse, J. Appl. Phys. **70**, 1270 (1991).
8.72. P.F. Miceli, C.J. Palmstrøm and K.W. Moyers, Appl. Phys. Lett. **61**, 2060 (1992).
8.73. N. Nakayama, T. Okuyama and T. Shinjo, J. Phys.: Condens. Matter **5**, 1173 (1993).
8.74. P.F. Miceli, in *Semiconductor Interfaces, Microstructures, and Devices: Properties and Applications*, edited by Z.C. Feng (Institute of Physics, Bristol, 1993), p. 87.

8.75. C.A. Ramos, M.O. Cáceres, and D. Lederman, Phys. Rev. **53**, 7890 (1996).
8.76. S. Hendricks and E. Teller, J. Chem. Phys. **10**, 147 (1942).
8.77. I.K. Schuller, M. Grimsditch, F. Chambers, G. Devane, H. Vanderstraeten, D. Neerinck, J.–P. Locquet, and Y. Bruynseraede, Phys. Rev. B **65**, 1235 (1990).
8.78. J. Kakinoki and Y. Komura, J. Phys. Soc. Jpn. **7**, 30 (1952).
8.79. E.E. Fullerton, D. Stoeffler, K. Ounadjela, B. Heinrich, Z. Celinski, and J.A.C. Bland, Phys. Rev. B **51**, 6364 (1995).
8.80. C.M. Schmidt, D.E. Bürgler, D.M. Schaller, F. Meisinger, H.-J. Güntherodt, and K. Temst, J. Appl. Phys. **89**, 181 (2001).
8.81. J. Birch, J.-E. Sundgren, P.F. Fewster, J. Appl. Phys. **78**, 6562 (1995).
8.82. M. Chládek, V. Valvoda, and C. Doerner, J. Magn. Magn. Mater. **172**, 218 (1997).
8.83. E.E. Fullerton, S.D. Bader, and J.L. Robertson, Phys. Rev. Lett. **77**, 1382 (1996).
8.84. D.L. Windt, Computers in Physics, 12, 360–370 (1998); http://cletus.phys.columbia.edu/~windt/idl/.
8.85. Commercial packages include the REFS from Bede Scientific and WinGixa for Philips Analytical.
8.86. J. Bai, E.E. Fullerton, and P. Montano, Physica B **221**, 411 (1996).
8.87. T. Bigault, F. Bocquet, S. Labat, O. Thomas, and H. Renevier, Phys. Rev. B **64**, 125414 (2001).
8.88. M.K. Sanyal, S.K. Sinha, A. Gibaud, K.G. Huang, B.L. Carvalho, M. Rafailovich, J. Sokolov, X. Zhao, and W. Zhao, Europhys. Lett. **21**, 691 (1993).
8.89. W.B. Yun, J. Kirz, and D. Sayre, Acta Crystallogr. A **43**, 131 (1987).
8.90. M. Sutton, S.G.J. Mochrie, T. Greytak, S.E. Nagler, L.E. Berman, G.A. Held, and G.B. Stephenson, Nature **352**, 608 (1991).
8.91. I.K. Robinson, I.A. Vartanyants, G.J. Williams, M.A. Pfeifer, and J.A. Pitney, Phys. Rev. Lett. **87**, 195501 (2001).
8.92. F. Pfeiffer, W. Zhang and I.K. Robinson, Appl. Phys. Lett. **84**, 1847 (2004).
8.93. B. Hu, P. Geissbuhler, L. Sorensen, S.D. Kevan, J.B. Kortright, E E. Fullerton, Synch. Rad. News **14**, 11 (2001).
8.94. F. Yakhou, A. Lëtoublon, F. Livet, M. de Boissieu, and F. Bley, J. Magn. Magn. Mater. **233**, 119 (2001).
8.95. K. Chesnel, M. Belakhovsky, F. Livey, S.P. Collins, G. van der Laan, S.S. Dhesi, J.P. Attane, and A. Marty, Phys. Rev. B **66**, 172404 (2002).
8.96. M.S. Pierce, R.G. Moore, L.B. Sorensen, S.D. Kevan, O. Hellwig, E.E. Fullerton and J.B. Kortright, Phys. Rev. Lett. **90**, 175502 (2003).
8.97. S. Eisebitt, M. Lörgen, W. Eberhardt, J. Lüning, J. Stöhr, O. Hellwig, C. Rettner, E.E. Fullerton, G. Denbeaux, Phys. Rev. B **68**, 104419 (2003).

Subject Index

MIX
Papier aus verantwortungsvollen Quellen
Paper from responsible sources
FSC® C105338

If you have any concerns about our products,
you can contact us on
ProductSafety@springernature.com

In case Publisher is established outside the EU,
the EU authorized representative is:
Springer Nature Customer Service Center GmbH
Europaplatz 3, 69115 Heidelberg, Germany

Printed by Libri Plureos GmbH
in Hamburg, Germany